Microbial Plant Pathogens-Detection and Disease Diagnosis

Microbial Plant Pathogens-Detection and Disease Diagnosis

Bacterial and Phytoplasmal Pathogens, Volume 2

P. Narayanasamy

Former Professor and Head, Department of Plant Pathology
Tamil Nadu Agricultural University, Coimbatore, India

Springer

P. Narayanasamy
Grand Parade Apartments
32 D Thilagar Street, R. S. Puram
Coimbatore 641002
India
pnsamyraj@gmail.com

ISBN 978-90-481-9768-2 e-ISBN 978-90-481-9769-9
DOI 10.1007/978-90-481-9769-9
Springer Dordrecht Heidelberg London New York

Library of Congress Control Number: 2010938430

Cover illustration:

Sesamum phyllody (Volume 2)
Symptoms of infection may be seen generally at the time of flowering. All floral parts are converted into green leaf-like structures. The infected plants become partially or totally sterile, depending on the time of infection.
(Courtesy of B. Sreenivasulu, Tamil Nadu Agricutlural University, Coimbatore, India)

Printed on acid-free paper

Springer is part of Springer Science+Business Media (www.springer.com)

Dedicated to the Memory of my Parents
for their Love and Affection

Preface

Existence of prokaryotes including bacteria has been recorded several billions of years ago. However, bacteria were demonstrated to be the causative agents of plant diseases only in the last quarter of the 19th century. Now numerous plant diseases are known to be due to the bacterial pathogens which induce symptoms distinct from those induced by fungal pathogens. Diseases caused by phytoplasmas were considered earlier to be due to viruses, because of the similarities in the modes of transmission of these two pathogenic groups. But visualization of plemorphic bodies without cell wall in the phloem tissues of the infected plants indicated the nature of the causative agents and provided evidence that phytoplasmas are related to bacteria and lack cell walls that are present in bacteria.

As the bacteria and phytoplasmas are much smaller than the fungal pathogens, more powerful light microscopes and electron microscopes have to be employed for the detection and differentiation of these disease-causing agents. As the morphological characteristics have limited value for the differentiation of bacterial species, several biochemical tests have to be performed for the detection and identification of bacterial species causing the newly observed disease(s). The phytoplasma, on the other hand, have not been successfully cultured in cell-free media and hence, their cultural characteristics could not be studied. The biochemical tests are time-consuming, labor-intensive and often yield inconsistent results. It is essential that rapid detection and precise identification of bacteria and phytoplasmas have to be achieved within short time to assess the extent of disease incidence. In order to successfully prevent the spread of the bacterial and phytoplasmal diseases, techniques with greater sensitivity, specificity, reproducibility and rapidity have to be applied. It has been acknowledged that rapid detection and accurate identification of the microbial plant pathogens up to subspecies/strain level form the basis for the development of short- and long-term strategies for management of diseases caused by them. Immunoassays and nucleic acid-based techniques have been demonstrated to be precise and reliable, in addition to being amenable for automation and providing the results much earlier when compared with the isolation-based conventional methods. Furthermore, these modern molecular techniques have higher level of utility where the conventional methods fail to detect the pathogens in asymptomatic plants and tissues containing low concentration of bacteria and phytoplasmas. Hence, the molecular methods have become the methods of choice for the researchers in the recent years.

This volume presents exhaustive information based on extensive literature search on various methods of detection of bacterial and phytoplasmal pathogens and diagnosis of diseases caused by them. Comparative effectiveness of different techniques is discussed critically to enable the researchers, teachers, extension specialists and graduate students to choose the suitable techniques for their investigations. Further, several protocols have been presented as appendix in appropriate chapters to meet this requirement. This volume is expected to provide the necessary platform for planning more critical studies that may improve the efficiency and reliability of the existing methods that have been employed for simultaneous detection of two or more kinds of pathogens, without compromising the sensitivity and specificity of detection of bacterial and phytoplasmal pathogens and diagnosis of the diseases caused by them.

Coimbatore, India P. Narayanasamy

Acknowledgement

With a deep sense of humility, gratitude and reverence I bow to my Alma Mater that continues to remain a source of inspiration for me for over five decades. I wish to place on record appreciation to my colleagues and graduate students of the Department of Plant Pathology, Tamil Nadu Agricultural University, Coimbatore, India for their suggestions and critical comments. I am thankful to Dr. T. Ganapathy, Professor of Plant Pathology, for providing technical assistance requiring enormous patience and skill. Permission granted by different copyright holders to reproduce the figures published in various journals is gratefully acknowledged.

It is with great pleasure, I thank profusely my wife Mrs. N. Rajakumari who showers her love and kindness, enabling me to devote my attention exclusively for the preparation of this book. Expression of abundant affection and endless encouragement of my family members Mr. N. Kumar Perumal, Mrs. Nirmala Suresh, Mr. T. R. Suresh and Mr. S. Varun Karthik has been the source of support for all my academic efforts during my career.

Contents

Table of Contents for Volumes 1 and 3

Volume 1

Volume 3

Chapter 1
Introduction

Abstract Bacteria have been recognized as a distinct group of plant pathogens capable of inducing significant losses in various crops cultivated in different countries all over the world. Various methods based on their morphological, biochemical/physiological, immunological and nucleic acid characteristics have been applied for their detection, differentiation and identification with different levels of sensitivity, specificity and reliability. Diagnosis of the diseases caused by the bacterial plant pathogens has been successfully accomplished using different biological methods that require different periods of time. Immunological and nucleic acid-based techniques have been demonstrated to provide more efficient and reliable diagnosis of diseases rapidly, especially when the plants and plant materials remain asymptomatic and the infection levels are low in the field conditions. The role of disease diagnostic centers, plant quarantine and certification programs in providing advice to the growers, prevention of introduction of new diseases and establishment of disease-free nuclear stocks is highlighted.

1.1 Microbial Plant Pathogens as a Major Limiting Factor of Crop Production

Microbial plant pathogens including bacterial and phytoplasmal pathogens have the ability to infect a few or a wide range of plant species causing varying magnitude of quantitative and qualitative losses in crops cultivated in different ecosystems. Global losses caused by crop diseases have been estimated to range from 9% to 14.2% of potential yield (Orke et al. 1994). Assessments of losses made later indicated that about 14.1% of produce may be lost due to crop diseases with a monetary value of $220 billions per annum, the developing countries suffering more losses compared with developed countries (Agrios 2005). The loss assessments have been made for different types of diseases with different levels of accuracy. However, irrespective of the levels of accuracy, the disease assessments emphasize the imperative need for measures to be taken urgently to avoid the losses to the extent possible. To achieve this aim, three principles of crop disease management viz.,

P. Narayanasamy, *Microbial Plant Pathogens-Detection and Disease Diagnosis: Bacterial and Phytoplasmal Pathogens, Vol. 2*, DOI 10.1007/978-90-481-9769-9_1,

exclusion, eradication and immunization form the basis for formulating short- and long-term disease management strategies (Narayanasamy 2002). The effectiveness of crop disease management systems depends heavily on the rapid, reliable and sensitive detection of microbial plant pathogens and accurate diagnosis of the diseases caused by the putative pathogen(s) detected in infected plants.

1.2 Discovery of Bacteria as Plant Pathogens

Various bacterial species have been demonstrated as the causative agents of plant diseases. Occurrence of diseases affecting cereals and wine crops has been mentioned in ancient scriptures. But they were attributed to supernatural elements, because of the prevailing superstitious beliefs, religious dogmas and faiths. Theophrastus (ca. 300 BC) and other Greek philosophers expressed grave concern over the drastic reduction in crop yields and speculated on the cause of the crop diseases (Orlob 1964). Plant diseases were considered as God-sent curses as punishment for sins committed, because of lack of scientific and analytical observations. However, scientific thoughts took roots based on the contributions of early researchers like Micheli (1729), Tillet (1755) and Prévost (1807). The remarkable contributions of Anton de Bary (1831–1888) formed the cornerstone for the future development of modern investigations on fungal, bacterial and viral plant pathogens that cause a wide range of diseases affecting crops all over the world. Hence, he is deservingly regarded as the father of Plant Pathology (Horsfall and Cowling 1977).

Fossil prokaryotes which include plant pathogenic bacteria have been recorded long ago (ca. 3.5 billion years) and the prokaryotes seem to be the only form of life on earth over 20 billion years (Schumann 1991). Bacteria as the causative agents of plant diseases was demonstrated about 2 decades after the fungus *Phytophthora infestans* was proved to be the cause of the destructive potato late blight disease by de Bary (1861). The fire blight disease of apple was shown to be caused by the bacterial species *Erwinia amylovora* by Burrill in 1878. Another bacterial species *Agrobacterium tumefaciens* was found to be the incitant of crown gall disease affecting several plant species by Smith and Townsend (1907). Later several species of bacteria were demonstrated to be the causative agents of diseases affecting various crops.

1.3 Detection of Bacterial and Phytoplasmal Pathogens and Disease Diagnosis

Taxonomic characteristics and guidelines for identification of bacterial species are described in the Bergey's Manual of Systematic Bacteriology (Kreig and Holt 1984). Bacterial cultures developed in appropriate media are examined for the cultural characteristics such as colony morphology, production of enzymes, toxins and other metabolites and pattern of utilization of carbon sources. The other biological characteristics such as pathogenicity, host range and types of symptoms induced in different host plant species are also studied for the detection, differentiation and identification of the bacterial species isolated from the plants exhibiting the newly observed disease(s). A set of differential host plant species or cultivars may be used for detection and identification of the strains of the bacterial species. Immunological properties of surface proteins on the bacterial cells or flagella have been used for the production of polycolonal and monoclonal antibodies (PAbs and MAbs). These antibodies have been employed for the detection, differentiation and quantification of bacterial pathogens in plant tissues. Nucleic acid-based techniques have been shown to be more precise, sensitive, specific and reliable, in addition to their capacity to provide the results rapidly. The nucleic acid based techniques are particularly very useful for the detection and differentiation of phytoplasmal pathogens which have not been cultured in the cell-free media. It is possible to analyze several samples simultaneously and to detect single or multiple infections by the bacterial and phytoplasmal pathogens simultaneously present in the same samples or in different samples by employing hybridization-based and/or polymerase chain reaction (PCR)-based techniques (Chapter 2).

The presence of bacterial pathogens in the soil, irrigation/rain water and air has been demonstrated by several studies. Application of techniques based on the biological, immunological and genomic nucleic acid characteristics of the bacterial pathogens in the environmental samples has resulted in the detection and identification of bacterial pathogens with different levels of sensitivity and specificity. Long time and large greenhouse space and inconsistent results have been the limiting factors for the continued application of biological methods, although they indicate the pathogenic potential of the putative bacterial pathogen isolated from the plants affected by the newly observed disease. The presence of the bacterial pathogens in the environmental samples and the role of additional host plant species capable of serving as sources of infection and the vector species involved in the spread of the disease within the field and to other locations has been established conclusively by applying immunological and nucleic acid-based techniques (Chapter 3).

Bacterial plant pathogens are capable of producing new strains/varieties to be able to infect newly developed cultivars incorporated with disease resistance gene(s). Application of chemicals indiscriminately may create a condition that adversely affects the survival of the bacterial pathogens. Under such situations, the bacterial pathogen has to form new strains that can overcome the effects of chemicals. Emergence of new strains with increased virulence (pathogenic potential) or resistance to chemicals has to be recognized immediately by applying sensitive and reliable technique(s) that can provide the results rapidly for the differentiation and identification of the new strains of the bacterial species. Application of immunological and nucleic acid-based methods has been shown to be effective to meet the demands of such critical situations. Determination of variability in bacterial and phytoplasamal pathogens by employing nucleic acid-based procedures has been found to be useful to make required changes in the measures to be adopted for effective management of diseases caused by the bacterial pathogens or their strains (Chapter 4).

Diagnosis of a newly observed bacterial and phytoplasmal disease in a location may be accomplished by fulfilling the steps of Koch's postulates. Primary pathogen and secondary invaders have to be precisely identified when more than one pathogen is involved as in the case of complex diseases. Disease diagnostic centers, plant quarantines and certification programs have to be highly competent to handle large number of samples of diseased plants or plant materials. This is essential to prevent the introduction and spread of the diseases caused by bacterial and phytoplasmal pathogens from one country to another country. The need for providing disease-free seeds and planting materials and proper advice for taking preventive measures by the growers has been well realized. Hence technical skill and equipments for detection, differentiation and identification of bacterial and phytoplasmal pathogens have to be improved for the effective functioning of the personnel manning these centers and programs (Chapter 5).

The information reflecting extensive literature search is presented in an easily understandable style. It is expected that the various aspects of detection, differentiation, quantification and identification of bacterial and phytoplasmal pathogens, as well as the diagnosis of the diseases caused by them presented in this volume, will be highly useful to the researchers, teachers and graduate students in the Departments of Plant Pathology, Microbiology, Plant Protection, Molecular Biology and Plant Breeding. In addition, the extension plant pathologists in disease diagnostic centers and personnel of plant quarantine and certification programs will find the information to have practical utility. Presentation of several protocols appended as appendices in appropriate chapters will assist in selecting the right procedures for reaching their research targets.

References

Agrios GN (2005) Plant Pathology, Fifth edition, Elsevier/Academic, Amsterdam.

Burril TJ (1878) Remarks made in a discussion. Trans Ill State Hortic Soc (NS) 11: 79–80.

de Bary A (1861) Uber die Geschlechtsorgane von Peronospora. Bot Z 19: 89–91.

Horsfall JG, Cowling EB (1977) The sociology of plant pathology. In: Horsfall JE, Cowling EB (ed) Plant Disease – An Advanced Treatise, Academic, New York, Vol 1, pp. 12–34.

Kreig NR, Holt JG (1984) Bergey's Manual of Systematic Bacteriology, Vol I. Williams & Wilkins, Baltimore, MD.

Micheli PA (1729) Nova Plantarum Genera. Florence, Italy.

Narayanasamy P (2002) Microbial Plant Pathogens and Crop Disease Management. Science Publisher, Enfield, NH.

Orke EC, Dehne HW, Schonbeck F, Weber A (1994) Crop Production and Crop Protection: Estimated Losses in Major Food and Cash Crops. Elsevier, Amsterdam.

Orlob GB (1964) The concepts of etiology in the history of plant pathology. Pflanzenschutz-Nachr Bayer 17: 185–268.

Prévost B (1807) Memore sur la cause immediate de al carie ou carbon des bles, et de plusieurs autres maladies des plantes, et sur les preservatifs de la carie. Paris [English translation by GM Keitt in Phytopathological Classics No. 6, 1939].

Schumann GL (1991) Plant Diseases: Their Biology and Social Impact. American Phytopathological Society, St. Paul, MN.

Smith EF, Townsend CO (1907) A plant tumor of bacterial origin. Science 25: 671–673.

Tillet M (1755) Dissertation sur la cause qui corrompt et noircit les grans de ble dans leseps; et sur les moyens de prevenir ces accidents. Bordeaux [English translation by HB Humphrey in Phytopathological Classics No. 5, 1937].

Chapter 2
Detection of Bacterial and Phytoplasmal Pathogens

Abstract Bacterial pathogens classified as prokaryotes are much simpler in structure and smaller in size and they have less complex structural features compared to fungal plant pathogens. As the morphological characteristics of bacterial cells are less variable, biological, biochemical, physiological, immunological and genomic characteristics have to be determined for reliable identification and meaningful classification of bacterial pathogens. Detection and identification of bacterial plant pathogens present in whole plants and in propagative plant materials have been possible by employing isolation on cultural media and metabolic fingerprinting methods. But they are labor-intensive and require long time. Results often are inconclusive. Isozyme analysis, direct colony thin layer chromatography and gel electrophoresis techniques have been successfully applied for the detection of some bacterial pathogens. Immunoassays and nucleic acid-based assays have become widely accepted techniques, providing more sensitive and specific detection and quantification of bacterial pathogens affecting a wide range of host plant species. However, the major limitation of the molecular techniques is their inability to discriminate living cells from the dead ones. Attempts have been made to overcome this limiting factor by using DNA binding dyes and estimating pathogen RNAs as an indicator of cell viability. The diagnostic methods have both merits and demerits and hence, appropriate method has to be selected based on cost-effectiveness and possibility of obtaining reliable results rapidly to suit the requirements of the investigation concerned.

Phytoplasmas are cell wall-less, nonculturable bacteria belonging to Mollicutes. Lack of cultural characteristics has made obligatory to study the morphological characteristics of phytoplasmas present in the phloem cells of infected plant hosts by observing under electron microscopes. As most of the phytoplasmas look alike in the ultrathin sections, the morphological characters have no diagnostic value. Histochemical methods using DNA binding dyes have been employed to localize the phytoplasmas in the host cells. Application of immunoassays and nucleic acid-based techniques has been effective in providing information for the identification and differentiation of phytoplasams. Polyclonal and monoclonal antibodies have been generated for detection of phytoplasmas infecting several plant species. Universal and species-specific primers and probes have been designed based on

P. Narayanasamy, *Microbial Plant Pathogens-Detection and Disease Diagnosis: Bacterial and Phytoplasmal Pathogens, Vol. 2*, DOI 10.1007/978-90-481-9769-9_2,

DNA sequences of the target phytoplasma(s) to be detected. Precise, rapid and reliable results obtained from molecular methods can be used for detection, quantification and classification of phytoplasmas infecting whole plants, as well as those present in propagative plant materials which form the primary sources of infection for the crops to be planted in the subsequent seasons.

Bacterial plant pathogens are smaller in size and have less defined structural characteristics compared with fungal plant pathogens. Fungi are placed under eukaryotes, whereas bacteria are primitive organisms classified as prokaryotes which include Mollbicutes enclosing phytoplasmal pathogens. Fossil prokaryotes have been recorded in rocks about 3.5 billion years of age and it appears that prokaryotes were the only form of life on earth over 2 billion years. Fossils of eukaryotic organisms have not been observed in rocks older than about 1.6 billion years (Schumann 1991). It is estimated that about 100 of 1,600 bacterial species are known to infect various kinds of plants (Agrios 2005). Most of the bacteria are strictly saprophytic and help decompose large quantities of organic wastes produced by industrial establishments or by dead plants and animals. Many bacterial species are beneficial to humans, because of their role in building up nutritional levels of the soil and as biocontrol agents (BCAs) that have the potential for reducing the activity of or inhibiting the development of bacterial plant pathogens.

The bacterial cell contains nuclear material which is not separated from the cytoplasm by nuclear membrane as in eukaryotes and there is no mitotic mechanism observed in higher life forms. The genetic material is present in the form of a single chromosome with double-stranded (ds) DNA in a closed circular form. The bacterial cell may also contain plasmids capable of replicating independent of chromosome. They have extra-chromosomal DNA governing some characteristics such as pathogenicity, resistance to chemicals and antibiotics and tumor formation. Plasmids can pass from one bacterial cell to another easily. No other organelle is present in the bacteria as in the case of eukaryotes.

The bacterial cells are much smaller in size compared with fungal cells and may measure about 1 μm in diameter, whereas the fungal spores on an average may have diameter varying from 50 to 200 μm. They are simple in structure with different shapes like spherical, ellipsoidal, rod-shaped, spiral, filamentous or comma-shaped. All plant pathogenic bacteria, except two species of filamentous *Streptomyces*, are rod-shaped. The rod-shaped bacterial cells are short and cylindrical, measuring 0.3–1.0 × 0.5–3.5 μm. In addition to the rod-shaped ones, a club, a Y or V shaped and branched forms may also be observed. Pairs of cells or short chains of cells may also be enveloped by a thin or thick slime layer made of viscous gummy materials. The slime layer (capsule) may be found as a larger mass around the cells. Most of the plant pathogenic bacteria are motile. The flagella, organs of locomotion, may be present either singly or in groups at one or both ends the bacterial cell or distributed over the entire cell surface. The cells of *Streptomyces* spp. consist of nonseptate, branched threads which usually have a spiral formation and produce conidia in chains on aerial hyphae.

Bacterial colonies show variation in shape, size, color, elevations, form of edges etc. and these characteristics may be useful in the identification of certain bacterial genera. The size of colonies may vary from 1 mm to several centimeters in diameter and they may be circular, oval or irregular with smooth, wavy or angular edges. The elevation of the colonies may be flat, raised dome-like or gray. The bacterial protoplasm is enclosed by a distinct cell wall that can be differentiated from the outer cytoplasmic membrane. The bacteria can be separated into two biologically different groups based on the ability to retain the stain developed by Hans Christian Gram. Gram-positive bacteria can retain the stain and have a relatively thick uniform cell wall. On the other hand, gram-negative bacteria that cannot retain the stain, have a thinner cell wall with an additional outer layer of polysaccharides and lipids. The lipids present in this layer prevent the absorption of certain substances like penicillin making them insensitive/resistant to such substances. None of the plant pathogenic bacterial species, except *Bacillus* spp. is known to produce endospores. Hence they are sensitive to desiccation.

Bacteria, unlike fungi, do not have complex life cycle. During asexual reproduction, a process known as binary fission is initiated by the growth of the cytoplasmic membrane toward the centre of the cell, forming a transverse membranous partition dividing the cytoplasm into two approximately equal parts. Two layers of cell wall material are laid down between two cytoplasmic layers. As the partition of the cytoplasm is in progress, the DNA which is in the form of a circular chromosome, condenses into a dumbbell structure. The DNA then divides into two equal pieces and gets distributed into the new daughter cells that are formed from the dividing single cell. This process can continue indefinitely. Under favorable conditions, the bacteria can divide once in every 20 min, reaching astonishing population levels rapidly.

The process known as conjugation considered similar to sexual reproduction, occurs in bacterial species. Two physiologically opposite cells come into contact side by side and a small fragment of DNA from the male cell (donor) is transferred to the female cell (receptor). The female cell then multiplies by binary fission, leading to the production of a population of cells possessing the characteristics governed by the DNA fragment of the donor cell. Apart from conjugation, changes in the genetic constitution of bacteria are possible in three ways. Mutation may occur in a very small proportion of bacterial cells leading to permanent changes in some of the characteristics of the bacteria. The genetic material in some bacteria is liberated from one cell either by secretion or rupture of the cell wall. A fragment of the DNA so released gains entry into a genetically compatible bacterial cell of the same or closely related species and the recipient cell becomes genetically different. This phenomenon is called transformation. A bacteriophage is involved as a vector in the phenomenon called transduction, for the transfer of the bacterial genetic material (DNA) from one bacterial cell to another cell. The vector (a virus) acquires a fragment of DNA of the infected bacterial cell which is lysed later, liberating the virus (bacteriophage). When the bacteriophage infects another bacterial cell, the DNA fragment acquired from the lysed cell DNA, gets integrated with the freshly infected bacterial cell. Thus, different characters can be transferred from one cell to another, resulting in changes in the genetic constitution of bacteria. However, such changes in the genetic constitution apparently occur rarely. The ability of bacteria to multiply very rapidly and reach high population levels within a short period of time makes them an important factor to be considered in any ecosystem in general and as pathogens of plants and animals in particular (Narayanasamy 2001).

2.1 Detection of Bacterial Pathogens in Plant Organs

2.1.1 Biological Methods

The taxonomic characteristics and guidelines for describing bacteria are included in Bergey's Manual of Systematic Bacteriology (Krieg and Holt 1984). The names proposed for bacteria should be in accordance with the International Code of Nomenclature of Bacteria. As the variations in morphological characteristics are insufficient to distinguish different bacterial species, several physiological and biochemical properties are also considered for classification of bacteria. Naming a new bacterial species requires the results of polyphasic tests which include nucleic acid analyses such as DNA–DNA and DNA–rDNA hybridization, chemotaxonomic comparisons such as cell wall composition, lipid composition, isoprenoid quinones, soluble and total proteins, fatty acid profiles, enzyme characterizations, in addition to biochemical and nutritional tests, so that determinative keys may be formulated (Young et al. 1992). The International Society of Plant Pathology established the criteria for use of the term 'pathovar' for those plant pathogenic bacteria that did not satisfy the criteria for species designation. The nomenclature of 'pathovar' is applied at the infrasub-specific level for bacteria distinguished chiefly based on the differences in pathogenicity on a set of plant species/cultivars. The plant pathogenic bacterial genera are classified as indicated in Table 2.1. As most bacteria do not possess

Table 2.1 Classification of plant pathogenic bacteria Kingdom: Procaryotae (Agrios 2005)

Division	Class	Family	Genus
Gracilicutes (Gram-negative)	Proteobacteria	Enterobacteriaceae	*Erwinia*
			Pantoea
			Serratia
			Sphingomonas
		Pseudomonadaceae	*Acidovorax*
			Pseudomonas
			Ralstonia
			Rhizobacter
			Rhizomonas
			Xanthomonas
		Rhizobiaceae	*Agrobacterium*
		Yet to be named	*Xylella*
			Candidatus Liberibacter
Firmicutes (Gram-positive)	Firmibacteria		*Bacillus*
			Clostridium
	Thallobacteria		*Arthrobacteria*
			Clavibacter
			Curtobacterium
			Leifsonia
			Rhodococcus
			Streptomyces

distinctive morphological characteristics, the taxonomy and systems of nomenclature are less stable than that of fungi. Bacterial strains that share certain phenotypic and genotypic similarities are placed in a bacterial species. The predominantly occurring strain is named as 'type strain', while other strains of the same species may differ to varying degrees from the type strain. Variations in strains may be observed in morphological, cultural, physiological, biochemical or type of effects induced in plants.

Plant pathogenic bacteria induce characteristic water-soaked lesions in the infected tissues at the initial stages of infection and these lesions may turn necrotic (brown) later. Formation of encrustations or bacterial ooze from infected tissues is another distinguishing feature of bacterial diseases. If the infection is localized, leaf spots, blights, scabs, cankers and tumors may be produced in the infected leaves, fruits, roots or stems. If infection becomes systemic as the bacterial pathogen is able to move to other plant tissues away from the site of infection, symptoms such as chlorosis, rotting and wilting of whole plants may be observed. Different kinds of symptoms caused by bacterial pathogens in some of the economically important crops are presented in Figs.2.1–2.3.

Some of the symptoms such as leaf spot and blight may have similarities with those caused by fungal pathogens. However, the absence any kind of spores usually produced by fungal pathogens and the presence of bacterial ooze from the cut ends of tissues may indicate the bacterial origin of the disease concerned. Isolation of the bacteria associated with the disease consistently and proof of its pathogenicity to the plant species under investigation, followed by reisolation of the bacteria and comparison of characteristics have to be accomplished to satisfy Koch's postulates. Application of Koch's postulates to establish the identity of disease-causing agent(s) may not be always possible, as in

Fig. 2.1 Symptoms of cassava angular leaf spot disease (Courtesy of International Development Centre, Ottawa, Canada)

Fig. 2.2 Symptoms of pepper bacterial wilt disease (Courtesy of Asian Vegetable Research and Development Center, Taipei, Taiwan)

Fig. 2.3 Symptoms of rice bacterial leaf streak disease (Courtesy of International Rice Research Institute, Manila, Philippines)

the case of xylem-limited bacteria (XLB) which includes *Xylella fastidiosa* (*Xf*), causing several economically important diseases such as grapevine Pierce's disease, citrus variegated chlorosis and leaf scorch diseases of plum, mulberry, pear, almond and coffee. These bacteria are difficult to isolate from the infected plant tissues in pure culture. The presence of *Xf* in the xylem vessels of infected plants has been visualized by electron microscopic observations. Aggregates of bacteria seem to be attached to the vessel walls by extracellular strands produced by the bacteria that are usually most abundant at the end of rod-shaped bacterial cells (Fig.2.4; Machado 2009). In addition, various tests have been performed to establish the identity of pathogen unequivocally.

In addition to the host plant species from which the bacterial pathogen was first described, one or more additional plant species may be infected under natural or glasshouse conditions. Determination of the host range for the pathogen may provide clues to its identification. Some plant species may be highly sensitive to certain bacterial pathogen(s), expressing the diagnostic symptoms early and such plant host species may be used as bioindicators for the presence of the pathogen in the infected plant or

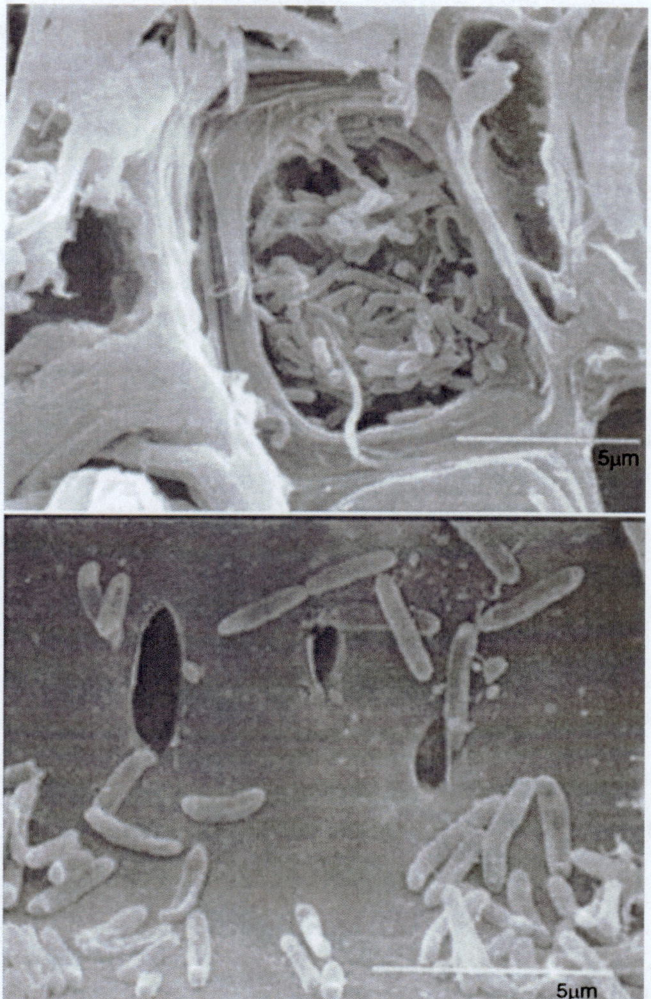

Fig. 2.4 Electron microscopy of citrus xylem tissue infected by *Xylella fastidiosa*. Note the presence of rod-shaped bacterial cells in the xylem vessels (Courtesy of Machado 2009; Agronomic Institute of Campinas, Citrus Research Center, Brazil)

plant materials. Phytopathogenic bacteria induce hypersensitive reaction (HR) in leaf mesophyll tissues, while saprophytic bacteria cannot induce such a response when inoculated on plants. Hence, production of visible necrotic lesions at the site of inoculation of a bacterial species, not only indicates the presence of the target bacteria, but also its pathogenicity on the plant species. Most gram-negative bacteria produce HR in tobacco, whereas gram-positive bacteria induce HR in four-o'clock (*Mirabilis jalapa*) plants. Another host plant species sedium (*Sedium hybridum*, Crassulaceae) was found to be responsive to gram-negaive bacteria such as *Erwinia amylovora*, *E. carotovora* pv. *atroseptica*, *E. chrysanthemi*, *Pseudomonas syringae* pv. *syringae*, *P. syringae* pv. *tomato*, *Ralstonia solanacearum*, *Xanthomonas axonopodis* pv. *malavacearum*, *X. axonopodis* pv. *vesicatoria X. maculogardenia*, and *X. campestris* pv. *oryzae*. In addition, the gram-positive bacteria *Clavibacter michiganensis* pv. *michiganensis* and an unidentified bacterium from irrigation water were able to induce HR in the leaves of sedium. All bacterial pathogens tested, were able to produce necrotic lesions within 24 h after infiltration of bacterial suspension at concentration of 1×10^6 cfu/ml. *S. hybridum* may be useful for detection of other bacterial plant pathogens also (Umesha et al. 2008). *Ralstonia solanacearum* race 4 causing bacterial wilt of edible ginger (*Zingiber officinale*) could be detected by using tissue-cultured edible ginger (TCG) plants which wilted completely within an average of 9 days at the highest (log 8 cfu/ml) level of inoculum and within 19 days at the lowest (log 4 cfu/ml) inoculum level. The bioindicator plants increased the pathogen population rapidly (100–1,000 fold), making the early detection possible (Paret et al. 2009).

Detection and identification of plant bacterial pathogens by biological methods involves isolation of bacteria in a suitable culture medium followed by a series of physiological and biochemical tests and adoption of Koch's postulates to prove the pathogenicity of the bacteria isolated. 'Pathovars' within a bacterial species are identified by their ability to infect one or more differential host plant species and the type/intensity of symptoms induced in the differentials. This approach has been found to be time-consuming and laborious and the results obtained are often inconclusive.

2.1.1.1 Use of Media for Culturing Bacteria

The main problem during isolation of bacteria is to separate the pathogen(s) concerned from the numerous saprophytes that may overgrow the presumptive pathogens on culture media. It is desirable to make judicious initial isolation from a series of symptoms in different stages of disease development on a general (nonselective) medium. A knowledge of basic characteristics of bacterial colonies may be useful in differentiating the target bacterial pathogen from the saprophytes. A general purpose nonselective, but differential medium containing tetrazolium choride is helpful for distinguishing a potential pathogen from the saprophytes. The concentration of tetrazolium chloride above 0.001% is likely to inhibit bacteria belonging to genera such as *Xanthomonas* (Schaad et al. 2002a). If the pathogen population is in sufficient numbers, it will be able to grow on nonselective media. Generally semiselective/selective media may be required for isolation of bacterial pathogens from the soil samples, but nonselective media may be sufficient for isolation of bacteria from plant tissues which can be surface disinfected [Appendix 1].

Suitable selective medium has to be identified for the target bacterial pathogen. Nutrient agar containing 5% saccharose (NASA), NASA + agar Bf were found to be most suitable for isolating and detecting *Pseudomonas syringae* pv. *phaseolicola* in bean (*Phaseolus vulgaris*) (Naumann et al. 1988). *P. viridiflora* causing bacterial streak and bulb rot of onions could be isolated using a semiselective diagnostic medium (T-5), in conjunction with a low temperature incubation (Gaitaitis et al. 1997). *Ralstonia solanacearum* was isolated by using a semiselective medium (PCCG) that supports the development of viable bacterial cells present in the soil (Ito et al. 1998). In the case of *Xanthomonas axonopodis* (*campestris*) *manihotis* causing cassava bacterial blight disease, Cefazolini trehalose agar (CTA) medium was shown to be useful for the detection of this pathogen both in infected cassava leaves and infested soil (Fessehaie et al. 1999). It may be possible to rapidly identify certain presumptive pathogens up to the genus level by using a minimal number of key diagnostic characteristics. Both differential and semiselective media may be used for the second stage of presumptive identification. The bacterial pathogens infecting tomato may be identified by examining the growth characteristics on media that differentially or selectively support the development of various bacterial pathogens (Schaad et al. 2002a; Alvarez 2004).

'*Candidatus* Liberibacter', a phloem-limited bacterium, causing the huanglongbing (greening) disease in citrus was considered as nonculturable. Three species '*Ca. Liberibacter asiaticus*', *Ca. Liberibacter africanus*' and '*Ca. Liberibacter americanus*' have been recognized. A new medium designated Liber A was formulated and successfully employed to cultivate all three '*Candidatus* Liberibacter spp.' This medium containing citurs vein extract and a growth factor sustained the growth of '*Ca.* Liberibacter spp.' for four of five single-colony transfers, before viability declined. The identity of '*Ca. L. asiaticus*' was confirmed by performing a 16S-based rDNA real-time PCR assay and sequencing. The colonies were irregular-shaped, convex and 0.1–0.3 mm after 3–4 days. The presence of fimbriae-like appendages (as in bacteria) was revealed by scanning electron microscopy observations. Pathogenicity of two strains of '*Ca. L. asiaticus*' and one of '*Ca. L. americanus*' grown on Liber A medium was proved by inoculating citrus trees and seedlings, followed by reisolation of the agent from inoculated plants. The identity of the reisolated agent was established by real-time polymerase chain reaction (RT-PCR) assay and sequencing as done for the cultured organism. This is the first successful attempt in culturing this bacterial species and proving its pathogenicity (Sechler et al. 2009).

Pathogenicity test has to be performed for the ultimate verification of the presumptive pathogen isolated from infected plant tissues or other substrates such as soil, water and air. However, it requires a long time, availability of susceptible plant species/cultivar, optimal in vitro (glasshouse) conditions and comparable controls. Leaf spotting pathogens may need relatively shorter incubation period to produce hypersensitive responses (HR) on specific plants such as tobacco. However, certain pathogens like *Clavibater michiganensis* subsp. *michiganensis* may require long incubation periods up to 4 weeks, depending on the strains of this pathogen, making the pathogenicity test to be impracticable (Alvarez 2004).

2.1.1.2 Use of Diagnostic Hosts

Host plant species that can rapidly produce symptoms of infection by bacterial pathogens may be used as diagnostic hosts. A simple, fast and reliable bioassay technique was developed for the detection of *Agrobacterium tumefaciens* (*At*) causing tumors in large number of plant species. Detached leaves of the higly susceptible *Kalanchoe tubiflora* was used as biological baits for trapping tumorigenic cells of *At* from soil as well as from tumor tissues of infected plants. A large and heterogenous microflora are known to be associated with soils and infected plant tissues. However, this bacterial pathogen was alone retrieved by the bait plants and the pathogen could be isolated in pure culture using the nutritional medium (Romeiro et al. 1999).

2.1.1.3 Use of Bacteriophages (Phage Typing)

Bacteriophages (bacterial viruses) infect specific species/strains of a bacterial species, causing lysis of susceptible bacterial cells leading to the formation of plaques (cleared areas) in the bacterial colony in the petridishes. Bacteriophages that infect phytopathogenic bacteria have been isolated from infected plant tissues, soil and irrigation water. *Xanthomonas campestris* pv. *oryzae* (*Xco*) causing rice bacterial blight disease is susceptible to several phages with tadpole-shape (polyheadral head and tail) and a filamentous phage Xf (Kuo et al. 1967). The possibility of detecting *Xco* and its quantification by employing OP_1 phage was reported by Wakimoto (1957). Based on the sensitivity of *Xco* isolates to phages could be detected and differentiated into 15 lysotypes. However, the sensitivity to phages of bacterial strains and their immunological properties did not show any relationship to one another and also to virulence of the strains (Ou 1985). On the other hand, the phage sensitivity of *X. campestris* pv. *malvacearum* (*Xcm*) was related to the virulence of two races causing bacterial blight of cotton. The phage sensitivity of these two races exhibited distinct differences. Race 1 was lysed by three or rarely four of the six phages used for typing, whereas race 2 was sensitive to all six pahges (Freigoun et al. 1994).

The phages CP115 and CP122 were isolated from canker lesions caused by *X. axonopodis* pv. *citri* (*Xac*) on grapefruit and sweet orange respectively. These two phages could be employed for detection and identification of *Xac* (Wu et al. 1993). When contamination of bacteria was excluded by passing the bacteria through a membrane filter (pore size 0.22 μm) and centrifugation of the solution or sample, the effectiveness of the phage in detecting *Xac* was significantly enhanced (Ebisugi et al. 1988). In order to investigate temporal and spatial dispersal of *Xac* on Unshiu orange from known inoculum sources, the presence of *Xac* was detected by employing phage technique. Detection of the pathogen overwintering in canker lesions in spring season was accomplished by applying pathogen-specific phage, as it is important to warn the growers about the possible disease incidence. Based on the assessment of pathogen population by phage technique, it was estimated that over 2% of bacterial pathogen overwintered in canker lesions. Symptoms of infections were first observed 1 month after detection by phage technique. Phages on symptomless leaves were detected and their population was closely related to disease incidence and severity (Myung et al. 2001).

Of the two lysotypes of *X. axonopodis* pv. *citri* (*Xac*), lysotype I was found to be predominant in Korea, as revealed by interaction of the phage CPK with *Xac*. The phage technique and ELISA methods were evaluated for their efficacy in detecting *Xac*. Both methods were equally effective. However, the phage technique provides a distinct advantage of detecting and quantifying only living bacterial cells which is not possible with ELISA method. Furthermore, phage technique is rapid, simple and inexpensive, as the phage lysis zones (plaques) become visible within 18–24 h of incubation. Phage technique is being practically applied in Korea for the detection of *Xac* on Satsuma mandarin fruits meant for exports (Myung et al. 2006). The possibility of employing phage technique for the management of Xanthomonas leaf blight on onion (*Allium cepa*) caused by *X. axonopodis* pv. *allii* was explored, because of the specific lytic activity of the phage on the pathogen. Phage population persisted for at least 72–96 h under field and greenhouse conditions. Weekly or biweekly applications of the phage reduced disease severity by 26–50% and the effectiveness of treatment was equal to or better than chemical application (Lang et al. 2007). The phage detection of bacteria has the potential for use in ecological and epidemiological investigations.

Four different kinds of bacteriophages viz., øRSL, øRSA, øRDM and øRSS, were isolated from the soilborne pathogen, *Ralstonia solanacearum*, causing bacterial wilt diseases of several crops. These pahges had relatively wide host ranges and produced large clear plaques with different host strains. The phage øRSA1 was able to infect all 15 strains of *R. solanacearum* of different races or biovars tested. Three host strains contained øRSA1-related sequences in their genomic DNAs, suggesting a lysogenic cycle of øRSA1. The results of this investigation suggest that the bacteriophages may be useful for both detection and control of bacterial pathogens infecting economically important crops (Yamada et al. 2007).

2.1.2 Biochemical Methods

2.1.2.1 Properties of Bacteria in Culture

Gram staining is an important test based on the results of which the bacteria are classified into two groups. The gram-positive bacteria retain the stain. The Gram's complex consisting of magnesuium-ribonucleoprotein which retains the stain, is not present in the ionic form in gram-negative bacteria whose cells are unable to retain the stain. Among the bacterial plant pathogens, coryneform bacteria including the genera *Arthrobacter*, *Clavibacter*, *Curtobacterium* and *Rhodococcus* are gram-positive, whereas those in other genera are gram-negative. Other characteristics useful for identification of bacteria include the size of bacterial cells, location of flagella, presence of food reserve materials (volutin, fat, glycogen and iogen), ability to reduce nitrates, production of hydrogen sulfide, ammonia and indole, utilization of carbon and nitrogen compounds, starch hydrolysis, lipolytic activity, action on litmus milk and gelatin liquefaction. In addition, tests for determining the activities of Kovoc's oxidase, Thornlely's arginine dehydrolase and tyrosinase are also recommended for gathering information that may be helpful for identification of the bacteria (Narayanasamy 2001).

2.1.2.2 Metabolic Fingerprinting

As the physiological tests involving the determination of different enzymatic activities or fatty acid analysis are time-consuming and labor-intensive, they are not suitable for large scale testing. Based on the differential utilization of 147 carbon sources, a technique was developed by API (API System, Monatiew-Vercelu, France). This system depending on the visualization of growth of target bacterium needs a long time as in *Xanthomonas* spp. which required 1 week for obtaining results. This technique was employed to identify *Pseudomonas syringae* pathovars (Gardan et al. 1984) and to distinguish *X. campestris* pv. pathotypes (Verniere et al. 1991). For the identification of strains of *Brennaria* (*Erwinia*) *quercina* causing bark canker and drippy bud and nut disease in *Quercus ilex* and *Q. pyrenaica*, the API System was applied. The API 20E and API 20NE systems showed that all Spanish isolates of *B. quercina* had similar biochemical characteristics and resembled the reference strain CFBP 1266 (Biosca et al. 2003).

A system of identifying gram-negative bacteria on the basis of metabolic fingerprints was developed by Biolog Inc., Hayward, California. The Biolog automated identification system, is based on the differential utilization of 95 carbon sources by different bacterial species/isolates/strains. A redox dye, tetrazolium violet is used to visualize the increased respiration of bacteria while utilizing a carbon source. Each bacterial species produces a metabolic fingerprint that can be compared to those of known bacteria whose profiles are entered into a database. The Biolog system is conducted in Biolog GP microplates with wells that are inoculated with a predetermined aliquots (150 µl) of target pathogen. The plates are incubated at 25°C for 24 h and they are evaluated on a microplate autoreader (Bio-tek EL 311) at a wave length of 590 nm and outputs are compared by the Micrолog Software (Biolog. release 3.5). The Biolog System provides a printout of identification choices that includes the identification of the target bacterial species followed by the next ten closest species. The Biolog System is relatively much less laborious than other earlier techniques and allows a faster identification than API assay. Biolog's identification system was employed to identify 39 American Type Culture Collection reference taxa and 45 gram-negative isolates from water samples. Within 4–24 h, 98% of the reference strains were identified to genus level (Klinger et al. 1992).

Metabolic fingerprints of 148 strains of *Xanthomonas axonopodis* pv. *citri* originating from 24 countries and associated with various forms of citrus bacterial canker disease (CBCD) were obtained by employing the Biolog substrate utilization system. Metabolic profiles were evaluated for strain identification. Only 6.8% of the strains could be correctly identified, when commercial Microlog 2N database was used alone. The Biolog metabolic fingerprinting technique was found to be useful for differentiating strains associated with CBCD and citrus bacterial spot disease (CBSD). None of the *X. axonopodis* pv. *citri* strains was identified as a strain of *X. axonopodis* associated with CBSD and vice versa. As the accuracy of identification based on Biolog system alone is questionable, it is desirable to combine metabolic fingerprinting with other diagnostic methods (Verniere et al. 1993).

Carbon source utilization patterns of 140 plant pathogenic coryneform bacterial strains were determined using Biolog System. All *Curtobacterium* strains were correctly identified up to the genus level. But identification of *Clavibacte*r strains correctly up to genus level varied from 27% to 77% depending on subspecies. Identification of *Rhodococcus* spp. required additional data. As such, the Biolog database cannot identify all phytopathogenic coryneform bacteria with high accuracy, although this system could reliably identify strains of the genus *Curtobacterium* (Harris-Baldwin and Gudmestad 1996). For the identification of bacteria isolated from spots of pepper plants, utilization patterns of 95 carbon sources on the Biolog GN Microplate were analyzed as per the manufacturer's protocol of Biolog Identification System. Positive and negative growth reactions of bacterial isolates were recorded after 28 h. Analysis of the results with BiologGN Database revealed that all strains from Macedonia were similar to that of *Xanthomonas axonopodis* pv. *vesicatoria*, group A (Mitrev and Kovačevic 2006).

2.1.2.3 Properties of Bacterial Cells

The components of bacterial cells and cellular contents may vary depending on the species of bacteria. The bacteria may be detected and identified based on the variations in the cellular composition and contents.

Direct Colony Thin Layer Chromatography

One loopful of bacterial colony (cell) suspension is applied directly on the origin line on a silica gel thin layer chromatography (TLC) plate (20 × 20 cm or 10 × 20 cm, 0.25 mm thickness, Merck Co., Si 60) and dried completely. The silica gel plate was developed with chloroform-methanol (CM, 2:1, v/v) in well-moistured glass tank at 25°C for 10 min, until the solvent front reached 6 cm line from the origin spots. The first run extracted the lipids from whole bacterial cells. After drying the plates well, bacterial cells on the plate were scraped out. The plate was developed again at the same direction using another solvent, chloroform-methanol-water (CMW, 60:25:4, v/v/v) for about 90 min at 25°C in an incubator. Ninhydrin was sprayed after drying the plate well, followed by heating at 100°C for 10 min for the development of colored spots and photographed. Several bacterial pathogens belonging to *Agrobacterium*, *Clavibacter*, *Erwinia*, *Pseudomonas* and *Xanthomonas* were identified by employing this technique. *E. chrysanthemi* and *E. carotovora* subsp. *carotovora* exhibited marked differences in their lipid profiles. The major types of pseudomonads were distinguished based on lipid profile differences (Matsuyama and Furuya 1993; Matsuyama et al. 1993). Later, the usefulness of the direct colony TLC method for rapid identification of different *Pseudomonas* spp. in ribosomal RNA (rRNA) homology group II was demonstrated by Matsuyama (1995). High performance liquid chromatography (HPLC) technique was employed to detect and differentiate *Clavibacter* and *Erwinia*. Distinct differences were recorded at species level in *Erwinia* (Matsuyama 1995). By applying HPLC technique in conjunction with direct colony TLC method, *Burkholderia gladioli* isolates were identified by distinct peak at Rt 6.2 min and they could be differentiated from the isolates of *B. glumae* and *B. plantarii* which did not show such a peak (Matsuyama et al. 1998).

Phytopathogenic bacteria can be identified and differentiated by thin layer chromatography (TLC) profiles using chloroform-methanol-NaCl or 2-propanol solvent system. Comparison of TLC of aminolipids extracted from 315 phytopathogenic and nonpathogenic bacteria showed that TLC profiles were characteristic at the genus or species level and the results were highly reproducible. For most gram-negative bacterial species, the upper most spot appeared at R_f 0.7 on the chromatogram and this spot was absent on the chromatograms of gram-positive bacteria *Clavibacter michiganensis*. The profiles of 96 isolates of *Ralstonia solanacearum* were identical, irrespective of their sources. In the case of *Erwinia carotovora* an intensive bench-mark spot appeared at R_f 0.64 which was absent on the chromatograms of pathovars of *E. chrysanthemi* and *E. herbicola*. The profile of pathovars of *Pseudomonas syringae* was identical and simple. On the other hand, distinct diversity was reflected in the profiles of *Xanthomonas campestris* and *X. oryzae*. The TLC profiles of aminolipids can be prepared easily and use for presumptive identification and differentiation of bacterial plant pathogens (Matsuyama et al. 2009).

The fatty acid composition of phytopathogenic bacteria has been demonstrated as a basis for their identification. The dendrogram based on fatty acid composition showed that all pathovars of *Pseudomonas syringae*, *P. viridiflora* and pear and radish strains were closely related. Lauric acid and palmitoleic acid were the major fatty acids present in these pathogens. Physiological and biochemical tests also provided similar results indicating that strains of *P. syringae* infected pear blossoms, whereas strains of *P. viridiflora* caused rotting of radish leaves (Khan et al. 1999). Qualitative and quantitative cellular fatty acid analyses were performed for the identification of *Brennaria* (*Erwinia*) *quercina*, causative agent of bark canker and drippy bud and nut disease of *Quercus ilex* and *Q. pyrenaica.* The fatty acid profiles of the Spanish isolates were similar to the strains from California. The fatty acid methyl ester (FAME) profiles of Spanish isolates and the Californian strain were very similar to the type strain NCPPB 1852, confirming the identification of Spainsh oak isolates as belonging to this species (Biosca et al. 2003).

Polyacrylamide Gel Electrophoresis

Manipulating and analyzing DNA are fundamentals in the field of molecular biology. Separating complex mixtures of DNA into different sized fragments by electrophoresis is a well established technique. The isolated DNA from target organisms are treated with restriction enzymes to generate pieces small enough to resolve by electrophoresis in agarose or acrylamide. In conventional electrophoresis involving constant electric field, DNA migrates to a distance that is inversely proportional to the logarithm of its length. For smaller molecules (<5 kb), relatively small differences in length result in larger differences in mobility. Owing to the logarithmic relationship, sensitivity significantly reduces with increasing size of DNA. Polyacrylamide gel electrophoresis (PAGE) is particularly useful to resolve the DNA fragments amplified by polymerase chain reaction.

Two major phenotypic groups could be identified by using PAGE and silver staining of sodium dodecyl sulfate (SDS)-lysed cells of *Xanthomonas campestris* pv. *vesicatoria* (*Xcv*). Broad dark gray bands with molecular weight (MW) of 32.35 and 25.57 kDa designated α and ß are present in different strains. The α-band is present in 192 or 197 tomato race 1 strains and the ß-band is seen in all 55 strains of tomato race 2. Moreover, race 1 strains expressing α-band cannot hydrolyze starch (Amy−) or degrade pectate (Pec−), whereas most strains of race 2 are Amy^+ and Pec^+. Silver staining of protein profiles and testing for amylolytic activity may be useful to differentiate strains of *Xcv* (Bouzar et al. 1994). It is difficult to differentiate the pathovars of *Pseudomonas syringae* merely by biochemical and O-antigen serological reactions. The envelope protein profiles of bacterial cells have provided a reliable basis of differentiation. Three protein bands 60, 65 and 150 kDa present in *P. syringae* pv. *pisi* strains differentiated them from 29 strains of *P. syringae* pv. *syringae* in which they were absent (Malandrin et al. 1996).

A two dimensional polyacrylamide gel electrophoresis (2-D PAGE) of acid ribosome-enriched proteins was used to detect and differentiate *Erwinia* spp. and soft- rotting *Pseudomonas* strains. Electrophoretic profiles of acid ribosome-enriched proteins of selected strains of *E. carotovora* pv. *atroseptica* (*Eca*), *E. carotovora* pv. *carotovora* (*Ecc*), *E. chrysanthemi*, *E. amylovora*, *E. herbicola* and *E. rhapontici* were compared. Seven major clusters were identified. Soft-rotting bacterial strains could be readily identified and differentiated by the 2-D PAGE technique. *Ecc* and *Eca* produced protein profiles that were consistent and distinct enough to be separated into two species. Other unknown soft-rotting isolates also could be reliably identified (Moline 1985).

Pulsed-field gel electrophoresis (PFGE) has been shown to be useful in overcoming the limitations of PAGE and this technique allows the electrophoretic separation of larger molecules in agarose gels. PFGE introduced by Schwartz and Cantor (1984) is capable of resolving extremely large DNA, raising the upper size limit of DNA separation from 30–50 kb to well over 10,000 kb. Application of PFGE are numerous and diverse and these include cloning large plant DNA, using yeast artificial chromosomes (YAC's) identifying restriction fragment length polymorphisms (RFLPs) and determining the number and size of chromosomes (Joppa et al. 1992). Currently PFGE has been applied widely and commercial pulsed field units have been manufactured for routine large scale use. It is now possible for cloning and analysis of a small number of very large pieces of a genome. The genomic analysis of *Erwinia amylovora* strains from the Mediterranean region and European countries was carried out. The PFGE patterns were determined by assaying the *Xba*1 digests of bacterial genomic DNA. The strains from Austria and Czechia were placed along with central European type (Pt1), whereas the strains from Eastern Europe and Mediterranean regions belonged to another group Pt2. The strains from Italy showed patterns of all three types (Zhang et al. 1998).

Acidovorax avenae subsp. *citrulli* (*Aac*) causes a serious disease, bacterial fruit blotch (BFB) in cucurbits in Israel. The isolates of *Aac* have been characterized in terms of their aggressiveness in different hosts by seed, seedling and fruit inoculations and according to their DNA fingerprinting profiles using PFGE and repetitive-polymerase chain reaction (rep-PCR) assays. The PFGE analysis of the isolates of *Aac* showed 23 unique DNA bands and five different profiles, each of which contained 9–13 bands. The isolates were classified into two groups within *Aac*, one including strains that were more associated with watermelon (group II) and the other enclosing strains that were usually found to be associated with nonwatermelon cucurbits (group I). Further, PFGE analysis indicated that the 12 analyzed isolates could be divided into five different haplotypes of which four were found to be new ones. The distinct advantage offered by PFGE over rep-PCR assay is the greater level of reproducibility of results. Further, it is possible to assess the genetic diversity of the bacteria by PFGE analysis more reliably (Burdman et al. 2005) [Appendix 2].

Conductimetric Assay

Detection of bacteria by measuring the conductance of the media in which the phytopathogenic bacteria are grown, has been possible in the case of metabolically active soft-rot pathogens infecting potato. The pectate medium containing pectate as the sole carbon source was used for the specific automated conductimetric detection of *Erwinia chrysanthemi* (*Ech*) and *E. carotovora* subsp. *atroseptica* (*Eca*) in potato peel extracts. The threshold of detection of *Eca* by this procedure showed variations between 10^2 and 10^3 cfu/ml of peel extracts respectively at 20°C and 26°C. It is desirable to confirm the results obtained from conductimetric assay by employing additional tests such as immunoassay or nucleic acid-based assay (Fraaije et al. 1996).

Isozyme Analysis

Bacterial pathogens may be identified on the basis of their isozyme profiles. The patterns of enzyme esterase (EST) and superoxide dismutase (SOD) of *Pseudomonas syringae* pv. *pisi* strains were studied for their usefulness for the purposes of identification and diagnosis of diseases caused by them. Two EST zymotypes specific for this pathogen could be used for their identification. Further, these two EST patterns were correlated to the race structure of this pathovar. Zymotypes corresponded to races 2, 3, 4 and 6, whereas zymotype 2 was correlated to races 1, 5, and 7 (Malandrin and Samson 1998). The applicability of isozyme analysis for the identification of pathovars of other bacterial pathogens has to be examined.

Use of Fluorescent Markers/Bioluminescent Strains of Pathogens

The fluorescent marker gentisic acid (GeA) was employed for the detection of the citrus greening agent in the alcohol extracts of leaves or stem bark by using silica gel thin layer chromatography (Schwarz 1968). This test was reevaluated for its reliability as a method of diagnosing greening disease. The severity of foliar symptoms was found to be correlated significantly with the amount of GeA in young and old bark tissues. Positive results were obtained only, if the GeA levels were more than 300 μg/g of tissue. However, it was suggested that it should be used only along with other diagnostic criteria under glasshouse conditions (Hooker et al. 1993). The presence of fluorescent dye carboxy fluorescein (CF) in galls induced by *Agrobacterium tumefaciens* (*At*) was used to monitor the symplastic movement of *At* by using epifluorescence microscope (Pradel et al. 1999).

Green fluorescent protein (GFP) assay has been applied to detect the presence and distribution of GFP-tagged organisms in their substrates. A bioluminescent strain of *Xanthomonas axonopodis* pv. *diffenbachiae* (*Xad*) causing bacterial blight disease of anthurium was used to detect the latent infection using autoradiophotography (Fukui et al. 1998). In a later study, a bioengineered *Xad* strain containing p519ngfp plasmid was employed to determine levels of resistance of anthurium cultivars to bacterial blight disease. Injection of GFP-tagged *Xad* strain into cut petioles resulted in 100% infection establishment in susceptible cultivars. The GFP assay requires only simple microscopic observation of representative portions of petioles and leaves and hence, it is possible to evaluate large number of plants from a segregating populations in a limited time in a cost-effective manner. Furthermore, the *gfp* gene remained stable for over a 4-year period during this investigation. The correlation coefficient between GFP-fluorescence and eventual death of inoculated anthurium plants was high ($r = 0.90$), indicating that the final death of individual plants can be reasonably predicted based on GFP-fluorescence data at 5 weeks after inoculation (WAI). By using the bioluminescent strain, the time required for screening for resistance could be reduced to 5 weeks from 28 weeks (time required for death of plants). The GFP data was used to detect the latent infection of plants early (Elibox and Umaharan 2007).

2.1.3 Immunoassays

Bacterial pathogens are less complex compared with the fungal pathogens which produce different spore forms that have different surface proteins which are antigenically important and distinct. The bacteria may have various immunodeterminants of capsular polysaccharide antigens, lipopolysaccharide (LPS) O and K antigens and murein lipoproteins that have been characterized in medically important genera of bacterial pathogens. But much less is known about membrane protein envelopes of plant pathogenic bacteria. The membrane proteins were reported to be useful for the identification of *Ralstonia solanacearum* (Schaad et al. 1978) and

Erwinia chrysanthemi (Yakrus and Schaad 1979). Characterization of the inner and outer membrane fractions of *Xanthomonas campestris* pv. *campestris* (*Xcc*) showed that a 44-kDa peptide of the outer membrane of *Xcc* was species-specific (Dianese and Schaad 1982). In a further study, a subspecies-specific antigenic determinant of membrane proteins of *Xcc* was identified and characterized. Sodium dodecyl sulfate–polyacrylamide gel electrophoresis (SDS-PAGE) of membrane proteins revealed the presence of six major peptide bands. Antiserum to purified 125-, 100- and 44-kDa peptides could be used to identify membrane proteins of *Xcc* by using Ouchterlony double-diffusion tests (Thaveechai and Schaad 1986). The sensitivity and specificity of polyclonal antisera can be increased by either diluting the cross-reacting antibodies or by cross-absorbing such antibodies in the antiserum. Marked increase in the specificity of serological reactions may be achieved by use of MAbs, leading to differentiation of strains/pathovars of a pathogen species.

Polyclonal antibodies (PAbs) generated against bacterial antigens posed problems such as inconsistent reproducibility of the reaction, loss of reactivity of a strain after subculturing and colony-type variants with different serological properties. Furthermore, cross-reactions with unrelated species of bacteria are common when PAbs are used, since polyclonal antisera contain mixtures of antibodies with multiple specificities. With the advent of monoclonal antibody (MAb) technology, some of these problems could be overcome. MAbs secreted by hybridoma lines exhibit defined specificity to a single epitope of the bacterial species to be detected, enabling reproducible results consistently in different laboratories. After characterization of MAbs, panels of MAbs may be combined to form a reactive group that can be employed to detect and identify up to genus, species, subspecies and pathovars of a bacterial pathogens (Alvarez et al. 1996).

MAbs produced against certain cellular or extracellular fractions have been used to differentiate serogroups of *Erwinia carotovora* subsp. *atroseptica* or *E. chrysanthemi* (De Boer and McNaughton 1987). A pathovar-specific MAb was employed in an ELISA test to identify the pathogen in an outbreak of bacterial blight in the United States (Jones et al. 1989). MAbs raised against two German isolates of *Xanthomonas campestris* pv. *campestris* (*Xcc*) were employed to subdivide the *Xcc* isolates into three serogroups S1, S2 and S3 (Rabenstein et al. 1999). Serological groups of *Pseudomonas syringae* pv. *coronafaciens* from otas were recognized based on the antigenic properties of the bacterial strains (Pasichnyk 2000). Serological techniques have been applied to detect latent infections that do not exhibit any visible symptoms, as in peanut (groundnut) plants infected by *Ralstonia solanacearum* (Shan et al. 1997) and in cabbage seedlings infected by *Xcc* (Shigaki et al. 2000).

The flagellar extracts of *X. campestris* pv. *campestris* (*Xcc*) were used to raise the PAbs and MAbs. The PAbs were employed to probe the immunoblots to identify bands of flagellar extracts separated by SDS-PAGE technique (Franken et al. 1992). The antibodies raised against flagella from strains representing most pathovars of *P. syringae* were used to differentiate flagella serotypes H1 or H2 (Malandrin and Samson 1999). The MAbs generated against the European potato strains of *Erwinia chrysanthemi* reacted with a fimbrial antigen present in all except two strains of *E. chrysanthemi* isolated from potato, demonstrating the usefulness of the

MAbs for detection of the pathogen not only in potato, but also in other host plant species (Singh et al. 2000).

Gram negative plant pathogenic bacterial pathogens produce lipopolysaccharides that form one of the structural components of the outer membrane of bacterial cells. Antisera specific to lipopolysaccharides have been effectively employed for the detection of *Agrobacaterium tumefaciens*, *Erwinia carotovora* subsp. *atroseptica*, *X. axonopodis* pv. *citri*, *Ralstonia solanacearum* and *Pseudomonas syringae* (De Boer et al. 1996; Ovod et al. 1999; Mc Garvey et al. 1999; Gallo et al. 2000). Many *Pseudomonas* strains elaborate phytotoxic lipodepsipeptides such as syringomycin, syringopeptins and related compounds. The antiserum raised against a macromolecular derivative of syringopeptins (KLH-SP_{25A+B}) detected free syringopeptins (SP_{22} and SP_{25}) and also syringopeptins present in the aqueous extracts of cotyledons of zucchini infected by *P. syringae* pv. *lachrymans* (Fogliano et al. 1999; Gallo et al. 2000). Coronatine (COR) another bacterial toxin is produced by *P. syringae* pv. *atroseptica* (Nishiyama et al. 1977) and *P. syringae* pv. *glycinea* (Mitchell and Young 1978). The MAbs generated against COR could be used to detect the presence of the toxin and quantitatively estimate its concentration (Jones et al. 2001).

Monoclonal antibodies have been produced to provide greater specificity and sensitivity to the immunoassays. Commercial test kits for detection and identification of phytopathogenic bacteria have been developed for application under different conditions. Of the 97 immunodiagnostic tests available during 2004, 66 tests were based on PAbs and 21 tests were based on MAbs, while ten tests incorporated both PAbs and MAbs (Alvarez 2004). Various tests, based on different kinds of serological reactions, have been employed for the detection of bacterial pathogens in different plant parts with varying degrees of success.

2.1.3.1 Agglutination Test

Agglutination test is the simplest of immunoassays, depending on the formation of visible precipitate due to the reaction between the antigen and antibody. This test was performed commonly by earlier workers, before the development of more efficient methods like enzyme-linked immunosorbent assay (ELISA). The agglutination test requires large volumes of antiserum, in addition to its low level of sensitivity. The agglutination test was employed to distinguish different species or strains of *Xanthomonas* spp. (Fang et al. 1950; Patel et al. 1951). A rapid slide agglutination test using PAbs conjugated to *Staphylococcus aureus* cells which have high concentrations of protein A on their surface was employed for detecting *Pseudomonas syringae* pv. *phaseolicola* and *P. syringae* pv. *pisi* infecting bean and pea respectively. Furthermore, this test was found to be useful also for detecting *P. gladioli* pv. *alliicola* and *Lactobacillus* sp. in rotted onion bulbs. The presence of specific strains of *Rhizobium phaseoli* in bean root nodules could also be detected by slide agglutination test (Lyons and Taylor 1990) [Appendix 3]. The bacterial pathogen causing banana blood disease was identified by employing an agglutination test and colony blot test using a monospecific antiserum (MSA) raised against a specific protein associated with the pathogenic agent. In these tests, specific reaction was observed between the MSA and all virulent strains, but only weak reaction with *Ralstonia solanacearum* and other bacterial species tested (Baharuddin et al. 1994). The cells of *S. aureus* were sensitized using the PAbs produced against *Erwinia amylovora*, causative agent of apple fire blight disease. It was possible to detect *E. amylovora* more reliably and efficiently in plant sap and bacterial slime of hawthorns (*Grataegus* sp.) infected by this pathogen by applying the improved agglutination protocol (Mráz et al. 1999). The presence of *X. campestris* pv. *campestris* (*Xcc*) in seed washings of *Brassica* could be detected effectively (Mguni et al. 1999). *Ralstonia solanacearum* (*Rs*), causing the economically important bacterial wilt diseases in several crops, required a rapid and reliable technique for detection and its identification precisely. The *Staphylococcus aureus* slide agglutination test was employed for direct detection of *Rs* in tomato and potato plants under field conditions and on bacterial cultures under laboratory conditions in Portugal. The results obtained indicated that this rapid technique could play a major role in the control programs by providing a sensitive tool for rapid detection of this pathogen (Lyons et al. 2001)

2.1.3.2 Gel Diffusion Test

Gel diffusion tests are based on the ability of the reactants (antigen and antibody) to diffuse towards each other in gel or agar medium. Precipitin lines are formed at the junction of the reactants in equivalent concentrations. Antisera to 23.4, 44 and 125 kDa peptides purified from membrane proteins of *X. campestris* pv. *campestris* (*Xcc*) reacted weakly with membrane proteins in Ouchterlony double diffusion tests. However, after the concentration of antisera by sixfold, intensity of precipitin lines was appreciably increased. All antisera to purified peptides resulted in reactions of identity with membrane proteins of homologous strains, but not with membrane proteins of heterologous strains of *Xcc* (Thaveechai and Schaad 1986). By employing the double diffusion test, the antisera raised against glutaraldehyde-fixed flagella of two strains UQM 551 of *Pseudomonas syringae* pv. *pisi* (*Psp*) and strain L of *P. syringae* pv. *syringae* (*Pss*) were shown to have high levels of specificity in the detection of respective strains. Further, the antisera specific to heat-killed cells of *P. syringae* could be used for distinguishing *P. syringae* from all other bacterial species and genera tested including strains of *P. fluorescens*, *Escherichia coli*, *Agrobacterium* and *Rhizobium* (Marzarei et al. 1992).

2.1.3.3 Enzyme-Linked Immunosorbent Assay

Polyclonal and monoclonal antibodies (PAbs and MAbs) have been widely employed for detection, differentiation and quantification of bacterial pathogens by using enzyme-linked immunosorbent assay (ELISA) and its variants. Generally MAbs generated for a large number of bacterial plant pathogens have been initially screened using an ELISA format. ELISA method has been demonstrated to be useful for testing large number of samples in quality assurance programs in several countries.

ELISA method is amenable for automation and often reproducible results have been obtained in large scale testing programs. The sensitivity of ELISA tests (10^3–10^6 cfu/ml) has been found to be sufficient for detection of bacterial pathogens in symptomatic plants. The sensitivity of ELISA method could be enhanced by tenfold by applying an extraction buffer containing EDTA and lysozyme which liberate LPS resulting in increased antibody-antigen reaction. Further, boiling the bacterial samples to destroy proteins or use of commercial extraction buffers like BEB (Agadia Inc., Indiana, USA) was also able to increase detection sensitivity. Detection kits have been developed by Agadia Inc., and they are capable of detecting more than one pathogen species. *Clavibacter michiganensis* subsp. *michiganensis* (*Cmm*) and *X. axonopodis* pv. *vesicatoria* (*Xcv*) can be detected simultaneously by ELISA format (Alvarez 2004).

The presence of *Xanthomonas campestris* pv. *oryzae* (*Xco*) was detected in leaves and seeds of 60 rice accessions (Zhu et al. 1998). The MAB (Xco-1) specific for *Xco* reacted positively with all 178 strains from diverse geographical locations, but not with *X. campestris* pv. *oryzicola* or other xanthomonads in modified ELISA tests. These strains were classified into groups I, II, III and IV on the basis of reaction with four MAbs. Two of these MAbs reacted positively with *X. campestris* pv. *oryzicola*. However, there was no reciprocal reaction between the MAb specific to *X. campestris* pv. *oryzicola* and any of the four groups of *Xco* strains (Benedict et al. 1989). Based on the serological reactivity of MAbs and PAbs, 63 strains of *Xco* were grouped into nine reaction types consisting of four serovars and seven subserovars (Huang et al. 1993).

An MAb specific to Florida citrus nursery strains of *X. axonopodis* pv. *citri* (*Xac*) was shown to be effective in demonstrating the existence of at least two distinct populations of this pathogen (Permar and Gottwald 1989). The MAb specific to *Pseudomonas andropogonis*, causative agent of carnation leaf spot disease, was employed to detect the presence of the pathogen in carnation cuttings (Li et al. 1993). The possibility of detecting and identifying *X. campestris* pv. *campestris* (*Xcc*) and *X. campestris* pv. *armoraciae* (*Xca*) infecting crucifers under field conditions as well as in seeds was reported by Alvarez et al. (1994). By employing MAbs in plate-trapped antigen (PTA)-ELISA format, *Xcc* serogroups were detected and identified (Rabenstein et al. 1999).

The sugarcane ratoon stunting disease caused by *Clavibater xyli* subsp. *xyli* was detected and identified by employing indirect ELISA protocol. The pathogen was also detected in asymptomatic sugarcane plants/cultivars that may be carriers of the disease (Viswanathan 1997a). The effectiveness of ELISA tests for the detection of *Xanthomonas albilineans* causing sugarcane leaf scald disease, was demonstrated. The ELISA tests revealed the bacterial infection in 99% of symptomatic stalks and in 14% of asymptomatic stalks (Comstock et al. 1998). Detection of *X. campestris* pv. *pelargonii* (*Xcp*) infecting geranium was accomplished by using an MAb specific for *Xcp* in an ELISA format. The intensity of response was moderately correlated ($r = 0.56$) with symptom severity induced by 14 strains of *Xcp* (Chitaranjan and De Boer 1997). The DAS-ELISA format was applied for the detection of *X. axonopodis* pv. *diffenbachiae* using PAbs, the detection limit of this test being 10^4 cfu/ml of the pathogen (Hseu and Lin 2000). The lucerne bacterial wilt pathogen *Clavibacter michiganensis* subsp. *insidiosus* could be detected and identified using PAbs in DAS-ELISA and PTA-ELISA formats (Kokošková et al. 2000).

Bacterial canker disease of citrus caused by *Xanthomonas axonopodis* pv. *citri* (*Xac*) is one of the most important diseases of quarantine concern in European countries and the USA. The ELISA test was applied for the detection of *Xac* on fruits and its efficiency was evaluated along with the bacteriophage test (BPT) that was used earlier. Positive results were obtained in the ELISA tests at bacterial densities of 3×10^5 cells/ml or more. ELISA test could detect *Xac* in all samples (100%), while successful detection of *Xac* by BPT was possible only in about 44% of the samples tested (Jin et al. 2001). Two strains of *Xac* with unique host specificity designated *Xac*-A (Asiatic strains) and *Xac*-A^w (Wellington strain) were tested using an indirect ELISA format and the commercially available MAb A1 specific for *Xac*-A strain. These two strains could be easily differentiated based on the ELISA results, since the MAb A1 reacted only with *Xac*-A strain, but not with *Xac*-A^w strain. Furthermore, these two strains had distinguishable fatty acid profiles, providing additional basis for differentiation of these two strains (Sun et al. 2004). *X. campestris* pv. *vesicatoria* (*Xcv*), causative agent of bacterial spot disease significantly reduces tomato and pepper production. Three specific MAbs 7AH10, 5HB3 and 4AD2 that reacted to lipopolysaccharides of *Xcv* were tested in three ELISA formats. Competitive ELISA format was the most sensitive one capable of detecting *Xcv* populations as low as 10^3–10^4 cfu/ml. The pathogen could be detected in artificially inoculated and field-collected samples, confirming the utility of ELISA test as a diagnostic tool for *Xcv* (Table 2.2) (Tsuchiya and d'Ursel 2004) [Appendix 4].

Strains of *Xanthomonas campestris* pv. *mangiferae-indicae* (*Xcm*), causing mango black spot disease were detected and differentiated into two groups based on pathogenicity, host origin, restriction fragment length polymorphism (RFLP), isozyme analysis and biochemical properties. MAbs were generated against these two groups of strains of *Xcm*. All the Japanese *Xcm* strains of group I reacted equally with MAb 1A7H12G3 which was the most specific for all but one world wide group I strains and to only one strain among group II. This MAb 1A7H12G3 was assayed

Table 2.2 Characteristics of three MAbs produced against stains of *Xanthomonas campestris* pv. *vesicatoria* (Tsuchiya and d'Ursel 2004)

Monoclonal antibodies	Titer (ABC-ABTS)[a]	Detection limit (cfu/ml)[a] Non-competitive ELISA ABC-ABTS	Non-competitive ELISA Anti-IgM-OPD	Competitive ELISA Anti-IgM-OPD
5 HB3	125,000	3×10^5	5×10^4	5×10^3
7 AH10	125,000	5×10^5	5×10^4	1×10^4
4 AD2	24,000	6×10^5	2×10^4	ND

ABTS 2,2′-azino-di-3-ethyl-benzothiazoline-6-sulfonic acid; OPD O-phenylene diamine dihydrochloride; ND Not determined quantitatively

[a] Avidin/biotin (ABC) or rabbit anti-mouse immunoglobulin M (IgM) peroxidase conjugate system was used as the detection system and ABTS or OPD as substrate

on selected leaf samples that were inoculated with *Xcm* strains from groups I and II. The reaction patterns of ELISA tests from infected leaf samples were in total concordance with that of pure cultures.The broad specificity of this MAb could be useful for the detection of the strains belonging to a predominant group of strains of *Xcm* occurring in Japan (Tsuchiya et al. 2004). Polyclonal antibodies generated against *X. axonopodis* pv. *vesicatoria* (*Xav*) were employed in indirect ELISA format using 96-well microtiter plates. Clear positive reactions were recorded for all eight strains from Macedonia and two reference strains of *Xav* (2524 and 1840) from California. *X. campestris* pv. *malvacearum* strain did not react positively with the antibodies specific *Xav*, indicating the specificity of the ELISA format. If the bacterial suspension was heated at 60°C for 30 min prior to performing ELISA test, the reactions could be seen more clearly and faster (Mitrev and Kovačevic 2006).

Clavibacter michiganensis subsp. *michiganensis* (*Cmm*), causing tomato bacterial wilt and canker disease, was detected by employing two polyclonal antisera and a monoclonal antibody in indirect sandwich ELISA format. All three antibodies reacted positively with the bacterial antigens. All *Cmm* strains gave a positive reaction ($OD_{405} > 0.4$) in indirect ELISA test, with the exception of two avirulent strains of *Cmm*. In addition, a strong reaction was observed with all subspecies of C. *michiganensis* and some phytopathogenic corynebacteria also (Dreier et al. 1995). *Erwinia chrysanthemi*, causing soft rot diseases in potato and other crops was detected using the MAb secreted by murine hybridoma cell line (6A6). This MAb reacted with a fimbrial antigen and fibrillin protein of *E. chrysanthemi*, as revealed by immunoglod electron microscopy and Western blotting respectively. In the ELISA test, the MAb reacted with all but two strains of *E. chrysanthemi* isolated from potato. The MAb also reacted with 20 out of 36 strains of *E. chrysanthemi* isolated from hosts other than potato. A triple antibody ELISA test using 6A6 MAb was effective in detecting *E. chrysanthemi* in infected potato stems as well as in tubers. However, the sensitivity was reduced to about 10^7 cfu/ml of pathogen cells (Singh et al. 2000; Appendix 5). *E. amylovora* (*Ea*), causing fire blight disease of apple and other rosaceous plants, is of quarantine importance in many countries where strict quarantine regulations are enforced. PAbs marketed by Neogen Europe Ltd. were employed in plate-trapped antigen (PTA)-ELISA for the detection of *Ea* in plant materials. The detection limit of the test was 10^5 cfu/ml of pathogen. Samples (63) from symptomatic plants yielded pure cultures of *Ea* and it was possible to detect *Ea* in these samples by PTA-ELISA format. However, there were cross-reactions (20%), when tested by PTA-ELISA format (Kokošková et al. 2007).

Ralstonia (*Pseudomonas*) *solanacearum* (*Rs*) causes bacterial wilt diseases in potato, tobacco and other crops. Latent infection of peanut (groundnut) by *Rs* could be detected by using ELISA method. The presence of the pathogens in the main root, hypocotyls, and stems of infected plants was revealed by ELISA method (Shan et al. 1997). PAbs generated against virulent strains of *Rs*, the cells of which are encapsulated with mucin, were employed. The ELISA format was efficient in detecting as few as 100 cells/ml. The PAbs specifically reacted with tomato isolate of *Rs*, but not with isolates from pepper (chilli) or aubergine (eggplant). The reactivity of the PAb in ELISA correlated well with the degree of infection in tomato seeds and plants (Rajeshwari et al. 1998). As the use of PAbs to *R. solanacerum*

(*Rs*) resulted in false-positive results due to the presence of cross-reacting saprophytic bacteria, attempts were made to produce recombinant single-chain antibodies (scFv) against the lipopolysaccharide of *Rs* biovar 2, race 3. The scFv antibodies were successfully selected by phage display from a large combinatorial antibody library. Antibodies with improved efficiency compared with PAbs were employed for detecting *Rs* causing brown rot disease in potato. Only a few cross-reactions with saprophytic bacteria were observed. Using ELISA test, as few as 5×10^3 bacteria in potato tuber extracts could be detected (Griep et al. 1998).

The ELISA technique was employed to establish the identity of the bacterial pathogen causing coffee leaf scorch disease. The leaf extracts from infected coffee plants reacted positively with the antiserum raised against *Xylella fastidiosa* (*Xf*) causing citrus variegated chlorosis (CVC) disease. The presence of the bacteria in the xylem of artificially inoculated coffee plants was detected by ELISA method, indicating the relationship between the pathogens causing diseases in coffee and citrus (de Lima et al. 1998). The MAbs specific to *Xf*, causative agent of pear leaf scorch disease, detected the pathogen in infected coffee plants. The MAbs showed no cross-reaction to 14 other bacterial strains belonging to nine genera (Leu et al. 1998). *X. fastidiosa* was also detected in oleander showing leaf scorch symptoms (Purcell et al. 1999). In the case of almond leaf scorch disease considered as a serious threat to almond production in San Joaquim Valley, California, ELISA procedure was found to be equally efficient as PCR assay in detecting *Xf* in asymptomatic almond trees during the field survey. The infected almond trees showed the presence of mixtures of strains of *Xf*, indicating the possibility of variations in the effectiveness of using certain antibodies (Groves et al. 2005).

Effective disease management relies critically on the accuracy of detection of pathogens prior to development of visible symptoms of infection. The efficiency of ELISA method in detecting *X. fastidiosa* (*Xf*) in whole tissue samples and xylem fluid samples from grapevine plants infected by Pierce's disease (PD) was assessed. ELISA method was more sensitive in detecting *Xf* in xylem fluids (40%) compared with detection in whole tissue samples (20.5%) in asymptomatic grapevine plants. ELISA and polymerase chain reaction (PCR) procedures had similar frequencies of detection of *Xf* (Bextine and Miller 2004). The incidence of PD within an American hybrid vineyards was assessed by employing ELISA test to detect *Xf* in the extracts of petioles of varieties Black Spanish, Blanc du Bois and Cynthia. Individual Cynthia vines had higher *Xf* populations as determined by ELISA tests and these vines had correspondingly lower yields ($p = 0.06$). A dramatic increase in bacterial population in Cynthia vines occurred over growing season. In contrast, *Xf* population levels in Black Spanish and Blanc du Bois varieties remained lower over all season, despite high disease pressure. The infection of six additional host plant species by *Xf* was also detected by ELISA tests (Buzombo et al. 2006). Grapevine plants were tested by ELISA for the presence of *Xf* after allowing infective smoke sharp shooter nymphs and apache cicada adults to feed on the test plants. Transmission of *Xf* by these insects to healthy grapevine plants was confirmed by ELISA tests (Krell et al. 2007). The ELISA technique was employed to detect *Xf* in grapevine plants at different periods after inoculation. *Xf* was detected in 96% of Thompson

Seedless and 100% of Redglobe inoculated shoots at 6 weeks post-inoculation and in 100% of inoculated shoots at 12 weeks post-inoculation. Detection of *Xf* in inoculated shoots in comparison with noninoculated shoots indicated that bacterial movement was both acropetal and basipetal with more acropetal movement detected earlier than basipetal movement (Krell et al. 2008).

Brennaria quercina, described earlier as *Erwinia quercina*, was isolated from *Quercus ilex* and *Q. pyrenaica* trees exhibiting bark canker, drippy bud and drippy nut symptoms. Physiological and biochemical characterization of the isolates from Spain, indicated the similarity with *B. quercina*. Fatty acid profiles also revealed that Spanish isolates were similar to *B. quercina* from California. The extent of relationship of *B. quercina* isolates was evaluated by indirect ELISA format. Only the antiserum for the Spanish isolate (1467a) reacted with all Spanish isolates of *B. quercina*. The Californian strain (CFBP 1266) from drippy nut exhibited a higher degree of relatedness with isolates from bark canker than with the Spanish isolates from corn and leaf bud exudations (Biosca et al. 2003).

2.1.3.4 Dot Immunobinding Assay

Dot immunobinding assay (DIBA) is quite similar to ELISA test in principle. The antigen is immobilized on nitrocellulose or nylon-based membrane instead of microtiter plates used for ELISA test. Detection of bacterial pathogens in asymptomatic plants is essential in order to restrict the incidence and spread of the disease(s) by eliminating infected plants/plant materials as in the case of citrus canker disease. The pathogen *Xanthomonas axonopodis* pv. *citri* (*Xac*) could be detected in 38.4% of asymptomatic citrus plants. The detection threshold was 1×10^4 cfu/ml. All isolates (26) of *Xcv* tested reacted positively providing reliability in the results obtained (Wang et al. 1997). Application of DIBA technique established that *X. fastidiosa* (*Xf*) was the causative agent of coffee leaf scorch disease. Antisera raised against the whole bacterial cells grown in artificial medium reacted positively with extracts from infected coffee plants. The antiserum cross-reacted with *Xf* isolated from citrus variegated chlorosis disease-infected plants, indicating the relationship between these two strains of the pathogen species. The threshold of detection of DIBA test was 5×10^5 bacteria/ml (de Lima et al. 1998).

2.1.3.5 Immunofluorescence Test

Immunofluorescence (IF) tests depend on the fluorescence emitted from the labeled antibodies that are able to react with epitopes present in the antigen or antigenic sites. *Xanthomonas albilineans* (*Xa*), causing sugarcane leaf scald disease was detected and differentiated by employing immunofluorescence test. The strains (215) of *Xa* were grouped into three serovars. Serovar I was the largest including the strains from Australia, Guadeloupe, India, Mauritius, South Africa and the United States, whereas serovar II included strains from Africa-Burkino Faso, Cameroon, Kenya and Ivory Coast. The serovar III enclosed strains from Asia (Sri Lanka), Caribbean Islands and

Oceania (Fiji) (Rott et al. 1994). The rhodamine-labeled oligonucleotide probe in conjunction with an indirect IF protocol based on specific MAb was applied. This MAb with a FITC-labeled conjugate was efficient in providing an improved detection of *Clavibacter michiganensis* subsp. *sepedonicus.* Bacterial cells labeled simultaneously with the oligonucleotide and antibody probes could be precisely identified by microscopic examination. This procedure may be useful especially when isolation and other methods of establishing the identity of the bacterial pathogens are difficult or not possible (Li et al. 1997). *Ralstonia solanacearum* was more efficiently detected in potato tissues by employing MAbs prepared from phage display library against the lipopolysaccharides of the pathogen (Griep et al. 1998).

Erwinia amylovora (*Ea*), the fire blight pathogen was detected by employing the PAb conjugated with an anti-rabbit FITC stain in immunofluorescence antibody staining (IFAS) technique. This method appeared to be very sensitive in detecting *Ea* in 36 samples from symptomatic plants. *Ea* was detected in 100 and 89% of samples at 1:100 and 1:1,000 dilutions respectively. IFAS technique was found to be more sensitive than PTA-ELISA procedure which was not able to detect *Ea* at a dilution of 1:1,000 of plant extracts (Kokošková et al. 2007).

The flagellins present in the flagella of *Pseudomonas syringae* pathovars were characterized by employing immunofluorescence staining, SDS-PAGE and immunoblotting techniques. Two serotypes were differentiated based on the nature of flagellins. The characteristics of flagellins were used as the basis of identification (Malandrin and Samson 1999). The MAbs generated against *Erwinia chrysanthemi* showed specificity in their reactions with the pathogen. In contrast, the PAbs cross-reacted with most strains of *E. carotovora* strains tested (Singh et al. 2000). Spanish isolates of *Brennaria* (*Erwinia*) *quercina* causing bark canker, drippy bud and drippy nut disease of *Quercus* spp. were analyzed by IF test. Most of the Spanish isolates reacted positively against the antisera, indicating the presence of many antigens in common, regardless of their geographical origin and host plant species. However, some antigenic heterogeneity was observed among them (Biosca et al. 2003).

For the detection of *Xylella fastidiosa* (*Xf*), causative agent of Pierce's disease (PD) of grapevine, two immunofluorescence (IF) procedures, were performed by tissue sectioning and membrane entrapment of *Xf* from inoculated grapevine hybrid selection of F8909-08. The tissue sections and bacteria-trapped polycarbonate membranes were incubated with specific polyclonal IgG and stained with fluorescein isothiocyanate (FITC)-conjugated IgG from rabbits to *Xf* cells. The stained preparations were observed by fluorescence microscopy. Detection and identification of *Xf* in thin cross sections of the petiole was possible at 3 weeks after inoculation. In addition, observations could be made at different periods after inoculation to study multiplication and invasion of xylem vessels by *Xf.* Using the membrane entrapment protocol, the pathogen at low concentrations could be isolated from asymptomatic plants (Buzkan et al. 2005). Indirect immunofluorescence (IF) microscopy procedure was applied for the detection of *Xf* infecting American hybrid vineyards in the Texas Gulf Coast. The infection of grape varieties Black Spanish, Blanc du Bois and Cynthia by *Xf* was detected by indirect IF format. The results of *IF* microscopy were similar to that of ELISA tests (Buzombo et al. 2006).

2.1.3.6 Immunolabeling and Electron Microscopy

Immunolabeling and electron microscopy techniques are very useful to visualize immunological reactions on electron microscope grids and to reveal the localization and distribution of various proteins of pathogen/host plant origin in different pathosystems. Enzyme-labeled antibodies may be applied after embedding and thin sectioning of plant tissues or they may be allowed to diffuse inside fixed cells and to interact with antigenic sites prior to thin sectioning. The monoclonal antibodies used for detection of *Xanthomonas campestris* pv. *oryzae* (*Xco*) were characterized by employing immunosorbent electron microscopy (ISEM) and immunofluorescent (IF) techniques (Benedict et al. 1989). The immunogold labeling technique has been shown to be useful for visualization of the bacterial pathogen per se or in situ in infected plant tissues. Anovich and Garnett (1989) demonstrated the effectiveness of immunogold labeling for rapid and specific detection of the causative agent of citrus greening disease. Immunogold labeling of whole cell lysates of *Pseudomonas andropogonis* was found to be useful for characterizing the epitopes located within the bacterial cell (Li et al. 1993). Some of the pectolytic *Erwinia* spp. possess nonflagellar appendages called as fimbriae that are considered to play role in the adhesion of bacterial cells to the plant surface/substrates. By employing fimbrial-specific MAbs, the presence of fimbriae in bundles in *E. chrysanthemi* was revealed by observations under electron microscope. A specific immunogold labeled MAb targeted a fimbrial epitope (Fig. 2.5). All

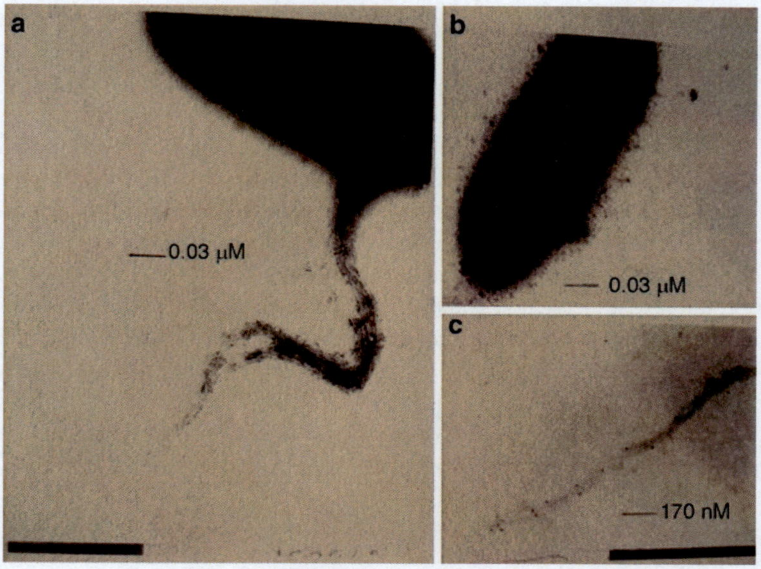

Fig. 2.5 Electron micrographs of *Erwinia chrysanthemi* treated with immunogold preparations containing monoclonal antibody (MAb). (**a**) Whole cell showing attached bundle of fimbriae; (**b**) whole cell showing single attached fimbria; (**c**) Magnified view of fimbria (Courtesy of Singh et al. 2000; The American Phytopathological Society, St. Paul, MN)

European potato isolates of *E. chrysanthemi* could be detected by using MAbs specific to fimbrial antigens (Singh et al. 2000; Appendix 6).

2.1.3.7 Immunomagnetic Fishing/Isolation of Bacterial Pathogens

Immunomagnetic 'fishing' procedure is useful to improve recovery ratio between the target and nontarget bacteria present together in samples. Many plant samples, particularly seeds and soils are contaminated commonly with many saprophytic bacteria which may frequently predominate, making the isolation of bacterial pathogen(s) difficult. The suspension containing the bacterial pathogen *X. campestris* pv. *pelargonii* (*Xcp*) and nontarget bacteria was incubated with rabbit PAb raised against *Xcp* for 1 h. Paramagnetic iron oxide particles coated with goat anti-rabbit IgG were mixed with the suspension and incubated. A polished surface of neodymium supermagnet (14 mm diameter) was placed at the air–water interface, so that magnetic particles–bacteria–antibody complex could be attracted to the magnet. After 'fishing out' all magnetic particles, the magnet was dipped into the sterile buffer to remove nontarget bacteria. The target bacteria firmly attached to the magnet were dislodged by gently rubbing over the surface of suitable nutrient agar medium that can support the development of target bacterial species. By following this procedure, the saprophytic bacterial population was reduced to 11.4% of the initial population, facilitating the isolation of target bacterial pathogen population (Jones and van Vuurde 1996).

Immunomagnetic separation (IMS) procedure involves selective separation of target bacterial pathogen species and cultivation of the separated pathogen in a selective medium that favors the pathogen development. *Erwinia carotovora* subsp. *atroseptica* (*Eca*), causing soft rot disease of potato tubers was detected by IMS technique. *Eca* was separated from the potato peel extract by applying a combination of advanced magnetics (AM)-protein A particles-anti-rabbit IgG particles and the selective medium, crystal violet pectate medium supplemented with 100 µg/ml of streptomycin. A streptomycin tolerant strain of *Eca* was included to reveal the effect of adverse effect of streptomycin, if any, on the target *Eca* strain. The potato soft rot bacteria could be enumerated in potato tuber extract consistently. The detection limit was 100 target bacterial cells/ml (Van der Wolf et al. 1996). A similar approach was made for identification of *Xylella fastidiosa* (*Xf*) present in the tissues of leafhopper vectors *Graphocephala coccinea* and *G. versuta*. IMS procedure was applied for the separation of *Xf* from insect tissue extract followed by nested PCR assay for detection and identification of the leaf scorch pathogen (Pooler et al. 1997).

2.1.3.8 Flow Cytometry

Flow cytometry (FCM) technique has been shown to be useful for rapid identification and quantification of cells or other particles as they move through a sensor in a liquid system. Bacterial cells can be identified by employing fluorescent dyes conjugated to specific antibodies. Multiple cellular parameters are determined simultaneously based on the cell's fluorescence and its ability to scatter light. The results may be confirmed by further confirmatory tests after sorting out the cells electronically or culturing the subpopulations of selected cells. Simple steps are involved in sample preparation. Large particles in the cell suspension are removed by filtering, followed by staining with fluorochrome-labeled antibodies. Vital stains are used as fluorescent markers to indicate cell viability. Carboxy fluorescein diacetate and Calcein AM are used for green fluorescent staining to indicate viable cells, whereas propidium and hexidium iodide are employed for red fluorescent staining to show dead cells (Alvarez 2001; Van der Wolf and Schoen 2004). Fluorescein diacetate (FDA), when absorbed into the cells, is cleaved by nonspecific esterases to release fluorescein, a polar fluorescent compound which is retained inside the cells. Thus, the cell viability can be correlated with the ability of the cells to accumulate fluroescein due to esterase activity. On the other hand, propidium iodide (PI) can pass through damaged cell membranes only and form complexes with ds-DNA and RNA. PI-stained cells are assumed to be nonviable (Chitaara et al. 2006).

Flow cytometry (FCM) was evaluated as a tool for rapid detection and quantification of *Xanthomonas campestris* pv. *campestris* (*Xcc*), causing black rot disease of crucifers. *Xcc* cells were labeled with a mixture of specific fluorescein isothiocyanate (FITC)-MAb in pure culture of *Xcc*, in mixed cultures of *Xcc* with either a common saprophyte *Pseudomonas fluorescens* (*Psf*) or a nonpathogenic *X. campestris* (*Xc*) isolate and in crude extracts. The *Xcc* cells labeled with FITC–conjugated MAbs (18G12, 2F4 and 20H6) were detected and quantified rapidly, the limit of detection being 10^3 cfu/ml in pure culture and in mixed culture with *Psf.* The presence of *Xc* cells did not affect the sensitivity of the test. FCM test was able to distinguish *Xcc* cells from other bacteria and small particles present in the seed extracts, because of the high intensity fluorescence of labeled cells (Chitaara et al. 2002).

The viability of cells of *Clavibacter michiganensis* subsp. *michiganensis* (*Cmm*), causing tomato bacterial canker disease, was assayed by employing the fluorescent probes Calcein acetoxy methyl ester (Calcein AM), carboxy-fluorescein diacetate (cFDA) and propidium iodide (PI) in combination with flow cytometry (FCM). Heat-treated and viable (untreated) *Cmm* cells labeled with Calcein AM, cFDA, PI or combinations of Calcein AM and cFDA with PI could be distinguished based on their fluorescence intensity in FCM analysis. Untreated *Cmm* cells exhibited relatively high green fluorescence levels due to staining with either Calcein AM or cFDA. In contrast, only *Cmm* cells damaged by heat treatment showed high red fluorescence levels due to staining with PI. Furthermore, FCM allowed a rapid quantification of viable *Cmm* cells labeled with Calcein AM or cFDA and heat-treated cells labeled with PI. The number of labeled cells detected by FCM analyses was always higher than the number of cfu/ml determined by the plate count procedure, implying that a large proportion of *Cmm* cells were enzymatically active, but only about 24–55% of stained cells were able to form colonies. However, both plate count and FCM results were positively correlated with the percentage of viable *Cmm* population. FCM analyses of Calcein AM stained cells may be used as a viability indicator, as this procedure was found to be superior to the plate count method, in addition to the shorter assay time required for FCM analyses (Chitaara et al. 2006; Appendix 7).

2.1.4 Nucleic Acid-Based Techniques

Nucleic acid (DNA)-based techniques have been demonstrated to be generally more sensitive, specific and reliable for the detection, identification and quantification of bacterial plant pathogens, than other methods. Closely related organisms share greater nucleotide similarity than those that are distantly related. A highly specific nucleotide sequence in the bacterial genome can be identified. Probes or primers based on the sequences of the nucleotides present in the DNA fragment may be employed as probes or primers for specific hybridization or amplification. Among the nucleic acid-based diagnostic techniques, polymerase chain reaction (PCR) assay or its variants have been used very widely for detection of bacterial pathogens in pure cultures or in single/multiple infections of plant hosts. One of the most important limitations of the nucleic acid-based techniques is their inability to differentiate live pathogen cells from the dead ones. The usefulness and applicability of different nucleic acid-based techniques under field conditions are discussed below.

2.1.4.1 Nucleic Acid Hybridization Methods

Development of specific DNA probes has been useful to rapidly detect, identify and quantify bacterial pathogens belonging to different species, subspecies, pathovars or strains, especially those bacterial pathogens transmitted through seeds, and other planting materials. The diagnostic DNA probes may show specificity at the genus, species, pathovar or race level. A probe that can be used for taxonomic comparisons is prepared by identifying a DNA fragment that is present only in the bacterial species to be detected/identified, but not in other closely related species. The DNA sequences of 16S ribosomal RNA of different bacterial species have been compared for selecting probes specific at genus level. DNA sequences of 16S ribosomal RNA of different bacterial species have been compared for selecting probes specific at genus level. A DNA probe for comparing partial sequences of 16S rRNA from 52 strains of bacteria including *Xanthomonas oryzae* pv. *oryzae* (*Xoo*) was developed. The rRNA molecule is present in large numbers (>10,000 copies/cell) in actively growing bacterial cells. The sensitivity of the probe derived from unique 16S rRNA sequences is greatly enhanced, because of the presence of large number of copies in a bacterial cell. A DNA sequence based on the 16S rRNA which hybridized only with plant pathogenic pathovars of *Xoo* could be identified (De Parasis and Roth 1990).

Variations in the sensitivity of dot blot hybridization procedure for the detection of bacterial pathogens may be noted depending on the probe and target bacteria and the minimum number of cells required for detection which may vary from 200 to 10^6 cfu/ml. Detection of *Pseudomonas syringae* pv. *phaseolicola*, causing bean halo blight disease was possible by using a semi-selective medium to support rapid multiplication of the pathogen, followed by concentration of bacteria before spotting on the nitrocellulose membrane (Schaad et al. 1989). *Xanthomonas axonopodis* pv. *vesicatoria* and *X. vesicatoria*, causative agents of bacterial leaf spots in tomato and peppers could be detected by employing a 1.75 kb fragment (KK 1750) that preferentially hybridized to both pathogens (Kuflu and Cuppels 1997). *Ralstonia solanacearum* race 3 biovar 2 was detected in potato tissues by using fluorescent *in situ* hybridization (FISH) protocol. The probe RSLOB was shown to have the potential for specific detection of *R. solanacearum* in pure cultures as well as in potato tissues which exhibit strong fluorescent signal (Wullings et al. 1998). Citrus greening disease pathogen, *Candiatus* Liberibacter was detected by employing pathogen-specific DNA fragment (0.24 kb) labeled with biotinylated nucleotides. This pathogen was detected in mandarins, tangerins, oranges and pummelos (Hung et al. 1999).

Clavibacter michiganensis subsp. *michiganensis* (*Cmm*), causative agent of bacterial wilt and canker disease of tomato has a pathogenicity gene (*pat-1*) and a gene for an endocellulase carried on two plasmids pCM1 and pCM2. Specific detection of *Cmm* was accomplished by employing Southern hybridization with DNA probes derived from the plasmid-borne *pat-1* and *celA*. The *celA* probe detected and differentiated *Cmm* from other subsp. *insidiosus*, *nebraskensis*, *tesellarius* and *sepedonicus*. On the other hand, the probe derived from *pat-1* gene distinguished virulent from avirulent strains of *Cmm* (Dreier et al. 1995). In another investigation, *C. michiganensis* subsp. *sepedonicus* (*Cms*), causing potato ring rot disease, was detected by using radio-labeled probes obtained from unique three single copy DNA fragments designated Cms 50, Cms 72 and Cms 85 isolated from CS3 strain of *Cms* by subtraction hybridization. These probes specifically hybridized to all North American strains of the pathogen tested by Southern hybridization. All strains including plasmidless and nonmucoid strains of *Cms* could be detected reliably (Mills et al. 1997). A tissue blot hybridization format was developed for the detection of sugarcane ratoon stunting disease pathogen *Clavibacter xyli* subsp. *xyli* (*Cxx*) by employing a 560-bp PCR amplified product as a probe. This probe was amplified from the intergenic region of the 16S/23S rDNA of *Cxx* with two universal primers. The protocol developed in this investigation may be suitable for wider application along with other diagnostic techniques (Pan et al. 1998).

Xylella fastidiosa (*Xf*) infects several economically important crops including grapevine and citrus. Fluorescence in situ hybridization (FISH) technique is based on the comibination of both specificity and adjustable to different phylogenetic levels with suitable oligonucleotide probes and detection and direct enumeration of the bacteria without the need for pure cultures. The probes were designed based on the sequences of the 16 rRNA gene of *Xf*. The probes S-S-X.fas-0067-a-A-18 and S-S-X.fas-1439-a-A-18 hybridized only with the DNA of 12 *Xylella* isolates and 15 closely related bacteria. These probes were able to detect and quantify *Xf* cells in sap from infected plants with an average of $2.9 \pm 0.3 \times 10^6$ cells/ml from three citrus orchards (Rodrigues et al. 2006).

2.1.4.2 Restriction Fragment Length Polymorphism

The genomic DNA of bacterial pathogens may be subjected to the restriction endonuclease activities leading to the production of unique banding patterns that can be visualized after gel electrophoresis. The restriction fragment length polymorphism

(RFLP) analysis of DNA fragments formed due to the cleavage of pathogen DNA by restriction endonucleases at specific sites, has been useful for the detection and differentiation of some bacterial species and their strains. The genomic DNAs obtained from 21 strains of *Xanthomonas axonopodis* pv. *citri* (*Xac*), 14 strains of *X. axonopodis* isolated from Florida citrus nurseries and ten strains of five pathovars of *X. campestris* were analysed by RFLP technique. Differentiation of *Xac* and *Xa* strains could be made out, indicating that the *Xa* strains occurring in citrus nurseries were not a form of strain of *Xac* (Hartung and Civerolo 1989). The cosmid clone PXCF 13–38 isolated from the genomic library of *Xac* covers almost the entire *hrp* genes cluster (involved in HR and pathogenicity). This clone was employed as a probe for RFLP analysis of xanthomonads. Based on RFLP analysis, a dendrogram different from the ones based on biochemical properties was generated and it was found to be useful for the identification of various xanthomonads (Kanamori et al. 1999).

Digestion of genomic DNA of *X. oryzae* pv. *oryzae* (*Xoo*), causing rice bacterial blight disease, with restriction enzyme *Pst*1 resulted only in a few high molecular weight DNA fragments which formed distinct patterns for different strains on electrophoresis. The strains of race 2 of *Xoo* commonly prevalent in the Philippines, could be distinguished by RFLP analysis. The strains of *Xoo* present in the United States were considered to be not closely related to the Asian strains, because of the differences in the RFLP patterns. It was suggested that rice bacterial blight disease observed in the US may be due to a distinct pathovar of *X. oryzae* (Leach and White 1991). RFLP analysis of ITS region (5S plus ITS1 and ITS2) of rDNA repetitive units amplified by PCR was useful in detecting and distinguishing *Burkholderia* spp. causing diseases in rice. *B. glumae*, *B. gladioli*, *B. plantarii* and *B. vandii* were identified based on distinct RFLP patterns. The DNA samples from rice leaf sheath tissues inoculated with *B. glumae* and *B. gladioli* and the cultured bacterial cells of the two species were subjected to digestion with *Hha*I and *Sau*3AI restriction enzymes followed by PCR-RFLP analysis. The patterns formed from the digests of bacterial cells in culture and tissue extracts were identical, indicating that pathogens present in naturally infected rice plant tissues can be reliably identified by RFLP technique (Ura et al. 1998).

2.1.4.3 Polymerase Chain Reaction

It may be difficult to detect by hybridization method, if the target bacterial species is contaminated with saprophytic bacteria. In such situations, polymerase chain reaction (PCR)-based assay may be effective for the detection of the target bacterial species concerned. In addition, sensitivity of detection is also enhanced significantly by employing PCR assay. For example, a PCR-based assay using the primers from DNA sequences of the phaseolotoxin gene was able to detect efficiently the pathogen *Pseudomonas savastanoi* pv. *phaseolicola*, causing bean halo blight disease, even in the presence of high populations of nontarget bacteria. The assay was highly sensitive with a detection limit of 1–5 cfu/ml of seed-soak wash (Prossen et al. 1991). Primer pairs based on the sequences of nucleotides of genes coding for virulence, enzyme production, or toxin production are employed to detect the bacterial pathogens concerned (Table 2.3).

Xanthomonas campestris pv. *phaseoli* (*Xcp*) was detected by employing primers from plasmid DNA, in PCR assay which had a detection limit of 10–100 fg of *Xcp* DNA (equivalent of 1–10 cfu) (Audy et al. 1994). A primer pair that amplified a 222-bp DNA fragment of *X. axonopodis* pv. *citri* (*Xac*) was employed for the first

Table 2.3 Nucleotide sequences for designing primers for amplification of pathogen-specific DNA

Bacterial pathogen	Target sequences	References
Acidovorax avenae subsp. *citrulli*	16S rRNA	Walcott and Gitaitis (2000)
Agrobacterium tumefaciens	T-DNA	Cubero et al. (1999)
Burkholderia spp.	ITS1 and ITS2 regions of rDNA	Ura et al. (1998)
Candidatus Liberibacter	16S rDNA	Coletta-Filho et al. (2005)
Clavibacter xyli subsp. *xyli*	Intergenic spacer region of 16S–23S rDNA sequences	Pan et al. (1998)
	16S rDNA sequences	Bereswill and Geider (1996)
Erwinia amylovora	50-mer Oligonucleotides	Merighi et al. (2000)
	pEA 71 primers	Taylor et al. (2001)
E. carotovora subsp. *atroseptica*	PEL (pectate lyase) encoding gene	Helias et al. (1998)
E. carotovora subsp. *carotovora*	16S-23S rDNA sequences	Toth et al. (1999)
E. herbicola pv. *gysophilae*	Cytokinins or IAA biosynthetic gene sequences	Manulis et al. (1998)
Pseudomonas savastanoi pv. *phaseolicola*	Phaseolotoxin (*tox*) gene	Prossen et al. (1991)
	argK gene sequences	Rico et al. (2003)
		Mosqueda-Cano and Herrera-Estrella (1997)
P. syringae pv. *atropurpurea*	Gene sequences associated pathogenicity and coronatine synthesis	Takahashi et al. (1996)
P. syringae pv. *syringae*	*syrD* gene sequences	Bultreys and Gheysen (1999)
Ralstonia (*Pseudomonas*)	16S rDNA sequences	Seal et al. (1999)
solanacearum	16S–23S rRNA sequences	Pastrik et al. (2002)
Streptomyces scabies	16S rDNA sequences	Lehtonen et al. (2004)
Xanthomonas axonopodis	IGS region between	Miyoshi et al. (1998)
pv. *citri*	16S and 23S rDNA	Taylor et al. (2002)
	rRNA operon 468-bp fragment of	Cubero and Graham (2002)
	plasmid DNA	Sun et al. (2004)
	rpf gene cluster	Colletta-Filho et al. (2006)
	pth gene	Cubero and Graham (2002)
X. axonopodis pv. *glycines*	DNA sequences encoding glycinecin	Oh et al. (1999)
X. campestris	*hrp* gene	Berg et al. (2005)
X. campestris pv. *carotae*	350-bp pathogen DNA fragment	Meng et al. (2004)
Xylella fastidiosa	16S-23S rRNA spacer region	Costa et al. (2004)

time to diagnose a field outbreak of citrus canker disease (Gillings et al. 1995). In order to standardize robust molecular diagnostic procedures for the identification of pathogens such as *Xac* that are considered to pose biosecurity risk to New Zealand, PCR analysis of *Xanthomonas*-like bacteria isolated from citrus was performed. Sequence analysis of the 16S and 23S ribosomal genes indicated that *Xanthomonas*-like organisms belonged to *X. campestris* subgroup and they were found to be common epiphytes on the citrus leaf surfaces (Taylor et al. 2002).

Polymerase chain reaction (PCR) assay was applied for the detection of 34 strains of *Agrobacterium tumefaciens* (*At*) infecting *Vitis* spp. The results of PCR assay were corroborated by the conclusions from pathogenicity tests and DNA slot blot hybridization technique (Dong et al. 1992). By employing two PCR primers designed from the sequences of *virD2* and *ipt* genes, a wide range of pathogenic strains of *At* were detected. The T-DNA-borne cytokinin synthesis gene was detected by primers corresponding to sequences of *ipt* gene in *At*, but not in *A. rhizogenes* (Haas et al. 1995). The *tms2* gene in T-DNA coding for indole acetamide amidohydrolase is required for the pathogenicity of *At*. Primers flanking a 220-bp fragment of one of the conserved regions of *tms2* gene were designed for PCR amplification which revealed the presence of T-DNA in infected plants and also in infested soils (Sachadyn and Kur 1997). PCR assay was successfully applied for the detection of *At* in more than 200 samples including naturally infected almond, apricot, chrysanthemum, grapevine, peach, raspberry, rose, tobacco and tomato. The pathogen was more effectively detected in crown and root tumors than in aerial tumors (Cubero et al. 1999). In a later investigation, the usefulness of including an internal control (IC) to prevent the false negatives in PCR assay was demonstrated. Addition of 0.15 pg of IC did not significantly affect the sensitivity of detection of *At*. Detection level for DNA from culture of C58 of *At* was from 10^2 to 10^3 cfu/ml, equivalent to 5–50 cfu per reaction (Cubero et al. 2002).

A PCR protocol for the reliable detection of tumorigenic agrobacteria in artificially inoculated chrysanthemum and peach plants was developed. The method of DNA extraction was improved to increase the yield and quality of the extracted DNA. The primers were designed based on the sequences of the intercistronic region between *virB* and *virG* of the *vir* region of pTi (tumor-inducing) plasmid. This PCR assay was highly sensitive, since the target sequence was detected in spiked samples containing only 10 ± 5 cfu/g of fresh tissue of *Agrobacterium*. The presence of *At* in the stems of chrysanthemum and peach plants was detected by the PCR assay, whereas the pathogen was not always detected by dilution plating procedure (Table 2.4). The PCR assay could detect *At* in the root and stem tissues of symptomless plants. Furthermore, this protocol allowed discrimination between pathogenic and nonpathogenic forms, because of the use of primers specific for the *vir* region of Ti plasmid of *A. tumefaciens* (Puopolo et al. 2007).

Erwinia carotovora subsp. *atroseptica* (*Eca*) was detected effectively in potato stem and also in tuber tissues by using primers capable of specifically amplifying the fragment of pathogen DNA. This PCR protocol was found to be more sensitive than ELISA employing MAbs (De Boer and Ward 1995). Later, a one-step PCR-based procedure was developed for the detection of all five species of *Erwinia* including

Table 2.4 Comparative efficacy of isolation and PCR methods for detection of *Agrobacterium tumefaciens* in inoculated plant tissues (Puopolo et al. 2007)

Detection method/ pathogen concentration	Chrysanthemum[a]			Peach		
		Stem			Stem	
	Roots	5–10 cm	15–20 cm	Roots	5–10 cm	15–20 cm
Isolation						
10^7 cfu[a]	12/12[b]	5/12	0/12	8/12	0/12	0/12
10^5 cfu	12/12	8/12	0/12	4/12	4/12	0/12
10^3 cfu	12/12	4/12	0/12	0/12	0/12	0/12
PCR						
10^7 cfu	12/12	12/12	12/12	12/12	12/12	12/12
10^5 cfu	12/12	12/12	12/12	8/12	12/12	11/12
10^3 cfu	12/12	12/12	12/12	9/12	12/12	12/12

[a]cfu – Pathogen concentration expressed as colony forming units (cfu/ml) used for in inoculation

[b]Number of samples showing positive reaction/total number of samples tested

subspecies *atroseptica* and *carotovora* and all pathovars/ biovars of *E. chrysanthemi* in micropropagated potato plants. The primers SR3F1 and SRIcR based on conserved region of the 16S rRNA gene amplified a DNA fragment of 119-bp from all strains (65) tested, when an enrichment step was used prior to PCR amplification, the sensitivity of detection was increased by about 200 times (Toth et al. 1999).

A rapid and sensitive PCR format was developed for the detection of *Erwinia amylovora* causing the destructive fire blight disease of apple, pear and rosaceous plants. This protocol relies on the amplification of 187-bp DNA fragment, probably of chromosomal origin. Two 30-mer oligonucleotide primers corresponding to the sequences adjacent to the termini of pEA 71 insert directed the synthesis of the 187-bp PCR product. This product was amplified from the DNA of 69 strains of *Ea*. Southern hybridization of PCR products from the 69 strains of *Ea*, using the pEA 71 insert as a probe, confirmed that the 187-bp product was specific to *Ea* and its strains tested. The PCR assay could detect *Ea* in samples containing as few as 10 cfu in culture and plant tissues. Furthermore, the PCR protocol developed in this investigation could detect *Ea* in apple flowers before the appearance of visible symptoms, indicating its potential for use in pre-sympotmatic disease management systems (Taylor et al. 2001).

Detection of *E. amylovora* (*Ea*) in asymptomatic plant materials is a major challenge, when applying the methods available earlier. Hence, a PCR assay was developed to overcome this problem by using the REDExtract-N-Amp™ Plant PCR Kit and by introducing several modifications such as dilution, addition of detergents, increase of extraction time and use of new primers. Based on the amplification of the *Pst*I fragment of the common *Ea* plasmid pEa 29, the detection limit was reduced to a single bacterial cell present in the plant extracts used for DNA extraction, when the PEANT1/PEANT2 primers were used for PCR amplification. The number of positive detection was nearly doubled, when 951 field samples were tested, using the new method compared with standard format. Of the 26 asymptomatic plants, only the new procedure detected *Ea* in 16 samples. A distinct advantage of this protocol was that extraction of DNA could be easily done in 96-well microplates for high throughput screening, allowing the analysis of more than 200 samples per person within a single day. In addition, the risk of contamination between samples is considerably reduced. The simple and rapid procedure together with high sensitivity makes this method, an ideal choice for routine large scale testing (Stöger et al. 2006; Appendix 8). In a later investigation, the *Ea* DNA was extracted using a DNeasy Plant Mini Kit (Qiagen) from the cultures and plant tissues. The sensitivity of the PCR assay using the primers EaF72 and EaR560 was less than 10^4 cfu/ml. The PCR assay was much more specific than PTA-ELISA or indirect fluorescent antibody staining methods employed for comparison of their efficacy in detecting the pathogen (Kokošková et al. 2007).

Suppression subtractive hybridization (SSH) is a PCR-based method that has been widely used to identify differences between prokaryotic genomes with differing phenotypes including those of pathogenic and nonpathogenic strains of the same species and between closely related species. SSH has also been used to analyze genetic differences between plant–pathogenic strains with varying host specificity. PCR-based subtractive hybridization was employed to isolate sequences from *Erwinia amylovora* strain Ea 110 which is pathogenic to apples and pears that were present in three closely related strains with differing host specificities. An *Ea* 110-specific sequence with homology to a type III secretion apparatus component of the insect endosymbiont *Sodalis glossindicus* was identified (Triplett et al. 2006).

Primers derived from the *pat-1* region of the plasmid, involved in the pathogenicity of *Clavibacter michiganensis* susbsp. *michiganensis* (*Cmm*), were employed in the PCR assay for the detection of *Cmm* in the homogenates of infected tomato plants and contaminated seeds (Dreier et al. 1995). The PCR primer pair SP1f and SP5r specifically amplified a 215-bp fragment of genomic DNA of *C. michiganensis* subsp. *sepedonicus* (*Cms*) from ring rot disease-affected potato stem or tubers. This primer pair did not amplify DNA from phenotypically and serologically related bacteria isolated from potato plants. The PCR assay was not only more sensitive than ELISA and immunofluorescence (IF) tests employing MAbs, but it could detect *Cms* also from infected plants which gave negative results with these two detection methods (Li and De Boer 1995). The oligonucleotide primers derived from the intergenic region between 16S and 23S rRNA genes amplified the specific DNA sequences in *C. xyli* subsp. *xyli*, causing sugarcane ratoon stunting disease. These primers, however, did not amplify the DNA from closely related *C. xyli* subsp. *cyanodontis*, indicating the specificity of the PCR assay (Fegan et al. 1998). *Ralstonia* (*Pseudomonas*) *solanacearum* (*Rs*), causes bacterial wilt diseases of several economically important crop plants. The primers PS96H and PS96I were able to specifically detect 28 strains of *Rs*. Sequence analysis revealed the presence of six different sequence groups in the region between these primers (Hartung et al. 1998). By employing the PCR amplification followed by electrophoretic analysis of amplicons, enabled the interception of a distinct pathotype of *Rs* race 1, biovar 1 infecting pothos (*Epipremnum aureum*) at the entry point (Norman and Yuen 1998).

PCR assay has been shown to be effective for the detection of bacterial pathogens in different organs/tissues of infected plants from which it may be difficult to isolate them in pure culture. The presence of *Xylella fastidiosa* (*Xf*), causing citrus variegated chlorosis (CVC) disease, was detected in all main sweet orange fruit vascular bundles and in seeds which exhibited no visible changes (Li et al. 2003). *X. fastidiosa* (*Xf*) infects grapevine, almond and oleander. A PCR-based method, involving the use of different sets of primers, was applied to detect *Xf* strains specifically in these host plant species. PCR assay performed with primers XF1968-L and XF1968-R amplified a 638-bp fragment from OLS strains infecting oleander, whereas PCR with primers XF2542-L and XF2542-R amplified a 412-bp from PD strains infecting grapevine. The PCR primers ALM1 and ALM2 specifically amplified a fragment of 521-bp from ALS strains infecting almond. Thus, by using appropriate primer sets *Xf* strains infecting three different host plant species could be efficiently detected and differentiated (Hernandez-Martinez et al. 2006).

A PCR assay was developed employing the primers designed based on the sequences of the 16S rRNA gene of *X. fastidiosa* (*Xf*) in the xylem sap of the infected citrus plants. The PCR amplicons (approximately 1,348-bp) were partially sequenced and the data revealed a 99% identity to the 16S rRNA gene sequences of *Xf*, confirming the presence of *Xf* in diseased plants (Rodrigues et al. 2006). A quick, simple and sensitive one-step PCR procedure was developed to detect *Xf* in the leaves of infected oleander plants. The DNA was extracted by using DNeasy Tissue Kit and Fast DNA Kit separately, followed by PCR employing 16S set C primers. *Xf* was detected in two inoculated plants showing leaf scorch symptoms, when DNeasy Tissue Kit was used for DNA extraction from leaves, but not in samples that were treated with materials of Fast DNA Kit for DNA extraction. This investigation reveals the need for employing suitable DNA extraction procedure that avoids the coextraction of compounds that may inhibit PCR amplification (Huang et al. 2006).

Grapevine accessions in a field Genebank of the National Clonal Germplasm Repository (NCGR), California, were screened for the presence of *Xylella fastidiosa* (*Xf*) by employing specific primers in PCR assay with high degree of sensitivity and specificity for all *Xf* strains. Basal leaf petioles were collected in the region where *Xf* was endemic from 60 different accessions in the NCGR representing a broad cross section of taxa and periods of time. Only Chardonnay grapevine previously inoculated with *Xf* showed the presence of the pathogen. It was concluded that although the budwoods were collected for NCGR from endemic areas of southeastern United States where *Xf* and PD were endemic, these accessions in NCGR were not considered as a potential reservoir of *Xf* capable of spreading the disease to areas around NCGR (Stover et al. 2008).

Citrus huanglongbing (HLB) or greening disease causes considerable losses in several countries. The effectiveness of the PCR assay depends primarily on the availability of pathogen DNA of high quality, specific primers and suitable polymerase enzyme for amplification of the unique DNA fragment present in the target pathogen. A non-phenol chloroform procedure for DNA extraction, three sets of primers (A, B and C) and two polymerase enzymes were evaluated for their efficacy in the detection of *Candidatus* Liberibacter, causative agent of HLB disease. The standardized PCR assay utilizing the nonphenol chloroform method of DNA extraction developed in this investigation, primer set C (amplifying a PCR product of 451-bp) and Klen *Taq* polymerase was effective for the detection of *Ca.* Liberibacter in citrus trees in India (Gouda et al. 2006). A new disease of glass-house-grown tomato and pepper occurring in New Zealand was investigated to determine the nature of the causative agent. The presence of a phloem-limited bacterium-like cells was observed under transmission electron microscope (TEM). Specific prokaryote PCR primers in combination with universal 16S rRNA primers were employed. Sequence analysis of the 16S rRNA gene, the 16S/23S rRNA spacer region and the *rpl*KAJL-*rpo*BC operon revealed that the bacterium shared high identity with '*Candidatus* Liberibacter' species. The bacterium infecting tomato and pepper was found to be distinct from the three citrus Liberibacter species and hence, the new bacterial pathogen was named '*Candidatus Liberibacter solanacearum*'. A specific PCR primer was developed for reliable detection of this bacterial pathogen species (Liefting et al. 2009).

Detection by Targeting Specific Gene Sequences

PCR assays may be developed based on the primers designed on the sequences of specific genes involved in the process of disease development like adhesion, production of phytotoxins, resistance to oxidative stress, plant cell wall degradation, secretory systems and interference and/or suppression of host defenses. Different kinds of enzymes and toxins produced by bacterial pathogens have a role in infection and disease development. A segment of *nec1* gene governing necrosis of tissues infected by *Streptomyces scabies*, causative agent of potato scab disease, is involved in the pathogenicity. Amplification of the segment of *nec1* gene using PCR assay was useful for reliable detection of *S. scabies* in potato tissues (Bukhalid et al. 1998; Joshi et al. 2007).

Pseudomonas syringae pathovars possess a large transmissible plasmid containing coronatine (COR)-producing genes. The sequences of the gene cluster have been used for designing primers for their amplification by PCR assays employed for the detection of *P. syringae*. However, some of the pathogenic strains are capable of producing COR, but they lack the coronatine gene cluster. It would not be possible to detect such strains using these primers (Bereswill et al. 1994). In a later investigation, *P. syringae* pv. *atroseptica* (*Psa*) was detected by employing primers from pCOR1 plasmid gene sequences associated with coronatine production and pathogenicity. Detection limit of the PCR assay was 0.1–1.0 cfu, when serially diluted pure cultures were used. *Psa* could be directly detected in infected tissues of Italian ryegrass showing halo blight disease symptoms (Takahashi et al. 1996). *P. syringae* pv. *maculicola* infecting leafy crucifers in Oklahoma was detected by employing primers to amplify the *Cfl* gene sequences from the gene cluster encoding coronatine toxin (Zhao et al. 2002).

Specific detection methods for the detection of toxin-producing pathovars of *Pseudomonas syringae* (*Ps*) have been developed, based on the genes required for their production. A PCR assay was developed for the detection of *Ps* pathovars producing tabtoxin, utilizing oligonucleotide primers derived from the coding sequences of the *tblA* and *tabA* genes. Thirty two strains of *Ps* and related species were examined.

Only a PCR product of 829-bp or 1,020-bp was produced in the PCR amplifications with the tblA or tabA primer sets respectively and cells from tabtoxin-producing pathovars of *Ps*. A single 829-bp amplification product was produced with tblA primers in reactions that contained cells from tabtoxin-producing strains of *P. syringae* pv. *tabaci* and *P. syringae* BR2R. A single 1,020-bp amplification product was present in reactions that contained cells from all the tabtoxin-producing strains of *P. syringae* tested. All known non-toxin producing bacterial species failed to produce amplification product with either primer set. The PCR protocol developed in this investigation was found to be simple, rapid and reliable for detection and identification of tabtoxin-producing strains of *P. syringae* (Lydon and Patterson 2001).

The xylem-limited bacteria (XLB), *Xylella fastidiosa* (*Xf*) is pathogenic to several economically important crop plants. Nearly complete nucleotide sequences of the 16S rRNA gene and partial sequences of the DNA gyrase (*gyrB*) were determined for 18 strains of *Xf* from different plant hosts. The specificity of *Xf* 16 rRNA-targeted primers was confirmed by amplification of a product of 645-bp (for set B) from the *Xf* DNA only. No PCR amplification occurred for the DNA of any of the 36 phylogenetically related microorganisms, endophytes or plant pathogens. Specific primers developed against the *gyrB* gene were also tested with DNA isolated from all strains as in the case of 16S rRNA-targeted primers. Amplification of the product of expected size (429-bp) was observed only when *Xf* DNA was used for amplification. These primer pairs were certified with 30 target and 36 nontarget pure cultures of microorganisms confirming 100% specificity (Rodrigues et al. 2003).

The type III secretion system (TTSS) of plant pathogenic bacteria is encoded by *hrp* gene clusters that are generally conserved. The type III secretion system delivers into plant cells, the pathogenicity factors, elicitors and avirulence proteins that largely determine the course of disease development. Primers designed based on DNA sequences related to *hrp* genes were employed for specific amplification of DNA of *X. campestris* pv. *vesicatoria* (*Xcv*) enabling the detection and identification of this pathogen (Leite et al. 1994). The primer pair DLH120 and DLH125 from the *hrpF* gene from *X. campestris* (*Xc*) amplified the 3′ end of *hrpF* gene. The primers amplified a 619-bp fragment of the *hrpF* gene only from *Xc*, but not from other *Xanthomonas* spp. indicating the primer specificity. By including primers targeting a 360-bp section of the ITS region from *Brassica* spp. in a multiplex PCR, the seedborne *X. campestris* pv. *campestris* (*Xcc*) could be detected even, if one seed in a lot of 10,000 was infected by *Xcc* (Berg et al. 2005).

The presence of *Xanthomonas campestris* pv. *campestris* (*Xcc*) in cabbage plants was detected by a specific and sensitive PCR assay using the primers XCF and XCR designed on the sequences of *hrpF* gene of *Xcc* homologous to *nolX*, host recognition protein. A specific fragment of 525-bp was amplified by the primers from the cabbage plants infected by *Xcc*. The specificity of the PCR assay was indicated by absence of amplification from the DNA of 40 isolates of *Xanthomonas* strains and other nontarget phytopathogenic bacterial species and *Escherichia coli* (Park et al. 2004). A sensitive and specific assay was developed for the detection of *Xanthomonas axonopodis* pv. *citri* (*Xac*), causing citrus bacterial canker disease. The primers XACF and XACR were derived from the sequences of *hrp*^{W} homologous to pectate lyase, modifying the structure of pectin in plants. These primers amplified a 561-bp DNA fragment from leaves and fruits of citrus infected by *Xac*. This specific PCR amplicon was produced only from *Xac* among 26 isolates of *Xanthomonas* strains, but not from *Escherchia coli* (0157:H7), *Pectobacterium carotovorum* subsp. *carotovorum* and other microorganisms tested. Further, the PCR protocol could be employed to detect *Xcc* in naturally or artificially infected leaves of citrus (Park et al. 2006).

The *hrpZPst* gene of *Pseudomonas syringae* pv. *tomato* (*Pst*) maps on a pathogenicity-associated operon of the *hrp/hrc* pathogenicity island. The primers based on the sequences of *hrpZPst* gene amplified a 532-bp product corresponding to the internal fragment of the *hrpZPst* from 50 isolates of *Pst*. Infection by *Pst* could be detected in leaf and fruit spots from plants under natural or controlled conditions and also in artificially contaminated seeds (Zaccardelli et al. 2005). The *rpf* genes regulate the expression of the pathogenicity factors in *Xcc*. In the case of *X. axonopodis* pv. *citri* (*Xac*), a specific genomic region was identified inside the *rpf* gene cluster between *rpfB* and *recJ* of strain JAPAR 306 (Da Silva et al. 2002). Computer analysis showed that *rpf* cluster in *Xac* was different from that of *Xcc*. Two primers Xaco1 and Xaco2 directed the amplification of a 581-bp fragment from DNA of *Xac* strains only. Limit of detection of *Xac* was as few as 100 bacterial cells in pure cultures and in extracts of both fresh and dried canker lesions and from washes of inoculated but asymptomatic leaf surfaces (Colletta-Filho et al. 2006).

The primer pair RST2/RST3 was designed based on the sequences of the *hrp* region of *X. campestris* pv. *vesticatoria* (*Xcv*). The primer pair amplified a unique fragment of 840-bp from the total genomic DNA of the strains tested. All strains pathogenic to pepper produced the expected amplicon (840-bp) as the reference strains from California. No amplification occurred from the DNA of the other *Xanthomonas* spp. tested, indicating the specificity of the PCR assay. The results indicated that strains of *Xcv* could be detected reliably and identified by employing primer pairs based on the sequences of *hrp* gene involved in the pathogenicity of the bacterial pathogen (Mitrev and Kovačevic 2006).

Pepper and tomato are infected by *Xanthomonas campestris* pv. *vesicatoria* (*Xcv*) and *X. axonopodis* pv. *vesicatoria* (*Xav*), causing indistinguishable bacterial spot disease in these two crops. The 20-mer oligonucleotide primer pair XCVF and XCVR based on the sequences of *rhs* family gene homolog in *Xanthomonas* strains was confirmed by dot blot analysis. *Xcv* strains exhibited high hybridization signal to the probe, whereas other *Xanthomonas* strains showed negative results in the dot blot analysis. The primer pair XCVF and XCVR amplified a 517-bp fragment from all six strains of *Xcv*, indicating the specificity of the PCR assay (Fig. 2.6). The results indicated that XCVF and XCVR primer pair can be used as rapid tool for the detection and identification of *Xcv* (Park et al. 2009; Appendix 9).

Xanthomonas fragariae infecting strawberry was detected by employing primers based on the sequences of *hrp* gene region. Amplification and sequencing of the *hrp* gene region allowed the selection of highly specific primers with high sensitivity for detecting *X. fragariae* in both symptomatic and asymptomatic strawberry plants. Specific primers based on *hrp* sequences have been shown to be particularly useful for the detection of pathogenic xanthomonads, since nonpathogenic strains lack

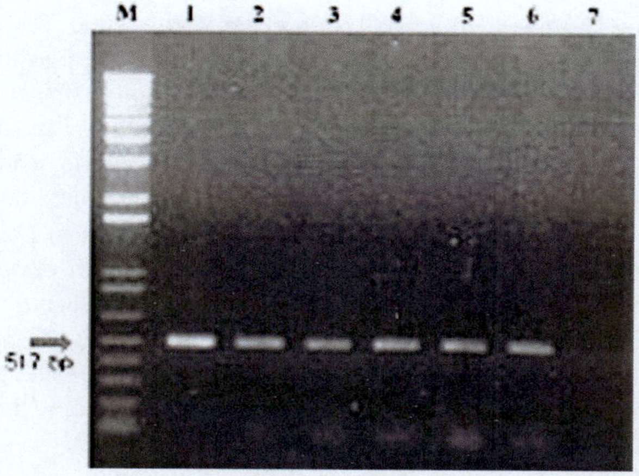

Fig. 2.6 Detection of *Xanthomonas campestris* pv. *vesicatoria* using pathogen-specific primers in PCR assay in artificially inoculated *Capsicum annuum* cultivars. M: 1 kb DNA ladder; Lane 1: *X. campestris* pv. *vesicatoria* KACC 11153; Lanes 2–6: Inoculated *C. annuum* cultivars; Lane 7: healthy *C. annuum* cultivar (Courtesy of Park et al. 2009; Elsevier Ltd, Oxford, UK)

these genes (Leite Jr. et al. 1994). *X. campestris* pv. *viticola* (*Xcv*) is the causative agent of bacterial canker, an important disease of grapevine. Primers based on the partial sequence of *hrpB* gene of *Xcv* were designed. The primer pairs Xcv1F/Xcv3R and RST2/Xcv3R were specific to amplify 243-bp and 340-bp fragments respectively from the DNAs of *Xcv* strains and also from the DNAs of *X. campestris* pv. *mangiferae-indicae* and *X. axonopodis* pv. *passiflorae* strains. However, the enzymatic products of PCR amplicons showed distinct differences for *Xcv* strains. None of the primer pairs amplified nonpathogenic bacteria from grapevine leaves and strains belonging to other genera of pathogenic bacteria. The limit of detection with Xcv1F/Xcv3R and RST2/Xcv3R was 10 and 1 pg of purified *Xcv* DNA respectively. By incorporating a second round of PCR amplification using the internal primer Xcv1F, the detection limit could be lowered from 10^4 to 10^2 cfu/ml of bacterial cells (Trindade et al. 2007).

Detection by Targeting Non-nuclear Nucleic Acids

Nonnuclear nucleic acids present in phytopathogenic bacterial cells may be used as targets for their detection and identification. Primers based on a promoter-like sequence from plasmid pCS1 were employed for the effective detection of *Clavibacter michiganenesis* subsp. *sepedonicus* (*Cms*). As certain *Cms* strains lack this plasmid, the usefulness of this approach is limited to the *Cms* strains with pCS1 plasmid (Schneider et al. 1993). However, another plasmid-derived *Cms*-specific primer set was demonstrated to be efficient and specific for the detection of *Cms* strains (Rademaker and Janse 1994). The effectiveness of plasmid-derived primers for reliable detection of *Xanthomonas axonopodis* pv. *citri* (*Xac*) strains (pathotype A) was indicated by Hartung et al. (1996). In another study, sets of primers based on sequence differences in the ITS and on a sequence from the plasmid gene *pthA* involved in pathogenicity were used for the detection of strains obtained from various geographical locations. In addition, pathotypes of *Xac* could be identified/differentiated. A new strain of *Xac* capable of causing canker on Key/Mexican lime (*Citrus aurantifolia*), designated *Xac-A*^{w} was differentiated by using plasmid-derived primers specific for *Xac* (Sun et al. 2004). *X. axonopodis* pv. *manihotis* (*Xam*) infecting cassava, was detected by employing primers designed from a plasmid fragment associated with the pathogenicity of *Xam.* In the PCR assay, a specific 898-bp fragment was amplified by the primers from 107 pathogenic strains of *Xam*, indicating the specificity of the PCR assay (Verdier et al. 1998). The nature and persistence of the target plasmid DNA in the bacterial pathogen to be detected, will be critical for the applicability of this approach.

The ribosomal DNA operon comprises of three functionally and evolutionarily conserved genes namely the small subunit 16S (rRNA) gene (*rrs*), the large subunit 23S rRNA gene and 5S rRNA gene interspersed with variable spacer regions (intergenic transcribed spacer) (Schmidt 1994). The PCR primers specific to 23S rRNA gene of *Erwinia amylovora* (*Ea*) were successfully employed for the detection of *Ea* strains in asymptomatic tissues (Maes et al. 1996). A single pair of primers was designed based on the sequences of the spacer region between 16S and 23S rRNA genes for five different *Clavibacter* subspecies. This primer pair specifically amplified a 215-bp fragment from *C. michiganensis* subsp.*sepedonicus* (*Cms*). This PCR format was more sensitive than ELISA and immunofluorescence (IF) tests in detecting (*Cms*) in naturally infected potato tissues (Li and De Boer 1995). *C. xyli* subsp. *xyli* (*Cxx*) was detected by PCR assay utilizing the oligonucleotide primers derived from the intergenic region between 16S and 23S rRNA genes which amplified the sequences specific to *Cxx.* The specificity of the PCR assay was indicated by the absence of amplification of the DNA of closely related subsp. *cyanodontis* (Fegan et al. 1998).

Selective PCR amplification of DNA sequences of *Ralstonia solanacearum* (*Rs*) biovar 3, 4 and 5 (in Division I) and biovar 1 and 2 (in Division II) was accomplished by employing different combination of forward and reverse primers derived from sequences of intergenic region between 16S and 23 S rDNA genes (Seal et al. 1999). The sequences of the 16S rRNA gene of a known strain of *Acidovorax avenae* subsp. *citrulli* (*Aac*) formed the basis for production of synthetic oligonucleotide primers for the detection of *Aac*. The synthetic primers amplified the DNA from all strains of *Aac* tested (Walcott and Gitaitis 2000).

Xylella fastidiosa (*Xf*), causing almond leaf scorch disease (ALSD) and grapevine Pierce's disease (PD) were considered to be due to a single strain of *Xf.* The usefulness of primers G1 and L1 for the amplification of the 16S–23S rRNA intergenic spacer region of *Xf* for the detection of *Xf* in many host plant species was demonstrated (Costa et al. 2004). In a later study, the strains of ALSD collected from two locations in California were identified as two genotypically distinct types of *Xf* strains. A conserved genomic region could be used for species recognition and a single nucleotide polymorphism (SNP) within the conserved region could be the basis for differentiating genotypes below the species level. An SNP-based PCR assay was developed to identify and differentiate *Xf* strains collected from ALSD-affected almond. The 16S rRNA gene sequences were used, because, they are highly conserved at species level and they are the most screened locus in bacteria. SNPs in the 16S rRNA gene of *Xf* were identified to characterize the population in infected trees. When genotype-specific SNPs were employed to design primers for multiplex PCR assays of early passage cultures, two genotypically distinct types of *Xf* strains, G type and A type were found to coexist in the same infected almond orchard. The RFLP analysis of a different genetic locus RST31–RST33 confirmed the findings of multiplex PCR assays. The coexistence of two different genotypes of *Xf* in the almond orchard appears to be of epidemiological significance for taking up measures for monitoring pathogen spread (Chen et al. 2005).

Candidatus Liberibacter (a phloem-limited nonculturable bacterium) infecting citrus was detected by employing the primers OI1 and OI2c designed using the sequences of 16S rDNA. The PCR assay using the primers detected the pathogen in 28% of samples analyzed during the survey undertaken in Brazil (Coletto-Filho et al. 2004). In a later investigation, the primers LSg2f and LSg2r developed using the consensus sequences of 16S rDNA fragments were able to detect *Ca.* Liberibacter in 38 samples that did not amplify with OI1 and OI2c. In addition, the Asian '*Ca.* Liberibacter' was also detected in six out of 53 samples tested (Coletta-Filho et al. 2005).

PCR assay has been employed to establish the identity of some bacterial pathogens that are not known to infect a plant species earlier as in the case of yellow vine disease of cucurbits. The presence of a bacterium-like organism (BLO) in the phloem of infected plants was observed. Based on the nucleotide sequence of the prokaryotic DNA isolated from symptomatic plants, three primer pairs YV1, YV2 and YV3 were designed. The primer pair YV1/YV2 amplified a product with 640-bp fragment, while YV1/YV3 primer pair amplified a 1.43-kb product. The phylogenetic analysis showed that the prokaryote causing yellow vein disease was a gamma-3 proteobacterium. This agent may cause the yellow vein disease in cantaloupe, squash and watermelon (Avila et al. 1998). Reliable evidence for establishing the identity of the causative agent of a new bacterial blight disease of leek (*Allium porrum*) occurring in California, was provided by a PCR-based assay. The DNA genomic fingerprints of leek isolate obtained by rep-PCR analysis were indistinguishable from those of known strains of *Pseudomonas syringae* pv. *porri.* The causative bacterial species was identified unambiguously by rep-PCR assay as *P. syringae* pv. *porri*, whereas fatty acid analysis did not provide a clean species identification that caused the new disease in leek (Koike et al. 1999).

Detection by Using Variants of PCR

Repetitive sequence-based (rep)-PCR genomic fingerprinting, used more frequently of the adaptations of standard PCR assay, is based on PCR-mediated amplification of DNA sequences located between specific interspersed repeated sequences in prokaryotic genomes. These repeated sequences are designated REP (repetitive extragenic palindromic sequences), BOX (DNA sequences of the BOXA subunit of the BOX element of *Streptococcus pneumonia*) and ERIC (enterobacterial repetitive intergenic consensus) elements. Amplification of the DNA sequences between primers based on these repeated elements generates an array of differently sized DNA fragments from the genomes of different strains. The resolution of these fragments on agarose gels yields highly specific DNA fingerprints. The rep-PCR genomic fingerprinting protocol may be carried out on whole cells (individual colonies of bacteria) bypassing the need for DNA extraction. This technique may also be applied directly to cell suspensions obtained from symptomatic plant tissues (Louws et al. 1994).

Rep-PCR technique was employed for the detection and differentiation of field isolates of *Xanthomonas fragariae* collected from the strawberry nurseries in California. There was good correlation between the results of rep-PCR fingerprinting procedure and pathogenicity tests that required much longer time for providing results. Further, rep-PCR technique was shown to be more sensitive than indirect ELISA format providing the results more rapidly and accurately (Opgenorth et al. 1996). Rep-PCR genomic fingerprint profiles from 33 isolates of *X. translucens* infecting asparagus were generated, in addition to those of 61 *X. translucens* reference strains infecting cereals and grasses. Amplified ribosomal restriction analysis profiles were prepared for most of these strains and they were compared with those in large *Xanthomonas* database using computer-assisted analysis. The isolates from ornamental asparagus (*A. virgatus*, tree fern) were identified as *X. translucens* pv. *undulosa.* A unique unamplified small subunit ribosomal gene *Msp*I/*Alu*I restriction profiles was found to be specific for all strains tested, including those strains pathogenic to asparagus, allowing the discrimination from other species of *Xanthomonas.* All hosts of *X. translucens* pathovars are known to belong to *Gramineae* and *Poaceae*. On the other hand, the novel asparagus isolates infect host plant species belonging to the phylogenetically distant *Liliaceae* (Rademaker et al. 2006). The isolates of *X. campestris* causing leaf spot disease of *Brassicas* and closely related pathovars were detected and differentiated by applying rep-DNA PCR-based fingerprinting technique. The leaf isolates were clustered separately from *X. campestris* pv. *campestris* isolates. Hence, it was proposed that *X. campestris* isolates that cause a nonvascular leaf spot disease on *Brassica* spp. should be identified as *X. campestris* pv. *raphani a*nd not as pv. *armoraciae* (Vicente et al. 2006).

The rep-PCR primers (BOX, ERIC and REP) were employed for genomic fingerprinting of *Clavibacter michiganensis* subsp. *michiganensis* (*Cmm*). The rep-PCR fingerprints identified at least four types A, B, C and D within *Cmm* based on the limited DNA polymorphisms (80% similarity in fingerprint patterns). This ability to differentiate individual strains may be useful for studies on the epidemiology and interactions of this pathogen and its hosts. In addition, a relatively large number of naturally occurring avirulent *Cmm* strains could be recovered from diseased tomato plants, with rep-PCR fingerprints identical to those of virulent *Cmm* strains (Louws et al. 1998). *Ralstonia solanacearum* (*Rs*) causes bacterial wilt diseases in ginger, curcuma and mioga. REP and BOX-PCR analysis was carried out using selected strains infecting these host plants in Japan, Thailand and Indonesia, Australia and China. In addition, representative Japanese race 1 and 3, biovar 2, 3 and 4 strains from tomato, eggplant, sweet pepper and statice were also examined. The DNA profiles obtained with BOX-PCR were highly reproducible. Two distinct types of DNA fingerprints based on the presence or absence of a few bands were observed among Japanese strains from plants belonging to Zingiberaceae family. Type 1 consisted of all curcuma strains, some ginger strains and all mioga strains, whereas the type 2 enclosed most of the ginger strains. However, neither DNA patterns was similar to the strains from hosts belonging to families other than Zingiberaceae (Horita et al. 2004; Tsuchiya et al. 2005).

Isolates of *Brennaria nigrifluens* (*Bn*), causative agent of shallow bark canker disease of walnuts were subjected to REP-PCR and two systems to determine metabolic activities of the pathogen viz., API 20E and Biolog analyses. The isolates whose REP-PCR fingerprint and the results of API20E and Biolog systems were very similar (87–100%) to those of *Bn* reference strains, were assigned to *Bn*. Four isolates of *Bn* among 28 gram-negative isolates obtained from bark cankers in walnut plantations on ten farms in Italy were identified. Pathogenicity tests confirmed that these four strains were pathogenic to walnut inducing the characteristic symptoms at 3 months after inoculation (Moretti et al. 2007).

Nested-PCR assay has been shown to be effective and efficient for the detection of some of the bacterial plant pathogens. In nested-PCR procedure, the sensitivity and specificity of detection of target pathogen are significantly enhanced by performing a second round of PCR using primers internal to the amplification product. Two oligonucleotide primers were designed based on the unique fragment of the plasmid pEA29 of *Erwinia amylovora* (*Ea*) causing fire blight disease. The nested-PCR was more sensitive (1,000 fold) than single-round PCR format. Even single cells of *Ea* could be detected in pure cultures by applying nested-PCR protocol. The presence of *Ea* in different plant organs such as leaves, axillary buds and mature fruit calyx samples was successfully detected by nested-PCR assay (McManus and Jones 1996). Nested-PCR assay in a single closed tube using two consecutive PCRs was developed for the detection of *Ea* in asymptomatic plant material. A higher annealing temperature that permitted amplification of only an external primer pair was maintained for the first PCR. The second standard PCR with internal primer pair amplified a specific DNA fragment from pEA29 of *Ea*. This nested-PCR format detected both endophytic and epiphytic populations of *Ea*. As this protocol is simple, specific, sensitive and rapid, it may be applied for routine detection of bacterial pathogens by quarantine programs. The sequences of the cytokinins (*etz*) or IAA-biosynthetic genes of *Erwinia herbicola* pv. *gysophilae* (*Ehg*) infecting *Gysophila paniculata* plants were employed for designing three primer pairs. It was possible by employing these primers to detect even single bacterial cell in pure cultures of *Ehg*, the limit of detection being at far lower concentration of *Ehg* compared to standard PCR assay (Manulis et al. 1998).

A nested-PCR technique was developed using a sequence-characterized amplified region (SCAR) marker for the detection of *X. axonopodis* pv. *diffenbachiae* (*Xad*) which is of quarantine importance for countries in European Union. A randomly amplified polymorphic DNA (RAPD)-based PCR technique was used to identify DNA fragments that were putatively specific to *Xad*. These fragments were further characterized as SCAR markers. One of the SCAR markers was used for detection of *Xad* by nested-PCR format in pure cultures and anthurium tissue extracts. The internal primer pair detected the expected product (785-bp) for all 70 *Xad* strains pathogenic to anthurium and for isolates originating from syngonium and not pathogenic to anthurium. There is a high level of relatedness between strains from anthurium and strains from syngonium. The DNA of unrelated phytopathogenic bacterial strains (98) did not yield the expected PCR product, confirming the specificity of the protocol developed in this investigation. Restriction enzyme analysis allowed differentiation of the two strains infecting anthurium and syngonium. The detection threshold was 10^3 cfu/ml of pure culture suspension or extracts of both symptomatic and asymptomatic contaminated plants. This protocol has the potential for application as a diagnostic tool for screening propagation stock plant material and monitoring international movement of this pathogen (Robéne-Soustrade et al. 2006).

Pseudomonas savastanoi pv. *savastanoi* (*Pss*) causing olive knot disease was detected by employing a nested-PCR in a single closed tube coupled with dot blot hybridization. The colorimetric detection of amplified products by dot blot hybridization increased the sensitivity of the test, facilitating accurate interpretation of results, saving time and avoiding the need for the use of gels stained with toxic ethidium bromide. The primer pair IAALF/IAALR as external primers and the internal primers IAALN1 and IAALN2 designed on the sequences of *iaaL* structural gene, were employed in the nested-PCR assay. A product of the expected size of 338-bp was amplified from the DNA of *Pss*. The designed probe, internal to the nested-PCR product hybridized only with amplified products from *Pss* strains. But no signal was obtained with the non-*Pss* DNA, indicating specificity of the nested-PCR assay. The nested-PCR assay developed in this investigation detected 1–10 cfu/ml of plant extracts with or without a pre-enrichment step showing tenfold improvement in detection sensitivity. In addition, the nested-PCR assay with a pre-enrichment step detected *Pss* in 9 of 30 asymptomatic seedlings in the nursery, whereas the isolation and conventional PCR methods did not detect the pathogen in any of the asymptomatic plants tested. This investigation showed that the nested-PCR assay coupled with dot blot hybridization would be very useful for the early detection of low levels of *P. savastanoi* in olive plants during plant propagation (Fig. 2.7) (Bertolini et al. 2003).

Pectobacterium carotovorum subsp. *carotovorum* (*Pcc*) causing bacterial soft rot of crown imperial (*Fritillaria imperialis*) was detected by the nested-PCR assay. The primer set EXPCCF/EXPCCR amplified a single fragment of 0.55 kb from the genomic DNA of all strains of *Pcc* tested by conventional PCR assay. In the nested-PCR format the primer set INPCCR/INPCCF amplified a single fragment of 0.4 kb from the PCR product of first PCR amplification. The strains infecting crown imperial were identified as *Pectobacterium carotovorum* subsp. *carotovorum* (Mahmoudi et al. 2007).

Although PCR assay is rapid, specific and sensitive, occurrence of false negatives limits its reliability. Hence, a competitive PCR technique that may eliminate the

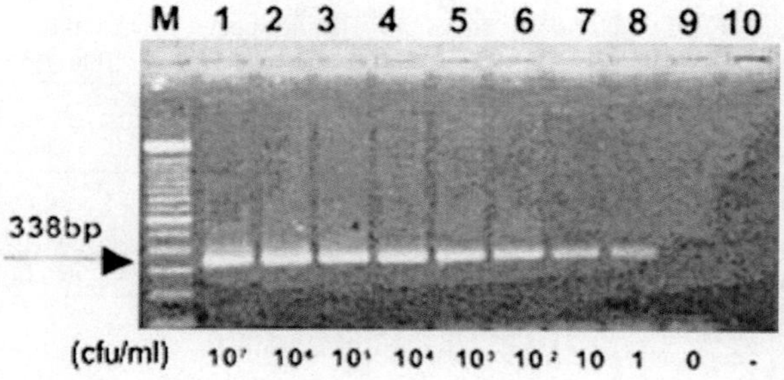

Fig. 2.7 Detection of *Pseudomonas savastanoi* pv. *savastanoi* (*Pss*) in extracts of infected plants at different concentrations using nested PCR technique employing primers targeting the *iaaL* gene of the pathogen. Lanes 1–8: 10^7 to 1 cfu/ml of plant extract respectively; Lane 9: non-inoculated plant extract; Lane 10: PCR negative control; M: DNA ladder (Court resy of Bertolini et al. 2003; Elsevier, Oxford, UK)

risk of false negatives has been found to be desirable for the detection of bacterial pathogens. For the detection and quantification of *Clavibacter michiganensis* susbsp. *sepedonicus* (*Cms*), a competitive PCR format was developed. An internal standard DNA template that served as control for all PCR assays was generated by the amplification of *Arabidopsis* genomic DNA under low annealing temperature with primers specific for *Cms*. The 450-bp product amplified from the internal standard DNA template was distinct from the 250-bp product characteristic of *Cms*. The ratio of PCR products amplified in the presence of a constant amount of internal DNA template increased with increase in the amount of *Cms* DNA. The PCR product ratios obtained from bacterial cells in cultures and in tissues of inoculated potato plantlets were corroborated by the cell numbers estimated by immunofluorescence antibody staining (IFAS) procedure. Competitive PCR assay was found to be tenfold more sensitive than IFAS method, as this PCR format was able to detect as few as 100 bacterial cells. As efficient primers for specific amplification of most genotypic and phenotypic variants of *Cms* may become available, competitive PCR format can be integrated into other PCR-based detection systems for elimination of possible false negative results (Hu et al. 1995).

Citrus huanglongbing (HLB) or citrus greening disease caused by *Candidatus* Liberibacter (CL) occurs in serious forms reducing the production levels significantly in many countries. The causative bacterial species occurs as Asian, African and American forms which are designated *Ca. L. asiaticus* (Las), *Ca. L. africanus* (Laf) and *Ca. L. americanus* (Lam) respectively (Jagoueix et al. 1994, 1996; Teixeira et al. 2005). Competitive PCR technique was developed for the detection and quantification of the pathogen in infected citrus tissues. Competitive DNA was produced from a DNA fragment amplified by rep-PCR with DNA extracted from infected citrus plants. The PCR primer sequence for primer pair OI1/OI2c was designed for specific amplification of 16S rRNA gene sequence of CL. The number of amplification cycle was increased to 45 for higher sensitivity. As the amplification of target DNA and competitor DNA occurred in the same tube, the inhibitory activity of compounds coextracted with host DNA was reduced to the maximum

Table 2.5 Quantification of *Candidatus* Liberibacter in huanglongbing disease-infected leaves of two citrus cultivars (Kawabe et al. 2006)

		Pathogen	
Cultivar	Number of leaves	f mol/m[a]	f mol/g[b]
'Cam sanh'	47	39 ± 26	35 ± 24
'Nam roi'	33	13 ± 10	6 ± 4
P (T ≤ t)[c]		<0.0001	<0.0001

[a] Molar concentration per length of leaf midrib
[b] Molar concentration per fresh mass of leaf midrib
[c] *t*-Test between average pathogen molar concentrations in two citrus cultivars

extent. The amounts of CL varied between citrus varieties as determined by competitive PCR assay (Table 2.5). This quantification system has advantages over the real-time PCR assay such as lower risk of error resulting from differences in amplification efficiencies, due to inhibitors and requirement of less expensive equipment and chemicals for competitive PCR format (Kawabe et al. 2006).

Incidence of citrus huanglongbing (HLB) in Florida during field survey was assessed by examining six citrus species using a quantitative PCR assay. The populations of *Ca. Liberibacter asiaticus* (*Las*) inferred from the distribution of 16S rDNA sequences specific for *Las* in leaf midribs, leaf blades and bark samples varied by a factor of 1,000 among samples and by a factor of 100 between two sweet orange trees tested. In naturally infected trees, the aerial plant parts had an average of 10^{10} *Las* genomes per gram of tissue. All root samples did not contain the pathogen genome. In symptomatic fruit tissues, highest levels of *Las* were found in the locular membranes and septa, while the mesocarp and pericarp tissues contained low levels (100-fold) of that found in locular membranes and septa. The quantitative PCR format may be useful for studying biology of *Las* in citrus plants providing clues for the management of the disease effectively (Li et al. 2009).

It is desirable to have pathogen detection systems that can provide the possibility of simultaneous detection and identification of as many pathogens as possible. This is required to minimize the time and cost of detection tests, since many pathogens can infect a crop either simultaneously or at different periods during crop growth. Infection of a plant species by many bacterial pathogens simultaneously is not as common as in the case of viral pathogens. Multiplex PCR assays have been shown to be effective in a few crops. Seeds or propagative plant materials may be infected simultaneously by many bacterial pathogens. Multiplex PCR assay was demonstrated to be effective for detecting four RNA viruses and the bacterial pathogen *Pseudomonas savastanoi* pv. *savastanoi* infecting olive trees (Bertolini et al. 2003).

Clavibacter michiganensis subsp. *michiganensis* (*Cmm*) and *Xanthomonas axonopodis* pv. *vesicatoria* (*Xav*) cause diseases in tomato with worldwide distribution. A multiplex PCR procedure that allows simultaneous detection of these two pathogens was developed. Concurrent amplification of *Cmm* and *Xav* with primer pairs RST2/RST3 and CMM-5/CMM-6 was accomplished. To optimize multiplex PCR conditions, annealing temperature was determined by gradient PCR and primer concentrations for each primer pair were tested. Primer concentration ratios of 0.2 and 0.4 μM in final concentrations respectively for RST2/RST3 and CMM-5/CMM-6 primer pairs were found to be optimal for multiplex PCRs. Annealing temperatures of 58.5–59.5°C were optimal for RST2/RST3 and CMM-5/CMM-6 primer pairs for multiplex PCR. The PCR products were clearly distinguishable for simultaneous detection of *Xav* and *Cmm* by the multiplex PCR format (Ozdemir 2005).

Integration of PCR with Other Detection Methods

In order to enhance the sensitivity and specificity of detection of bacterial pathogens, PCR assay has been integrated with other detection methods. Detection of *Ralstonia solancearum* (*Rs*) in plant tissues was accomplished by integrating the rapid self-replicating ability of bacteriophages with quantitative PCR (qPCR). Six bacteriophages were tested for their ability to specifically infect and lyse 63 strains of *Rs* and 72 isolates of other bacterial species. *Rs* strain GW-1 infecting ginger and the phage MDS1 were selected based on the susceptibility of host *Rs* strain and replication speed and reproductive burst sizes for the phage. The protocol was optimized using the primers based on the phage genome for detection of *Rs* from a number of substrates. In pure *Rs* culture, the procedure detected approximately 3.3 cfu/ml after an incubation of one hour with 5.3×10^2 pfu (plaque forming units)/ml of MDS1. The presence of *Rs* in infected potted ginger plants was detected by this protocol, the limit of detection being near 10^2 cfu/0.1 g of leaf tissues (Kutin et al. 2009).

A highly sensitive nested-PCR assay was developed based on a sequential nested amplification of PCR of a region of plasmid DNA that is highly conserved in *Xanthomonas axonopodis* pv. *citri* (*Xac*). The amplified PCR products were detected colorimetrically by DIANA (detection of immobilized amplified nucleic acid) method which uses labeled primers to allow amplification product capture and detection in a microtiter plate. Predicted amplification products were produced from all strains of *Xac* and also from four of six strains of *X. axonopodis* pv. *aurantifolii* (*Xaa*), but not from *X. axonopodis* pv. *citrumello*, incitant of citrus bacterial spot disease and other xanthomonads. This protocol was 50–100 times more sensitive than standard PCR assay. Immunomagnetic separation (IMS) of target pathogen prior to amplification by PCR was used to concentrate and recover *Xac* from the samples. The sensitivity of the assay could be further improved by immunocapture (IC)-PCR by concentrating the *Xac* from dilute plant extracts, by over 100-fold over nested-PCR format (Table 2.6). This integrated PCR format may be particularly useful for plant quarantine and certification programs (Hartung et al. 1996; Appendix 10).

Xylella fastidiosa (Xf), a xylem-limited bacterium (XLB) causes grapevine Pierce's disease (PD) in addition to diseases in citrus, almond and coffee plants. In order to enhance the sensitivity of detection of *Xf* and to reduce the effects of inhibitors of PCR, a pressure chamber was used to exude xylem fluids from grapevine shoots. *Xf* could be isolated on PD3 solid medium from the xylem fluids. Because of the presence of inhibitors in the xylem fluids no amplification of *Xf* DNA occurred. Hence, a purification procedure involving immunocapture (IC) was adopted prior to PCR. The *Xf* suspension (mixed pellets of *Xf* and antibodies) were

Table 2.6 Assessment of sensitivities of nested (N)-PCR and immunocapture (IC)-N-PCR for the detection of *Xanthomonas axonopodis* pv. *citri* (Hartung et al. 1996)

cfu/ml of cell suspension	Nested PCR	IC-N-PCR Experiment 1	IC-N-PCR Experiment 2
0	–	–	–
10	ND	–	+
10^2	–	+	ND[b]
10^3	+	+	+
10^4	+	+	ND
10^5	+	+	+
10^6	+	+	ND
10^7	+	+	ND
10^8	+	+	ND

[a] In experiment 1, the capture antibody used at 10 µg/10^6 beads; in experiment 2 the capture antibody used at 0.1 µg/10^6 beads
[b] ND–Not determined; presence (+) or absence (–) of an amplification product of the expected size in agarose gels

used as templates. Two target fragments of 511- and 733-bp were amplified from various xylem exudates samples. In a weakly infected grapevine plant, the 733-bp fragment was detected, indicating that IC-PCR protocol could be employed to detect Xf in the xylem exudates from plants before symptoms are expressed (Guo and Lu 2004; Appendix 11). *Xylella fastidiosa* (*Xf*) was detected by IC-PCR technique involving the capture of *Xf* cells by a specific antibody-coated magnetic immunocapture beads, followed by amplification of *Xf*-specific DNA fragment by PCR. Detection efficiencies of IC-PCR and ELISA techniques were similar in detecting *Xf* from naturally infected grapevine plants. As ELISA test was easier, less expensive and less-time consuming than IC-PCR technique, the ELISA test was recommended for use under field conditions for the detection of *Xf* on grapevines (Costa et al. 2004).

The bacterial blight disease in anthurium caused by *X. axonopodis* pv. *diffenbachiae* (*Xad*) could be detected effectively by IC-PCR procedure. A genus-specific MAb was employed for capturing *Xad* cells followed by PCR amplification with *Xad*-specific primers. The pathogen could be detected in both symptomatic and latently-infected anthurium plants. The IC-PCR format was found to be more sensitive than both standard PCR and indirect ELISA techniques used for comparison of efficiencies for the detection of *Xad* (Khoodoo et al. 2005). *Acidovorax avenae* subsp. *citrulli* (*Aac*), incitant of watermelon bacterial fruit blotch disease, was detected by IC-PCR technique. A polyclonal antibody specific to *Aac* was used to concentrate bacterial cells followed by standard PCR for amplification of *Aac* DNA by specific primers WFB1 and WFB2 designed on the sequences of 16S rRNA gene. The expected 360-bp DNA fragment was amplified only from four strains of *Aac* (Xiao et al. 2007).

Application of PCR in tandem with an immunoassay has been demonstrated to be more sensitive, rapid and effective for the detection of some bacterial plant pathogens. The immuno-enzymatic detection of PCR amplified products as in PCR-ELISA technique helps in circumventing the need for performing elelctrophoresis, image capture and other steps required. *Erwinia amylovora* (*Ea*) was detected reliably by using PCR–ELISA combination. The PCR primers were designed to amplify sequences of the ubiquitously present cryptic plasmid pEA29. Amplicons were labeled with 11-digioxigenin (DIG)-dUTP during amplification reaction, captured by hybridization to a biotinylated oligonucleotide in streptavidin-coated ELISA microplates, followed by detection with anti-DIG-Fab -peroxidase conjugated antibodies. Strains of *Ea* from various host plants and geographical locations could be detected by this PCR-ELISA protocol confirming the specificity and reliability of the procedure. The colorimetric and chemiluminescent detection assays had detection thresholds of 30 and 3 cfu/reaction tube respectively in pure cultures. This assay was able to detect *Ea* in both artificially inoculated and naturally infected plant organs. The results obtained with PCR-ELISA format were comparable to standard PCR assay (Merighi et al. 2000).

The sensitivity of PCR assay may be significantly enhanced by immunomagnetic capture (IC)-/separation (IMS)-PCR or enrichment-involved BIO-PCR techniques. The BIO-PCR procedure was developed by Schaad et al. (1995) based on an enrichment of the causal bacterial species using an appropriate nutrient medium, followed by enzymatic amplification of PCR targets by employing specific primers. *Erwinia herbicola* pv. *gysophila* (*Ehg*) was detected by BIO-PCR procedure in the program to establish disease-free nuclear stock of mother plants of gysophila (Manulis et al. 1998).

Pre-enrichment step when included prior to PCR, the assay was found to be effective for the reliable and sensitive detection of *Pseudomonas savastanoi* pv. *savastanoi* (*Pss*) infecting olive trees. A nonselective King's medium and a semi-selective PVF-1 medium were used for enrichment step prior to PCR amplification. The sequence of the gene *iaaL* of *P. savastanoi* EW2009 was used for designing primers for PCR amplification. The *iaaL*-derived primers directed the amplification of a 454-bp fragment from genomic DNA isolated from 70 strains of *P. savastanoi*, but not from any of the 93 non-*P. savastanoi* isolates tested. A pre-enrichment in King's medium did not improve the efficiency of detection of *Pss* by PCR in olive knot tissues from naturally infected plants. In contrast, bacterial pre-enrichment in the semiselective liquid medium PVF-1 improved the PCR sensitivity level significantly, allowing detection of the pathogen at concentration of 10–100 cfu/ml of plant extract. The enrichment-PCR protocol developed in this investigation could be employed for the detection of *Pss* from inoculated and naturally infected plants of different olive varieties. Furthermore, detection of *Pss* in symptomless stem tissues from naturally infected olive plants was successfully accomplished by the enrichment-PCR technique (Table 2.7). The enrichment-PCR technique was shown to be more sensitive and less cumbersome than isolation method and more sensitive than the standard PCR assay (Peñyalver et al. 2000).

Preenriching the pathogenic bacteria on agar media (BIO-PCR) to enhance the sensitivity of PCR was adopted for the detection of *Pseudomonas syringae* pv. *phaseoli* (*Psp*) in washings of leaves and seeds of bean plants (*Phaseolus vulgaris*),

Table 2.7 Detection of *Pseudomonas savastanoi* pv. *savastanoi* in asymptomatic stem tissues of olive varieties by enrichment-PCR technique and isolation method (Peñyalver et al. 2000)

	Number of positive samples/total tested				
Treatments	Blanqueta[a]	Gordal	Manzanilla	Picual	Picudo
Without enrichment					
A	1/6	2/3	2/6	5/20	0/3
B	0/6	0/3	0/6	4/20	0/3
Pre-enrichment with King's B medium					
A	0/6	0/3	3/6	4/20	1/3
B	1/6	3/3	4/6	6/20	2/3
Pre-enrichment with PVF-1 medium					
A	0/6	1/3	0/6	4/20	1/3
B	3/6	2/3	5/6	10/20	3/3

A – Isolation method; B – PCR and restriction analysis
[a] Olive varieties

using the classical PCR assay. A high throughput 96-well membrane BIO-PCR technique was developed to reduce the labor needed for the earlier BIO-PCR format. The primers were designed from the sequences of a tox-argK chromosomal cluster of the *Psp*-specific phaseolotoxin gene. Samples (1.2 ml) were filtered under vacuum in the 96-well membrane plates followed by incubation on soft agar medium for 48–52 h. After washing each well with sterile water, the samples were subjected immediately to nested (two-step) PCR or real-time PCR procedure. The standard PCR and BIO-PCR assays did not detect *Psp* in leaf washings. In contrast, the membrane real-time BIO-PCR format detected *Psp* at a concentration of 80 cfu/ml of bacterial cell suspension. The sensitivity of detection from leaf washings was lower compared with seed washings (Schaad et al. 2007).

Xanthomonas campestris pv. *viticola* (*Xcv*) causing grapevine bacterial canker disease was detected by PCR assay employing the primer pairs Xcv1F/Xcv3R and RST2/Xcv3R based on the partial sequences of *hrpB* gene. When the primers RST2 and Xcv3R were used for detection of *Xcv* in artificially inoculated petioles, amplification of *Xcv* DNA was positive, only when a pre-enrichment step was included. The pathogen could not be detected when suspensions were prepared from macerated tissue and added directly to the reaction mixture. By including the enrichment step prior to PCR amplification, only DNA from viable bacterial cells was detected. Detection of *Xcv* by BIO-PCR format was possible only in samples from symptomatic petioles of inoculated grapevine plants. The PCR protocol for detection of *Xcv* required only 3–4 days, while isolation method provided results after at least a period of 10 days. Hence, BIO-PCR format could be an effective tool for monitoring *Xcv*-contaminated plant materials effectively (Trindade et al. 2007).

Erwinia amylovora (*Ea*) infecting apple was detected by employing two 30-mer oligonucleotide primers corresponding to the sequences near the termini of the plasmid pEA71 insert. A 187-bp product was amplified from 69 strains that hybridized with total *Ea* DNA. Using the direct DNA extraction method, involving the use of

GeneReleaser™, as few as 10 cfu of *Ea* per tissue sample could be detected. However, detection at this level of *Ea* population was not always consistent. When the selective medium CCT was used to recover *Ea* from plant tissues, the pathogen could be detected consistently at *Ea* population concentration of 10 cfu/ml. Likewise, the sensitivity of PCR detection of *Ea* in inoculated apple flowers was improved by the enrichment-PCR method. The limit of detection of *Ea* in inoculated apple flowers was 10 cfu/ml at 6 days after inoculation with different concentration of *Ea*. Using the enrichment method, *Ea* was detected in samples of 100 flowers inoculated with a concentration of 10 cfu/ml of the pathogen (Taylor et al. 2001).

The crown gall pathogen *Agrobacterium tumefaciens* (*At*) causes tumors in infected peach plants. Enrichment of bacterial cells in selective liquid medium was carried out prior to PCR amplification. Sample extracts were incubated for 48 h in a liquid medium selective for biovar 2 of *At*. The DNA was extracted from the liquid broth. PCR products 172-bp for *At* DNA and 373-bp for internal control DNA were obtained for most of the samples analyzed (Cubero et al. 2002). *Clavibacter michiganensis* subsp. *michiganensis* (*Cmm*) causing tomato bacterial canker disease is a quarantine organism in Europe. Using the species-specific primers CMM5 and CMM6, the presence of the pathogen in infected tomato seedling was detected by PCR assay based on the amplification of a 614-bp product. An optimized BIO-PCR procedure was developed for the detection of *Cmm* in tomato transplants during the early latent stage of infection. Yeast glucose mineral agar (YGMA) medium was used for pre-enrichment with an incubation period of 2–3 days. It was possible to detect *Cmm* at a concentration of 4×10^2 cfu/ml in tomato seedling tissues. This protocol allowed detection of *Cmm* at a distance of 8 cm farthest from inoculation site in 95% of the samples, after 3 days of incubation in YGMA medium. The results indicated the possibility of detecting *Cmm* in very early stages of infection of tomato seedlings (Burokiene 2006).

The PCR amplicons are subjected to random fragment length polymorphism (RFLP) analysis for the detection and identification of some bacterial pathogens. *Erwinia carotovora* (*Ec*) subspecies could be detected by PCR-RFLP analysis based on a pectate lyase-encoding gene (*pel*), since pectate lyases have important role in the development of soft rot diseases. By adopting a 48-h enrichment step prior to PCR amplification, the sensitivity of the PCR-RFLP analysis was substantially improved. The presence of *E. caraotovora* subsp. *atroseptica* (*Eca*) in wash water, leaves, stem and tuber peel extract could be detected reliably (Helias et al. 1998). When the strains of *E. carotovora* subsp. *carotovora* (*Ecc*) were examined for the *pel* gene with RFLP using *Sau*3AI, three patterns (1, 2, 3) were observed (Seo et al. 2003). RFLP pattern 3 of *pel* gene contained only type 2 strains from mulberry. Seven additional mulberry strains were subjected to RFLP pattern analysis. Three strains of *E. carotovora* from mulberry with RFLP pattern 3 of the *pel* gene, produced the expected size of ca. 430-bp product. The PCR amplification of these strains allowed them to be assigned to *E. carotovora* (Seo et al. 2004).

Detection and identification of *Xanthomonas campestris* pv. *viticola* was accomplished by employing specific primer pairs Xcv1F/Xcv3R and RST2/Xcv3R in grapevine plants exhibiting symptoms of bacterial canker disease.

By incorporating the enrichment step prior to the PCR amplification, bacterial colonies could be recovered from symptomatic petioles of inoculated plants, but not from inoculated asymptomatic petioles. PCR assay was performed for aliquots of plate wash obtained from symptomatic tissues and using cell suspensions from a single suspect colony. Amplifcations occurred in both cases. The identity of the bacterium was established by *Hae*III digestion of PCR products. The RFLP analysis revealed the similarity of the restriction profiles of the samples to that of *Xcv* (Trindade et al. 2007).

Xanthomonas axonopodis pv. *manihotis* (*Xam*) causing bacterial blight of cassava is transmitted through planting materials and contaminated seeds, offering distinct scope for preventing the introduction of the disease by quarantine regulations. The presence of a specific 898-bp PCR fragment generated from all 107 pathogenic strains of *Xam* was revealed by PCR-RFLP technique. The DNA of five nonpathogenic strains of *Xam* or several closely related xanthomonads or cassava-associated saprophytes did not yield the expected product following amplification by PCR. Hence, this approach offers the possibility of early detection of *Xam* for effective prevention of disease introduction in new areas (Verdier et al. 1998).

2.1.4.4 Real-Time Polymerase Chain Reaction

The polymerase chain reaction (PCR) and its variants have been employed to detect the bacterial pathogens affecting various crops. However, it has not been possible to detect all strains of a bacterial species such as *Xanthomonas axonopodis* pv. *citri* (*Xac*) (Hartung et al. 1996). Further, the post-PCR steps involving the use of hazardous chemicals have to be performed extending the period needed for obtaining the results. In order to reduce the time required and to make the assay system portable, facilitating its use under field conditions away from the laboratory, an improved detection system had to be developed. Real-time PCR technology has been shown to be sensitive, reliable and rapid for detection and identification of microbial plant pathogens including bacterial pathogens. Real-time PCR, compared with standard PCR, is simpler to perform, less labor-intensive and much faster providing the results in less than 60 min, if appropriate sampling method is adopted.

A fluorogenic (TaqMan) PCR assay was developed for the detection of *Ralstonia solanacearum* (*Rs*), causative agent of bacterial wilt disease of several crop plants. Two fluorogenic probes were utilized in a multiplex reaction: one broad-range probe (Rs) detected all biovars of *Rs* and a second more specific probe (B2) detected only biovar 2A. Amplification of the target was measured by the 5′ nuclease activity of *Taq* DNA polymerase on each probe, resulting in emission of fluorescence. TaqMan PCR was performed with the DNA extracted from 42 genetically or serologically related strains of *Rs* to demonstrate the specificity of the assay. Detection limit of the assay was $\geq 10^2$ cells/ml in pure cultures. When inoculated potato tissue extract was tested, the sensitivity of the assay was reduced. The fluorogenic probe (COX) designed with potato cytochrome oxidase gene sequence was used as an internal control and it detected potato DNA in an RS-COX multiplex

TaqMan PCR with infected potato tissues. The specificity and sensitivity of the TaqMan assay showed high speed, robustness and reliability. This protocol offers potential advantages in routine indexing of potato tubers and other plant materials for detecting the presence of *R. solanacearum* (Weller et al. 2000).

Ralstonia solanacearum (*Rs*) race 3 biovar 2 is a subspecific taxon within the species complex that infects solanaceous plants at low temperatures and causes brown rot of potato tubers. A mutiplex TaqMan method specific for the detection of all strains of *Rs* as well as race 3 bv. 2 permitted analysis in real-time PCR within a closed tube system, reducing the risk of cross contamination between samples within the laboratory (Weller et al. 2000). Diagnostic PCR methods used for regulatory purposes have to provide adequate provision for validating negative results as well as confirming positive results. The negative results, in a later investigation, were validated through the use of a reaction control plasmid, designated pRB2C2 which was designed to generate a 94-bp product using the same amplimers targeting the primary diagnostic 68-bp sequence in *Rs* race 3 bv. 2 DNA. SYBR Green was included in the reaction mix to facilitate the identification of postreaction products using a melt peak analysis. The reaction control (94-bp) and diagnostic target (68-bp) amplicons had melt peak temperatures of about 90°C and 83°C respectively. Thus the positive results could be easily confirmed and distinguished from reaction control product. The modified TaqMan assay procedure, developed in this investigation, successfully detected *Rs* race 3 bv. 2 in infected asymptomatic tomato stems and leaves of tomato, as well as in potato tubers and stems (Smith and De Boer 2009).

A sensitive and reliable SYBR Green real-time PCR assay primers designed to amplify conserved regions of an essential pathogenicity gene was developed for the detection of *Xanthomonas axonopodis* pv. *citri* (*Xac*) and its strains present in the field-collected plant samples. Primers were selected to amplify relatively short segments from highly conserved regions of all members of the *pthA* gene family. A portable, field-hardened RAPID 7200 system was used for performing real-time PCR assay. Single lesion sampling methods were adopted, as they needed minimal handling and allowed complete real-time diagnosis in a total time of 4 h and with an apparent sensitivity of less than 10 cfu of target pathogen cells from diseased lesion samples. Significant improvement in sensitivity was achieved by the use of $CaCO_3$, minimizing the amount of lesion tissue sampled, soaking or swiping but not grinding the lesions and using appropriate primer designs. The real-time PCR assay proved to be highly reproducible, quantitative and very sensitive, using minimum labor and time (less than 4 h). The high level of sensitivity of the real-time PCR protocol was revealed by this molecular detection for the first time of *Xac* in a herbarium sample from a 1912 canker outbreak in Florida (Fig. 2.8) (Mavrodieva et al. 2004) [Appendix 12].

A quantitative real-time (QRT)-PCR protocol was developed by using the point in time when amplification of the target is first detected to estimate bacterial concentration. Fluorescence can be generated by nonspecific dyes such as SYBR green that binds to ds-DNA or by fluorogenic probes. During amplification, a specific DNA probe conjugated with a reporter and quencher dye hybridizes with target sequence and is degraded by *Taq* polymerase. After degradation, reporter and

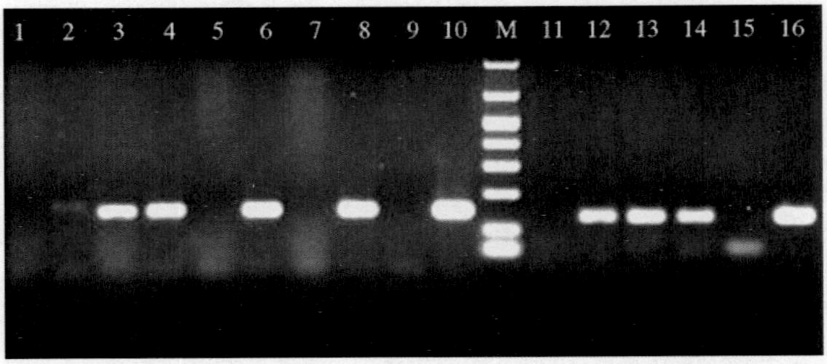

Fig. 2.8 Detection of *Xanthomonas axonopodis* pv. *citri* in direct tissue extracts of infected plants by real-time PCR assay. Note the bands of expected size after loading the real-time PCR products on agarose gels (Courtesy of Mavrodieva et al. 2004; The American Phytopathological Society, St. Paul, MN)

quencher are separated resulting in emission of a fluorescent signal of which the accumulation is monitored during the consecutive PCR cycles. The cycle at which the fluorescent signal exceeds a certain background fluorescence level, known as the threshold cycle (Ct), is directly proportional to the amount of target DNA present in the sample (Gibson et al. 1996).

Citrus bacterial canker (CBC) is caused by *X. axonopodis* pv. *cirti* (*Xac*) that produces type A citrus canker and by *X. axonopodis* pv. *aurantifolii* (*Xaa*) that causes types B and C citrus canker (Schubert et al. 2001) CBC disease is under international quarantine regulations, while citrus bacterial spot (CBS) disease due to *X. axonopodis* pv. *citrumelo* is not regulated by quarantine enforcements. A quantitative real-time (QRT)-PCR was employed for detection, identification and enumeration of bacteria in citrus plant samples infected by *Xac* and pv. *citrumelo*. Three sets of primers based on the pathogenicity gene *pth* in *Xac*, a ribosomal gene in *X. axonopodis* pv. *citrumelo* and the leucine-responsive regulatory protein (*lrp*) gene in both pathovars were combined with TaqMan probes and applied for specific strain detection and quantification. Calibration curves were generated for *Xac* and pv. *citrumelo* based on the assumption that the concentration of bacteria was directly related to DNA concentration. Specificity for CBC diagnosis was ensured, because the gene *pth* is present only in *Xac*, but not in other bacteria belonging to the same genus attacking the same host. The sensitivity of the QRT-PCR was improved, as evidenced by apparent overestimation of bacterial cells, compared with number of colonies detected in a semiselective medium. However, in most of the samples, colony counts coincided with the data obtained by QRT-PCR assay. Although, QRT-PCR format is suitable for quantification, DNA detection does not necessarily correspond with the presence of viable bacterial cells. However, QRT-PCR procedure, reduced the risk of false positives and enhanced the reliability of detection and differentiation of strains of *Xanthomonas* by allelic discrimination (Cubero and Graham 2005).

Xylella fastidiosa (*Xf*), a xylem-limited bacterial (XLB) species causing grapevine Pierce's disease demands more sensitive detection method(s), because of low

Table 2.8 Relative efficiencies of detection of *Xylella fastidiosa* using 16S primers and probe in real-time PCR assay under field and subsequent laboratory conditions (Schaad et al. 2002b)

	Number of samples positive/samples tested		
	16S[a]		
Vineyard	Field	Laboratory	Agreement of 16S field and laboratory[b]
Exeter	3/5	2/5	3/4
Bakersfield	3/10	1/8	4/8
McFarland	5/9	4/9	8/9
St. Helena	9/18	2/15	8/15
Total	20/42 (47%)	9/37 (24%)	23/36 (63%)

[a]Comparison between PCR conducted with 16S primers and probe on-site under field conditions and later in the laboratory
[b]Only those samples tested by both 16S field and 16S laboratory were considered

concentrations of bacteria in asymptomatic plant tissues and its irregular distribution in infected plants. A direct real-time PCR assay was developed for the detection of *Xf* in plant sap and samples of macerated chips of secondary xylem from trunks of vines without extraction of DNA. Using two different sets of primers and probe, *Xf* was detected in seven of 27 vines (26%) from four of six vineyards sampled at 10–12 days after budbreak in four counties of California. The results of direct real-time PCR assay were confirmed by standard PCR procedures (Table 2.8). As this protocol could detect *Xf* in asymptomatic vines just coming out of dormancy, it has potential for application for rouging out infected vines to prevent the spread of the disease from such foci of infection (Schaad et al. 2002b).

The generic detection of *X. fastidiosa* (*Xf*) was achieved by applying a dual purpose conventional PCR and quantitative PCR (TaqMan™) system. A unique region common to the sequenced genomes of four *Xf* strains was amplified by the primers HL5 and HL6. A 221-bp fragment from strains associated with grape Pierce's disease, almond leaf scorch and oleander leaf scorch diseases and also citrus variegated chlorosis was amplified by the HL5 and HL6 primers. Serial dilutions of *Xf* in water, grape extracts and insect cells were used for preparing the standard curves/regression curves to calculate the correlation coefficients. In quantitative PCR format, Ct values ranged between 20 and 36 cycles for 5–10^5 bacterial cells per PCR. *Xf* could be detected reliably and rapidly in grapes and almonds as well as in the insect vectors by applying the quantitative PCR protocol developed in this investigation which showed high level of sensitivity and specificity (Francis et al. 2006).

Strains of *Ralstonia solanacearum* (*Rs*), causative agent of bacterial wilt diseases affecting several economically important crops were detected by employing PCR assay and quantified by applying real-time PCR technique. PCR using the *Rs* universal primer pair 759/760 amplified the expected single 280-bp fragment from all the 15 strains tested. However, none of the 15 strains contained the DNA fragment that was amplified by the race 3-specific primer set 630/631 which indicated that none of the Florida isolates belonged to the quarantined subgroup R3B2. A quantitative real-time PCR format using the universal *Rs* primer/probe set amplified the expected *Rs* target DNA sequence from the strains tested. Real-time PCR format using the biovar 2-specific B2 primer/probe also amplified the expected target sequence from the biovar 2 reference strains as well as biovar 1 strain (potato isolate from Peru). Accurate race-specific detection of *Rs* is essential for crops susceptible to both race 1 and race 3 strains such as geranium, since strict government quarantine and eradication measures are adopted, because of the 'zero tolerance' level applied to geranium (Ji et al. 2007).

A real-time PCR assay was developed based on a 5′ nuclease and minor groove binding (MGB) probe to provide a fast, sensitive and reliable method for the detection of *Xylophilus ampelinus* (*Xa*), infecting grapevine, is a slow-growing bacterium belonging to the Betaproteobacteria (Comamonadaceae). Based on MGB-probe chemistry and real-time PCR product detection, this protocol avoids laborious post-PCR handling and the associated high risk of cross-contamination. This protocol could amplify all isolates of *Xa* from a variety of geographical locations. The target sequence appeared to be highly conserved in *Xa* isolates. The specificity of the assay was revealed by the absence of specific signal with other bacterial pathogens of grapevine or grapevine plant tissues with their microflora. Used in combination with the DNeasy plant mini kit, the sensitivity of *Xa* detection was approximately 100 cells from tissue extracts, surpassing the sensivity of nested-PCR method used earlier by ten times. In the case of field samples, a high correlation was observed between real-time PCR cycle threshold (Ct) values obtained and *Xa* isolation on artificial media. The real-time PCR procedure is a reliable and sensitive test for detection of *Xa* and it can be complementary to isolation or other detection methods and suitable for relative quantification of *Xa* in grapevine tissues of different cutivars (Dreo et al. 2007).

A combination of standard PCR and real-time PCR targeting repectively the putative DNA polymerase and 16S rDNA sequence of *Candidatus Liberibacter asiaticus* (*CLa*) was adopted for the detection of *CLa*, causing huanglongbing (HLB) disease of citrus. In addition to detection of *CLa*, its distribution in various tissues such as bark, leaf midrib, roots and floral and fruit parts of infected citrus was determined, using this combination of two techniques. Quantification of analysis of *CLa* showed that it was distributed unevenly in planta and ranged from 14 to 137,031 cells/μg of total DNA in different tissues. A relatively high concentration of *CLa* was observed in fruit peduncles (Tatineni et al. 2008)

Pantoea stewartii subsp. *stewartii* (*Pss*) (Syn. *Erwinia stewartii*) causing Stewart's bacterial wilt of maize is transmitted through the seeds at a very low frequency and overwinter in mature corn flea beetles (*Chaetocnema pulicaria*). Although ELISA test was commonly employed earlier for the detection of *Pss* in infected plants, TaqMan® real-time PCR assay was developed, because of its higher level of sensitivity and specificity and the possibility of obtaining results rapidly. This assay was based on the primers and probe designed on the sequences of *cpsD* gene for detection and identification of *Pss* in pure cultures, infected leaf or maize seeds. The *cps* gene cluster is one of the two genes that play an important role in the pathogenicity and virulence of *Pss* under optimal conditions, the selected primers and probe were specific for the detection of all 14 reference *Pss* strains by real-time PCR assay. The non-*Pantoea* and eight other *Pantoea* strains tested negative. The protocol could detect 1 pg of purified DNA and 10^4 *Pss* cfu/ml (ten cells per reaction) in pure cultures. Direct processing of leaf lesions by real-time PCR detected ten cells per reaction. This assay eliminated false negative results and considerably reduced the time required for providing the results. Further, this protocol was found to be robust and rapid and it could be automated with high sample throughput potential, permitting analysis of up to 96 samples within 24 h of receiving the samples (Tambong et al. 2008).

Real-time PCR technique was successfully applied for the detection of *Leifsonia* (= *Clavibacter*) *xyli* subsp. *xyli* (*Lxx*) causing sugarcane ratoon stunting disease (RSD).The youngest, fully expanded leaves of three sugarcane cultivars sampled at biweekly interval from the field nurseries were tested. Positive detection of the pathogen was possible even in 1 month old plant, indicating the high level of sensitivity of the assay, when the bacterial concentration was expected to be low. Real-time PCR assay was more sensitive than the standard PCR procedure, in addition to being capable of providing quantitative estimation of pathogen populations based on which the levels of resistance to *Lxx* could be ranked. Further, the results of real-time PCR assay were coroborated by tissue-blot immunoassay, when the assessments were made in tissues of 7–9 months old sugarcane stalks. As the real-time PCR format could be used for testing the resistance of the cultivars at the young stage, appreciable time would be saved for obtaining the results, facilitating screening of additional batches of sugarcane genotypes for their resistance to RSD within a season (Grisham et al. 2007).

2.1.4.5 Loop-Mediated Isothermal Amplification Technique

The DNA amplification method designated loop-mediated isothermal amplification (LAMP) is based on auto-cycling strand displacement DNA synthesis by a DNA polymerase which has high strand displacement activity and a set of specially designed inner and outer primers. Amplification is completed within 30 min using a simple affordable water bath which is maintained at 65°C constantly. The target bacterial pathogen may be readily detected by employing LAMP technique even in under-equipped laboratories of extension centers and quarantines.

The LAMP technique was applied for the detection of citrus greening (CG), also known as huanglongbing (HLB) disease, caused by *Candiatus* Liberibacter (*CL*). The 16S rDNA and 16S/23S intergenic regions of *CL* was not suitable for LAMP, because they share homology with plant genomic DNA. Hence, the *nus*G-*rpl*KAJL-*rpo*B gene cluster was selected, because of its specificity. Thermal asymmetric interlaced (TAIL)-PCR assay was performed to amplify the uncharacterized regions adjacent to the *nus*G-*rpl*KAJL-*rpo*B gene cluster of *CL* isolates from different locations in Japan and Indonesia. Standard PCR was employed to amplify the internal *nus*G-*rpl*KAJL-*rpo*B gene cluster of *CL* isolates and the complete sequence of the 6.1 kb fragment was determined. A LAMP assay based on the conserved sequence of the *nus*G-*rpl*KAJL-*rpo*B gene cluster was developed for the detection of *CL*. The LAMP product was detected on nylon membranes by staining with AzurB with a detection limit of about 300 copies of the target fragment. This assay does not require thermocycler and electrophoresis equipments as in standard PCR format. Detection of colored products on the nylon membrane is not only simple, but also very rapid and widely applicable for possible elimination of infected plants or planting materials (Okuda et al. 2005).

The uneven distribution and low concentration of '*Candidatus Liberibacter asiaticus*' (Las), African (Laf) and American (Lam) forms has made detection of the pathogen consistently associated with HLB or greening disease affecting citrus plants very difficult. Hence, the conventional PCR, LAMP and TaqMan real-time PCR systems were evaluated for their relative accuracy, sensitivity and specificity of detection of the pathogen. With primer sets OI1/OI2c or OAI/OI2c, the PCR assay could detect '*Ca*. Liberibacter' spp. in 2–10 μl of 100 μl of the original DNA extracts obtained from mixed samples made of 1 g of healthy midribs with a minimum of 20 mg of infected midribs. The lower detection limit of the protocol developed in this investigation was equivalent to 0.01 mg of infected midribs per reaction, showing improvement over earlier protocols. The TaqMan real-time PCR format successfully detected '*Ca*. Liberibacter' spp. from the equivalent of 20 ng of midribs from symptomatic leaves showing a 100-fold increase in the sensitivity of detection. The higher sensitivity of TaqMan real-time PCR assay may be due to increased amplification efficiency and its lower vulnerability to PCR inhibitors from citrus plants. Among the 204 plants, 126 showed foliar symptoms and 78 were asymptomatic. Of the 17 HLB symptomatic greenhouse-grown inoculated plants, the pathogen was detected by conventional PCR and TaqMan real-time PCR assays. But the pathogen was not detected by LAMP in three of these symptomatic plants. All 63 asymptomatic greenhouse-grown plants tested negative for '*Ca*. L. asiaticus' by conventional PCR, but TaqMan real-time format could detect this form in one of these plants. However, LAMP gave positive results for six of these asymptomatic plants. The results indicated that real-time PCR system was a more efficient tool for early diagnosis of HLB pathogen compared with conventional PCR assay (Li et al. 2007).

2.1.4.6 Isothermal and Chimeric Primer-Mediated Amplification of Nucleic Acids (ICAN) Technique

A detection system comprising of amplification of 16S rDNA by isothermal and chimeric primer-mediated amplification of nucleic acids (ICAN) and detection of amplified products with cycling probe (cycleave) technology was developed for the detection of '*Candidatus Liberibacter asiaticus*'. Cycleave ICAN technology is suitable for bacterial pathogens for which information on the genomic DNA as in the case of '*Ca. Liberibacter asiaticus*' is not entirely available, since this technique requires only two DNA–RNA chimeric primers based on 16S rDNA sequences. The cycling probe is a chimeric DNA–RNA probe that hybridizes to an amplified target sequence, but not to a nonspecific product, primer dimers. Once the probe hybridizes, the RNA part of the probe is cleaved by TliRNaseH. The fluorescer ROX and quencher Eclipse on each side of the probe are separated and red fluorescence is emitted. The red fluorescence is produced rapidly from positive samples confirming the presence of target DNA and the occurrence of false-positives can be prevented.

The Cycleave ICAN technique and the conventional PCR assay were evaluated for their efficacy, using the PCR primers for '*Ca. Liberibacter asiaticus*' (CLA) 16S rDNA. The PCR assay positively detected a 1,160-bp rDNA fragment in the positive templates from 40 ng to 102 pg. In the Cycleave ICAN technique, red fluorescence was detected from 40 ng to 0.512 pg. Weak fluorescence was observed at 0.102 pg concentration. No fluorescence was observed in healthy samples. The sensitivity of the Cycleave ICAN technique was about 125 times greater than that of the conventional PCR assay. Of the 95 samples tested, CLA was detected in 35 and 39 samples by conventional PCR and Cycleave ICAN protocols respectively. Cycleave ICAN procedure has distinct advantages over conventional PCR assay for the detection of CLA. The cycling probe enables the reaction to be performed in one tube, avoiding possible cross-contamination across the samples. Further the results can be obtained rapidly, as there is no need for electrophoresis for resolving PCR amplicons. As the time required is short, this technique has the potential for large scale application for diseases like citrus huanglongbing (HLB) or greening disease for which early eradication of infected plants forms one of the important component of disease management system (Urasaki et al. 2008).

2.1.4.7 DNA Array Technology

DNA array technology based essentially on a reverse dot blot technique has emerged as a potential technique for rapid detection and identification simultaneously of several microbial pathogens present in a sample. An array of species-specific oligonucleotide probes representing different target pathogen is built on a solid support such as nylon membrane or microscope slide. They can be probed readily with labeled PCR amplicons from plant samples. Conserved primers were employed to amplify common bacterial pathogen DNA fragment from extracts of potato tissues that might contain bacterial pathogens. The presence of DNA sequences indicative of target pathogen species would be revealed by hybridization to species-specific oligonucleotide probes within the array.

DNA array method was applied for the detection and identification of the bacterial pathogen *Erwinia amylovora* and fungal pathogens *Botrytis cinerea*, *Penicillium expansum*, *Podosphaera leucotricha* and *Venturia inaequalis* infecting apples. The oligonucleotide probes were spotted on nylon membrane by an amine-modified linker arm and arranged in a precise pattern to form an array. All pathogens were detected and identified accurately. The probe sequences for EA-H1 and EA-H4C, designed from the ITS region of *Erwinia amylovora* (*Ea*) were specific to two of the three *Ea* isolates and they did not react with the third isolate (G-5) from pear. However, the isolate G-5 positively reacted with the probe EA-H3d that was designed for the detection of both *Ea* and *E. pyrifoliae*, infecting Asian pear (Scholberg et al. 2005).

A PhyloChip high density 16S rRNA gene microarray and 16S rRNA clone library methods were applied to assess the diversity of bacteria present in the leaf midrbs of citrus plants affected by huanglongbing (HLB) disease. PhyloChip analysis indicated that 47 orders of bacteria in 15 phyla were present in the citrus leaf midribs, while 20 orders in 8 phyla were observed with cloning and sequencing procedure. PhyloChip arrays indicated that nine taxa were significantly more abundant in symptomatic midribs than in asymptomatic midribs. '*Candidatus Liberibacter asiaticus*' was detected at a very low level in asymptomatic plants, but was over 200 times more abundant in symptomatic plants. The results of PhyloChip analysis were further confirmed by sequencing 16S rRNA gene clone libraries which indicated the dominance of '*Candidatus Liberibacter asiaticus*' in symptomatic leaves. These findings lend support to the view that '*Candidatus Liberibacter asiaticus*' may be the causative agent of HLB disease (Sagaram et al. 2009).

2.1.4.8 Multilocus Sequence Typing System

Multilocus sequence typing (MLST) procedure is useful to detect and identify strains of a bacterial species based on the differences in the nucleotide sequences of a small number of genes. In this system, each allele of a gene is allotted a number and each strain characterized (for *n* loci) is represented by a set of *n* numbers defining the alleles at each locus. The sequence type (ST) is charactaeized by these numbers. A seven-locus MLST data is considered conventionally and this set of data indicate the basis of strain discrimination effectively. A single-nucleotide difference always produces a new allele in an MLST data set which distinguishes this method from other methods like multilocus enzyme electrophoresis (MLEE) and pulsed field gel electrophoresis (PFGE) which require greater number of substitution for discrimination (Peacock et al. 2002).

Strains (25) of *Xylella fastidiosa* (*Xf*) from grapevine (PD strains), oleander (OLS strains), oak (OAK strains), almond (ALS strains) and peach (PP strains) were detected and differentiated by using an initial set of sequences of ten loci (9.3 kb). Nineteen different STs (ST1–ST19) were recognized from the allelic profiles of the *Xf* strains. Based on the allelic profiles, six clonal complexes were formed to include PD, OLS, ALS, OAK and PP strains. The MLST approach, because of its simplicity, compared with phylogenetic investigations, offers a distinct advantage by providing required information for rapid recognition of the incidence of an unusual or new isolate of *Xf* in a given location for initiating damage control process. In addition, MLST is also useful for efficiently cataloguing the genetic diversity within a bacterial species (Scally et al. 2005).

2.2 Detection of Phytoplasmal Pathogens in Plant Organs

Phytoplasams, earlier designated mycoplasma-like organisms (MLOs), belong to the class Mollicutes which include minute, cell wall-less organisms that primarily inhabit phloem sieve elements of infected plants. Aster yellows (AY) disease described by Kunkel (1926) was considered to be due to a virus, because of the similarities in the modes of transmission through graft and leafhopper vectors. However, the plant-infecting causative agents were recognized as a distinct group of plant pathogens and designated mycoplasm-like organisms (MLOs), following the studies on certain yellows type of plant diseases by Doi et al. (1967) and Ishiie et al. (1967). The trivial name 'phytoplasma' was officially adopted in 1994 to replace the existing name 'mycoplasma-like organism'. Phytoplasma-associated plant diseases are primarily transmitted by leahoppers, planthoppers and psyllids. In plant hosts, the phytoplasmas are restricted to phloem tissues, whereas they occur in insect hosts in haemolymph. They have to enter the salivary glands for transmission to the healthy plant.

None of the phytoplasmas has been cultured in a cell-free medium so far and hence information on their cultural characterisitics is lacking. Based on the electron microscopic observations made on the phytoplasmas in phloem tissues, wide variations in size and shape (pleomorphic) and absence of cell wall could be visualized. In place of the cell wall, they have a triple-layered unit membrane and they are insensitive to penicillin, but sensitive to tetracyclines. Bacteria, in contrast, have distinct cell wall and are sensitive to penicillin. Because of the absence of a rigid cell wall, phytoplasmas are able to pass through pores of 220 nm diameter, even though the diameter of a viable cell may be more than 300 nm (Musetti and Favalli 2004). *Spiroplasma citri*, causative agent of citrus stubborn and corn stunt diseases, has been isolated in pure cultures and its pathogenicity was established using insect vectors (Davis and Lee 1982). On the other hand, all attempts to culture phytoplasmas in pure culture have been unsuccessful so far. Hence, precise identities of the phytoplasmas could not be established depending on the morphological and biochemical characteristics as in the case of bacterial pathogens.

During the last 2 decades, diagnostic methods based on serological and nucleic acid sequence characterisitics have been instrumental in gathering information useful for the detection, identification and quantification of phytoplasmal pathogens infecting various plant species, as well as their insect vectors. Poly- and monoclonal antibodies (PAbs and MAbs), cloned phytoplasma DNA probes and especially phytoplasma-specific universal (generic) or phytoplasma group-specific PCR primers designed based on the sequences of highly conserved 16S rRNA gene have been employed for detection and identification of a wide range of phytoplasmas that are presumably the cause of the diseases affecting various plant species (Lee et al. 1998a).

Restriction fragment length polymorphism (RFLP) analysis of PCR-amplified 16S rDNA sequences using several restriction endonucleases such as *Alu*I, *Hae*III or *Rsa*I was employed to identity and differentiate phytoplasmas.This resulted in characteristic RFLP patterns for different strains which could be resolved by agarose or acrylamide gel electrophoresis (Lee et al. 1993; Schneider et al. 1993). This procedure was found to be simple, reliable and practical. The phytoplasma groups identified based on comprehensive RFLP analyses with enough restriction enzymes have been considered with phylogenetic groups delineated based on sequence data. Based on extensive RFLP analyses of phytoplasma 16S rDNA sequences, phytoplasmas have been classified into 14 groups and 38 subgroups (Lee et al. 1998b). However, this method did not provide clear resolution of all 16Sr

Table 2.9 Classification of selected strains of phytoplasmas based on in silico RFLP analysis (Lee et al. 2007)

16S rRNA group	Subgroup	Phytoplasma (strain designation)
16SrI: Aster yellows (AY)	I-B	'*Candidatus Phytoplasma asteris*'
16SrII: Peanut witches' broom group	II-A	Peanut witches' broom
	II-B	'*Ca. P. aurantifolia*'
	II-D	Papaya yellow crinkle
16SrIII: X-disease group	III-A	Western X-disease
16SrIV: Coconut lethal yellows group	IV-A	Coconut lethal yellowing
16SrV: Elm yellows group	V-A	'*Ca. P. ulmi*'
	V-B	'*Ca. P. ziziphi*'
16SrVI: Clover proliferation group	VI-A	'*Ca. P. trifolii*'
16SrVII: Ash yellows group	VII-A	'*Ca. P. fraxini*'
16SrVIII: Loofah witches' broom group	VIII-A	Loofah witches' broom
16SrIX: Pigeonpea witches' broom group	IX-A	Pigeonpea witches' broom
	IX-D	'*Ca. P. phoenicium*'
16SrX: Apple proliferation group	X-A	'*Ca. P. mali*'
	X-B	'*Ca. P. prunorum*'
	X-C	'*Ca. P. pyri*'
	X-D	'*Ca. P. sparti*'
16SrXI: Rice yellow dwarf group	XI-A	'*Ca. P. oryzae*'
16SrXII: Stolbur group	XII-A	Stolbur
	XII-B	'*Ca. P. australiense*'
	XII-D	'*Ca. P. japonicum*'
	XII-E	'*Ca. P. fragariae*'
16SrXIII: Mexican	XIII-A	Mexican periwinkle virescence
16SrXIV: Bermudagrass white leaf group	XIV-A	'*Ca. P. cyanodontis*'
16SrXV: Hibiscus witches' broom group	XV-A	'*Ca. P. brasiliense*'
16SrXVI: Sugarcane yellow leaf syndrome group	XVI-A	'*Ca. P. graminis*'
16SrXVII: Papaya bunchy top group	XVII-A	'*Ca. P. caricae*'
16SrXVIII: American potato purple top wilt group	XVIII-A	'*Ca. P. americanum*'
16SrXIX: Japanese chestnut witches'broom group	XIX-A	'*Ca. P. castaneae*'
16SrXX: Buckthorn witches' broom group	XX-A	'*Ca. P. rhamni*'
16SrXXI: Pine shoot proliferation group	XXI-A	'*Ca. P. pini*'
16SrXXII: Nigerian coconut lethal decline (LDN) group	XXII-A	Nigerian coconut lethal decline
16SrXXIII: Buckland valley grapevine yellows group	XXIII-A	Buckland valley grapevine yellows
16SrXXIV: Sorghum bunchy top group	XXIV-A	Sorghum bunchy top
16SrXXV: Weeping tea tree witches' broom group	XXV-A	Weeping tea witches' broom
16SrXXVI: Mauritius sugarcane yellows D3T1 group	XXVI-A	Sugarcane phytoplasma D3T1
16SrXXVII: Mauritius sugarcane yellows D3T2 group	XXVII-A	Sugarcane phytoplasma D3T2
16SrXXVIII: Havana derbid phytoplasma group	XXVII-A	Derbid phytoplasma

groups and usually does not resolve subgroups clearly without the use of a wide range of restriction endonucleases (Wei et al. 2007). Hence, other methods including heteroduplex mobility assay (HMA) based on the 16–23S ISR have been used to differentiate 16SrI subgroups (Wang and Hiruki 2005). Terminal restriction fragment length polymorphism (T-RFLP) technique based on the 23S rRNA gene sequences which distinguished the various 16Sr groups have been used for identification of phytoplasmas (Hodgetts et al. 2007). In addition, by using less-conserved sequences such as ribosomal protein gene clusters or the 16S–23S rRNA intergenic spacer region, subgroups identified through these approaches were shown to be consistent with genomic subclusters delineated earlier based on DNA sequence homology studies with probe hybridization (Davis and Sinclair 1998) and later based on in silico RFLP analysis (Lee et al. 2007) (Table 2.9). As the descriptions of organisms cultured in vitro are required for naming of species in the class Mollicutes, phytoplasma lineages are being referred to as '*Candidatus* Phytoplasma species'. For example, two provisional species have been named: '*Candidatus Phytoplasma aurantifolia*' (Zreik et al. 1995) and '*Candiatus Phytoplasma australiense*' (Davies et al. 1997).

Phytoplasmas induce certain characteristic symptoms, though some symptoms may be similar to those caused by viral pathogens. Phytoplasmas cause general stunting or dwarfing of whole plants or plant parts. Chlorosis and smalling of leaves are also observed frequently in infected plants. Antholysis of floral parts is the most characteristic feature of many phytoplasmal diseases, resulting in virescence, phyllody and proliferation of floral parts that are transformed into leaf-like structures. Partial or total sterility of diseased plants may be commonly seen. Proliferation of axillary buds and formation of a large number of thin shoots may be evident. Different symptoms induced by phytoplasmas in some host plant species are presented in Figs. 2.9 and 2.10. Periwinkle is susceptible to several phytoplasmas and the type of symptom induced may help in tentative identification of phytoplasma concerned. The possibility of forumulating a molecular classification of phytoplasmas associated with four symptom types on periwinkle was explored. Apart from inducing virescence in periwinkle by the phytoplasmas belonging to aster yellows (AY) group, there was no consistent relationship between symptom type and phytoplasma group identity (Schneider and Seemüller 1994).

2.2.1 Histochemical Methods

Diene's stain has been shown to reveal the presence of some phytoplasmas in the infected tissues. Infected phloem cells appear blue, following staining with Diene's stain as in the case of eggplant (brinjal) little leaf and sandal spike diseases (Srinivasan 1982) and coconut root (wilt) disease (Solomon and Govindan Kutty 1991). *Spiroplsama citri*, causing citrus stubborn disease, phytoplasmas causing aster yellows (AY), pear decline (PD), tomato big bud (TB) diseases and clover club leaf (CCL) agent were visualized in the sieve elements of infected plants by treating the midribs

Fig. 2.9 Symptoms of sugarcane grassy shoot disease (Courtesy of Dr. S. Edison, Tamil Nadu Agricultural University, Coimbatore, India)

with macerating enzymes (cellulase and macerozyme) and separating the sieve elements. The helical spiroplasma, pleomorphic phytoplasmas and slender rod-shaped CCL agent could be seen under dark-field microscope. The morphological and other characteristics of viable cells of the disease-causing agents could be studied in situ in the infected host plant tissues (Lee and Davis 1983).

A fluorescent DNA-binding dye 4,6-diamidino-2-phenylindole-2HCl (DAPI) was used in a histochemical method devloped for detection of phytoplasmal infections in *Catharanthus roseus* (Hiruki and da Rocha 1986). A rapid compression procedure was applied for compressing firmly the excised leaf midrib sections, followed by crushing the compressed tissue with a small spatula. Using fine forceps, the vascular tissue was gently removed and fixed in Karnovsky's fixative for 20 min. The fixed tissue was rinsed in phosphate buffer for 5 min and then placed into DAPI solution for 3–5 min. The tissue was mounted in a drop of DAPI solution on a glass slide and examined under a fluorescence microscope. The presence of AY-phytoplasma in aster, blueberry stunt phytoplasma and *Spiroplasma citri* in periwinkle was detected by applying DAPI fluorescence stain (Dale 1988).

The usefulness of DAPI staining method for detecting infection of various plant species by phytoplasmas was demonstrated later. The DAPI fluorescence test was found to be superior to Diene's staining method for histological detection of AY phytoplasma in lilac witches' broom diseased plants (Sinclair et al. 1989). In a later investigation, DAPI fluorescence test was applied for the detection of ash yellows phytoplasma in white ash trees. The detection of DAPI fluorescence test was as consistent as the dot hybridization technique employing DNA probes (Sinclair et al. 1992).

Fig. 2.10 Symptoms of peanut witches' broom disease (Courtesy of International Crops Research Institute of the Semi-Arid Tropics, Patancheru, India)

Detection of phytoplasmas in six species of ash and lilac in 13 locations in the United States was efficiently accomplished by employing DAPI fluorescence test (Griffiths et al. 1994). The possible involvement of a phytoplasma as a component of sudden death syndrome of mango was explored. The presence of the phytoplasma was detected by DAPI fluorescence test only in one of 90 samples tested, suggesting that the phytoplasma may not have a causative role in the mango sudden death syndrome in Pakistan (Kazmi et al. 2007).

Histochemical and histological examination of sections of petioles from coffee plants affected by crispiness ('crespera') disease of uncertain etiology was taken up to establish the nature of the cause(s) of this endemic disease occurring for over 60 years in Colombia. The sections stained with Diene's stain showed the presence of deep blue coloration in phloem cells indicating the presence of a phytoplasma. On the other hand, such stained cells of the phloem of healthy plants could not be seen. In the DAPI fluorescence test, petiole sections from diseased leaves were fixed with glutaraldehyde followed by staining with DAPI. Characteristic fluorescence was observed in the sections of the diseased petioles, whereas no reaction was evident in comparable healthy petiole sections. Further, ultrathin sections of glutaraldehyde-fixed petioles of healthy and diseased plants were examined under transmission electron microscope (TEM). The presence of pleomorphic bodies in the phloem cells of infected pertiole tissues indicated the infection of coffee plants by a phytoplasma. The phytoplasma has a size range of 300–400 nm length (Galvis et al. 2007). These evidences indicated the possible involvement of a phytoplasma as the cause of coffee crispiness disease which was confirmed by the polymerase chain reaction [Appendix 13].

The health status of tissue-culture-generated mulberry plants grown under greenhouse conditions was verified by applying DAPI fluorescence test for the presence of the mulberry dwarf phytoplasma. The regenerated mulberry plants were found to be apparently free from infection to an extent of 70–90%. The results of DAPI fluorescence test were comparable to that of PCR assay (Dai et al. 1997). In another study, a total of 200 samples of Arizona or velvet ash (*Fraxinus velutina*) cv. Modesto were tested by DAPI fulorescence test. Positive reaction was noted in the majority of the trees tested. All apparently symptomless trees that were DAPI-positive developed typical ash decline (AD) symptoms within 1–2 years. The type of tissue samples affected the test results. The greatest percentage of positive DAPI tests was obtained from stem samples that had leaves with tip and edge necrosis (80%) or chlorosis (78%). Stem samples with healthy leaves also gave high percentage (73%) of DAPI positive reaction (Bricker and Stutz 2004). Although the histochemical methods are simple, rapid and relatively sensitive, they can be applied for preliminary diagnosis with caution when the antiserum or PCR primers are not freely available.

2.2.2 *Immunoassays*

Characterization of phytoplasmas based on cultural and morphological properties has not been possible, because they have not been successfully cultured on cell-free media so far. Further, characterization based on the type of symptoms induced, host range and vector-phytoplasma relationships was also found to be unreliable. Hence, the need for development of sensitive and specific diagnostic methods was realized by researchers. Serological studies have shown that in many phytoplasmas, a single reasonably abundant surface membrane protein is immunodominant (Clark et al. 1989). In other phytoplasmas several proteins may be strongly antigenic. In the case of Fluorescence dorèe and elm yellows phytoplasmas, three and four immunodominant proteins respectively were identified which are serologically related (Seddas et al. 1996). Immunoassays employing poly- or monoclonal antibodies have been shown to be more reliable and sensitive in detecting the phytoplasmas in plants and for establishing relationships among phytoplasmas (Clark et al. 1989; Lin et al. 1993).

2.2.2.1 Rapid Slide Test

Performing slide test is the easier way of detecting the microbial pathogens compared to other immunoassays involving several steps and requiring longer time to obtain the results. *Spiroplasma citri* has spiral forms based on which the generic name was proposed. In the presence of specific antibodies, the spiral forms loose this characteristic and are reduced to round forms, indicating positive reaction between the antibodies and antigenic determinants of the pathogen. A drop of the mixture containing the antiserum produced against *S. citri* (California strain) and the 72-h old *S. citri* broth culture (diluted with growth medium appropriately) is placed on a microscope slide and covered with a cover slip, followed by observation under the microscope. The extent of reduction of spiral forms is calculated. Positive reaction is indicated by reduction percentage of 50 or more. The antiserum against the California strain reacted positively with Morocco strain. *S. citri* was shown to be antigenically related to corn stunt spiroplasma, but not to aster yellows (AY) phytoplasma (McIntosh et al. 1974).

2.2.2.2 Enzyme-Linked Immunosorbent Assay

Both polyclonal and monoclonal antibodies (PAbs and MAbs) have been generated against the phytoplasmas for their detection and differentiation. Monospecific PAb capable of reacting with aster yellows (AY)-associated protein was employed for the detection of AY-phytoplasma. In partially purified preparations from AY phytoplasma-infected plants, proteins associated with the phytoplasma were detected by Western blotting. The PAbs produced against AY-phytoplasma recognized a specific protein (23 kDa) in infected, but not in healthy plants. The antibodies specific for phytoplasma-associated protein were purified by trapping them on AY-phytoplasma protein obtained by electrophoresis of infected plant extracts and then they were transferred to a nitrocellulose membrane. The monospecific antibodies eluted from nitrocellulose reacted specifically with the AY-phytoplasma-associated proteins. The monospecific PAbs reacted positively with AY isolates from carrots and lettuce, but not with other phytoplasma and spiroplasma tested, indicating the specificity of the PAbs (Errampalli and Fletcher 1993).

The spiroplsama that causes brittle root symptoms in horseradish plants was identified as a strain of *Spiroplasma citri* by employing a PAb in ELISA tests (Davis and Flectcher 1983). By employing MAbs specific to *Spiroplasma citri* in indirect ELISA format, specific reaction of nine clones of hybridoma cell lines with different strains of *S. citri* was demonstrated. It was possible to distinguish *S. citri* from corn stunt spiroplasma by employing MAbs, whereas PAbs were unable to differentiate the two strains (Lin and Chen 1985). By using specific PAbs in ELISA tests, *S. citri*, causative agent of citrus stubborn disease, was detected in citrus plants and also in the leafhopper vector *Circulifer haematoceps* (Najar et al. 1998). The corn stunt disease seemed to be a complex caused by mollicutes. The involvement of *Spiroplasma kunkelii* as one of the causative agents was demonstrated by employing specific $F(ab')_2$ in protein A ELISA format (Henriquez et al. 1999). The presence of a surface protein p89 in *S. citri* was detected by electrophoresis and Western blotting. By using ELISA, *S. citri* was detected in the leafhopper vector *Ciruculifer tenellus*. The surface protein may have a role in the adhesion of *S. citri* to vector cells (Yu et al. 2000).

The relative sensitivities of PAbs and MAbs generated against AY phytoplasma were determined using indirect ELISA procedure. The MAbs were more sensitive and specific in their reaction with AY phytoplasma present in infected plants, compared to PAbs. Further, MAbs could differentiate AY phytoplasma from other phytoplasmas (Lin and Chen 1986). Peanut (groundnut) wiches' broom phytoplasma could be detected by employing the polyclonal antiserum in proteinA indirect ELISA format. This phytoplasma was detected in crude extracts of leaves, stems and pegs of infected peanut plants. The specificity of the assay was revealed by the absence of positive reaction with extracts of tissues infected by eggplant (brinjal) little leaf, *Catharanthus roseus* witches' broom and *Datura* witches' broom diseases of presumed phytoplasma origin (Hobbs et al. 1989). Indirect ELISA format by coating directly either the whole antigen or $F(ab')_2$ fragments of the IgG was found to be effective for the detection of the phytoplasma associated with faba bean (*Vicia faba*) (Saeed et al. 1993). Likewise, sesamum phyllody phytoplasma was consistently detected in the infected sesamum plants by using indirect ELISA procedure (Srinivasulu and Narayanasamy 1995).

Different ELISA formats were applied for the detection of sugarcane grassy shoot disease (GSD) phytoplasma. Indirect ELISA protocol was found to be more efficient than other formats, in detecting GSD phytoplasma in different sugarcane tissues. The antiserum against sugarcane white leaf disease (WLD) occurring in Taiwan cross-reacted with the extracts from GSD-infected plants, indicating a possible serological relationship between these two phytoplasmas infecting sugarcane in India and Taiwan (Viswanathan 1997a). X-disease phytoplasma was detected using an MAb produced by employing an enriched antigen from an infected chokecherry (*Prunus virginiana*) plant. This MAb reacted in ELISA with five of seven tested phytoplasmas included in the X-disease phytoplasma cluster and one of the four phytoplasmas (pigeonpea witches' broom phytoplasma) outside the cluster. This ELISA protocol using the MAb has the potential for commercial diagnosis of X-disease phytoplasma infection in stone fruit nurseries and orchards and for screening germplasms for breeding programs (Guo et al. 1998). The MAbs generated against tagetes witches' broom (TWB) disease agent detected the causative agent in artificially inoculated *Catharanthus roseus* and *Tagetes* and also in naturally infected *Tagetes* plants. When ELISA and DIBA tests were used, cross-reaction was seen with a phytoplasma isolated from grapevine infected by yellows disease (Loi et al. 1998).

By employing DAS-ELISA, PCR and RT-PCR techniques, sugarcane yellow leaf syndrome was demonstrated to be due to combined infection by sugarcane yellows (ScY) phytoplasma and *Sugarcane yellow leaf virus* (ScYLV) (Aljanabi et al. 2001). An immunodominant membrane protein (IMP) associated with apple proliferation (AP) phytoplasma was detected using specific antibodies. The same protein (P-318B) was also detected by an antiserum raised against antigen preparation from AP-infected plants (Berg et al. 1999). AP phytoplasma could be detected in apple roots, stems and leaves by employing two MAbs in ELISA and immunofluorescence (IF) tests (Loi et al. 2002).

The sensitivity and specificity of immunoassays depend primarily on the purity of the immunogen. As the phytoplasmas have not been successfully cultured on cell-free media, they have to be purified from the extracts of tissues of infected plants. A differential filtration technique was developed for the purification of sandal spike phytoplasma (SPP). The PAbs generated against the purified SPP were employed to detect the pathogen using indirect ELISA test. The presence of SPP was detected in 24 sandal plants, exhibiting spike symptoms. Indirect ELISA format was found to be more sensitive compared to dot immunobinding assay (DIBA) (Thomas and Balasundaram 2001).

2.2.2.3 Immunofluorescence Technique

Detection of phytoplasma in situ in infected plant tissues can be accomplished by adopting immunofluorescence (IF) technique. For the detection of aster yellows (AY) phytoplasma, the acetone-fixed sections of midribs of AY-infected lettuce plants were stained with fluorescein isothiocyanate (FITC) conjugated antimouse IgG. Use of MAbs resulted in specific binding of the MAbs to AY phytoplasma bodies in the sieve tubes, whereas application of PAbs led to fluorescence throughout the sections in both healthy and diseased plants, due to nonspecific binding of PAbs to cell wall and membranes. The specificity of MAbs in detecting AY phytoplasma in plant tissues was clearly revealed by the IF tests (Lin and Chen 1986). Immunofluorescent staining procedure was shown to be effective for detecting, identifying and establishing genetic relatedness of phytoplasmas associated with geographically diverse grapevine yellows disease (Chen et al. 1994). Indirect immunofluorescence procedure was applied for the detection of sugarcane grassy shoot disease (GSD) phytoplasma in different tissues such as leaf lamina, midrib, leaf sheath, nodes and roots. All tissues tested, exhibited characteristic apple green fluorescence, indicating infection by GSD phytoplasma (Viswanathan 2000).

2.2.2.4 Immunosorbent Electron Microscopy and Gold-Labeled Antibody Decoration

Electron microscopy has enabled in situ visualization of microbial pathogens that are below the resolution power of light microscopes. A classical approach to identifying phytoplasmal diseases is thin sectioning and electron microscopy of the phloem tissues, within which the phytoplasmas are present. The morphological characteristics as observed under electron microscope could be used for identification of phytoplasmas and distinguishing them from bacterial pathogens. Localization of phytoplasma bodies is revealed by transmission electron microscope (TEM), when stained/labeled ultrathin sections are examined. Blackcurrant plant showing the severe Russian (R) form of the blackcurrant reversion disease (BCRD) was found to contain phytoplasma bodies measuring 530–750 nm. The phytoplasma infection was confirmed by PCR assay employing the universal primer pair (Špak et al. 2004).

As the phytoplasmas do not have differentiating morphological characteristics, different types cannot be distinguished based only on the observations made under electron microscope. Hence, electron microscopy in conjunction with immunological approach has to be employed for reliable detection and identification of phytoplasmas. Derrick and Brlansky (1976) demonstrated that corn stunt phytoplasma could be trapped using specific antibodies and visualized by immunosorbent electron microscopy (ISEM) technique. Later, similar approach was made for trapping phytoplasmas using specific antibodies by ISEM (Milne 1992). By using the homologous antiserum, aster yellows (AY) phytoplasma bodies could be trapped from partially purified extracts of AY-infected aster plants (Sinha and Benhamou 1983). AY- and peach X- phytoplasmas could be detected by ISEM in individual vector leafhoppers *Macrosteles fascifrons* and *Paraphelpsius erroatus* respectively. Some of the leafhoppers that had access to infected plants did not transmit the phytoplasma and they were also ISEM-negative (Sinha 1988). The phytoplasma causing Flavescence dorée (FD) disease of grapevine was detected by ISEM technique in extracts of both infected grapevine and leafhopper vector *Euscelidius variegatus* (Caudwell et al. 1982). The antiserum specific to FD-enriched material from inoculative leafhoppers was employed to detect the phytoplasma in plants, whereas the antiserum to material from infected *Vicia faba* plants was used to detect the phytoplasma in leafhoppers to prevent the effects of host-directed antibodies (Lherminier et al. 1990).

Methods for immunotrapping and gold labeling of phytoplasmas in crude sap extracts from infected plant tissues have been developed. The aster yellows (AY) phytoplasma bodies were trapped from the extracts of infected periwinkle plants or infected leafhoppers (*Macrosteles quadripunctulatus*) by using electron microscope grids coated with the F(ab′)2 portion of specific rabbit IgG conjugated with gold particles (5 or 15 nm diameter). The grids were negatively stained with 0.5% ammonium molybdate. The European AY phytoplasma could be distinguished from serologically unrelated tomato big bud phytoplasma, which is morphologically indistinguishable from AY phytoplasma (Veera and Milne 1994).

Pre- and post-embedding immunogold labeling procedures were evaluated for detecting and distinguishing serologically differing phytoplasmas in ultrathin sections of infected plants. Antisera against primula yellows (PY), tomato big bud (TBB) and bermudagrass white leaf (BGWL) phytoplsama and an MAb against PY were tested with the three serologically unrelated phytoplasmas. Labeling was specific for each phytoplasma and was localized to the outer surfaces of the phytoplasma bodies. The antisera were found to be effective, when applied for labeling at pre- and post-embedding steps. The MAb reacted well in pre-embedding conditions, but no labeling occurred with post-embedding treatment. Glutaraldehyde fixation reduced levels of labeling in post-embedding conditions. The results indicated the usefulness of immunolabeling techniques for detection and differntiatiation of phytopalsamas (Milne et al. 1995).

2.2.3 *Nucleic Acid-Based Techniques*

Immunoassays employing either polyclonal or monoclonal antibodies have been shown to provide more reliable results for the detection and identification of phytoplasmas, compared to microscope-based staining techniques. However, cross-reactions with other strains of a phytoplasma and inconsistency in the presence of phytoplasma-specific membrane epitopes at all stages of the life cycle of the target phytoplasma have been major limitations of immunoassays, restricting their applicability for routine detection of phytoplasma concerned. Nucleic acid (NA)-based techniques, on the other hand, have the potential for rapid, specific, sensitive and reliable detection of various phytoplasmas affecting a wide range of host plant species. However, isolation of high quality DNA from the target phytoplasma has been a formidable problem. Further, the presence in plants of inhibitors of polymerase chain reaction (PCR), the most frequently applied NA-based technique, is another difficulty faced during the application of NA-based techniques for the detection, identification and differentiation of phytoplasmas.

Specific tissues such as phloem sieve tube elements or haemolymph or salivary glands of insect vectors may be used to extract phytoplasma DNA, since the concentration of the pathogen is higher in such tissues. The phytoplasmas are usually detected by extracting the DNA from infected plants or inoculative insects and using a specific DNA probe in the dot blot hybridization technique. Several methods have been followed for the extraction and enrichment of phytoplasma DNAs. Enrichment procedures such as cesium chloride (CsCl) buoyant gradients (Nakashima et al. 1991) and pulsed field gel electrophoresis (PFGE) (Marcone and Seemüller 2001) were applied for the extraction of DNAs from rice yellow dwarf and European stone fruit yellows phytoplasmas respectively. But the pathogen DNA was still contaminated with host plant DNA. Later, new methods such as rolling circle amplification (RCA), suppressive subtractive hybridization (SSH) and mirror orientation selection (MOS) were employed to obtain DNA with high phytoplasma : host ratio for the genome analysis of *Candidatus Phytoplasma australiense* (Tran-Nguyen and Gibb 2007).

2.2.3.1 Dot Blot Hybridization Assay

Taxonomic classification of phytoplasmas is based mainly on the nucleic acid sequence of the 16S rDNA which is so far the only stretch of the DNA that have been sequenced for nearly all known types of phytoplasmas. Extensive studies on the 16S rDNA of phytoplasmas have identified variable regions and permitted the construction of the group-specific and strain-specific oligonucleotides for polymerase chain reaction (PCR). Phytoplasma-specific DNA probes have been produced from chromosomal or plasmid (extrachromosomal) DNA of the phytoplasmal pathogens. These probes are frequently labeled with either radioactive ^{32}P or nonradioactive biotin or digoxigenin. Labeled oligonucleotide probes have been efficiently employed as strain-specific probes in the dot blot hybridization of PCR-amplified phytoplasma DNA and in localization of phytoplasmas in plant cells using electron microscopy. Fluorescently labeled 16S rRNA-targeted oligonucleotide probes have been successfully used for the detection of individual species in bacterial biofilms using confocal laser scanning microscopy.

Western X phytoplasma was reliably detected by employing cloned DNA probes in infected plants and leafhopper vectors (Kirkpatrick et al. 1987). Biotinylated DNA probes were employed for the detection of alfalfa witches' broom, clover yellow edge, X-disease, clover phyllody, eastern and western aster yellows phytoplasmas in different host plants and also in leafhopper vector *Macrosteles fascifrons* for aster yellows phytoplasma (Davis et al. 1988). Likewise, the presence of ash yellows phytoplasma was detected by using dot blot hybridization procedure, in various host plant tissues such as inner most phloem at the trunk base, roots, twigs and leaves of white ash trees. The DNA probe was equally consistent as the fluorescence test using DAPI dye in detecting phytoplasma (Sinclair et al. 1992; Davis et al. 1992). Freedom from infection of micropropagated plants was verified by performing dot blot hybridization test (Bertaccini et al. 1992). Expression of symptoms of infection by phytoplasma may take a long time in perennial plants like fruit trees and palms. Phytoplasma causing lethal yellowing (LY) in true date (*Phoenix dactylifera*), cliff date (*P. rupicola*), Chinese fan (*Livistona chinensis*) and five coconut palm cultivars was successfully detected by employing specific probes in dot blot hybridization procedure. *Caryota rumphiana* and *L. rotundifolia*, two palm species, which were not known to be infected earlier by LY phytoplasma, also revealed the presence of this pathogen (Harrison et al. 1992).

Detection of Western-X phytoplasma was accomplished by using a ^{32}P-labeled single-stranded RNA probe (riboprobe) with plasmid vector pS64. This riboprobe was more sensitive and reliable than cDNA probe in detecting the pathogen in infected plant tissues. At higher concentrations of cDNA probe a nonspecific hybridization signal was observed with nucleic acid from healthy plants and from plants infected by other phytoplasmas limiting the specificity and reliability of the assay. On the other hand, sensitivity of detection with complementary riboprobe was increased at higher concentrations of the probe (Lee and Davis 1988). By employing ^{32}P labeled probes, it was possible to recognize the aster yellows (AY)-related phytoplasma strain cluster. In addition, the strains of this cluster could be distinguished by using these probes (Lee and Davis 1988; Davis et al. 1990).

Improvement in sensitivity of detection is necessary for the detection of phytoplasmas, since they are generally found in low concentrations and also restricted to phloem tissues. As many phytoplasmas contain extrachromosomal DNA, probes for extrachromosomal DNA that give stronger hybridization signal, have been employed. Further, the sensitivity of detection may possibly be enhanced, because of the presence of multiple copies of extrachromosomal DNA in phytoplasmal cells. The amount of plant tissue needed for detecting maize bushy stunt phytoplasma using extrachromosomal DNA probe was reduced to 0.02 g from 0.3 g of plant tissue required for detection by using chromosomal DNA probe (Davis et al. 1988).

Depending on the nature of host tissue, about 15–30 ng of phytoplasma DNA may be required for detection of phytoplasamas by dot blot hybridization (Goodwin and Nassuth 1993).

Non-radioactive labels are preferred, because of hazards associated with the use of radioactive labels and their short half-life. Biotinylated cloned DNA probes were found to be effective in detecting aster yellows (AY) phytoplasma in infected plants and also in the leafhopper vectors (Davis et al. 1990). In the case of ash yellows phytoplasma, hybridization signals were most consistent and intense, with samples from innermost phloem at the trunk base of infected plants, when biotin-labeled probes were employed for hybridization. It is possible that these tissues might have high concentration of the phytoplasma in that tissue (Sinclair et al. 1992). In another investigation, Davis et al. (1992) showed that the presence of ash yellows phytoplasma could be reliably detected in leaves, twigs, trunk phloem and roots of white ash trees.

Use of digoxigenin-labeled DNA probes has been demonstrated to be effective for the detection of the sweet potato witches' broom phytoplasma in infected sweet potato and periwinkle plants. Majority of the probes tested hybridized with serologically related peanut witches' broom phytoplasma. The probes could detect sweet potato witches' broom phytoplasma DNA at 10 and 0.39 ng of DNA from periwinkle and sweet potato respectively, indicating that this phytoplasma might reach higher concentrations in sweet potato than in periwinkle (Ko and Lin 1994). Polymorphism among the isolates of rice yellow dwarf and sugarcane white leaf phytoplasmas collected from the same field was recognized by examining the restriction enzyme digests of the extrachromosomal DNAs of these phytoplasmas. Homology in the seqeuences of the chromosomal DNA of these phytoplasmas was indicated by the hybridization tests. Genetic relatedness was inferred by the extent of homology in the DNA sequences. Sesame phyllody and aster yellows type phytoplasma showed little or no similarity in their sequences of their extrachromosomal DNA (Nakashima and Hayashi 1995). The phytoplasma causing decline syndrome in coconuts was detected by dot blot hybridization by using two probes from palm lethal yellowing (PLY) phytoplasma (Tymon et al. 1997).

In situ hybridization using short labeled oligonucleotide probes has been shown to be a powerful, sensitive and specific tool for the detection of phytoplasmas in infected plant tissues. Protocols were developed using 20- to 24-mer oligonucleotides, originally designed as primers for PCR, as hybridization probes for the nonradioactive detection of Italian clover phyllody (IcPh) phytoplasma in *Chrysanthemum carinatum* plants. In situ hybridization of paraffin embedded tissue sections was performed using oligonucleotide 5 end-labeled with either Cy5 (far-red emitting fluorochrome), biotin, or digoxigenin. The Cy5-labeled oligonucleotide probes that hybridized to phytoplasma present in plant tissues were visualized by confocal microscopy. The biotin- and digoxigenin-labeled probes could be detected in plant and insect vector tissues, using a chromogenic alkaline phosphatase-nitroblue tetrazoliun chloride/5-bromo-4-chloro-3-indolyl phosphate reaction. The fluorescence signal of hybridized Cy5-labeled oligonucleotide probes could be readily detected in the phloem tissue of all sections obtained from the three

IcPh-infected chrysanthemum plants tested. All IcPh-specific probes exhibited a similar localization of a signal within the phloem tissue. The specificity of the assay was revealed by the absence of hybridization in uninfected chrysanthemum plants, as well as with pear decline-specific oligonucleotide probe. As the labeled oligonucleotide probes developed in this investigation have the potential for detection and differentiation of phytoplasmas, they may be useful for epidemiological and vector ecology investigations (Webb et al. 1999).

Non-radioactive nucleic acid probes have been shown to be effective and reliable for the detection of *Candidatus Phytoplasma aurantifolia* belonging to 16SrII group, causing bunchy top disease (BTD) of papaya in Cuba. Total DNA from more than 200 plant and insect *Empoasca papayae* samples were tested by nonradioactive nucleic acid hybridization technique. Very well defined and strong hybridization signals were observed in 172 of 177 symptomatic papaya plant samples, 37 of 53 asymptomatic plant samples and 63 of 67 *E. papayae* insect samples. Further, PCR amplification products of phytoplasma DNA from selected samples were purified, labeled with alkaline phosphatase. Chemiluminiscent detection for confirmation of the infection by BTD phytoplasma was followed (Arocha et al. 2008).

2.2.3.2 Restriction Fragment Length Polymorphism

Digestion of phytoplasmal DNA with restriction endonucleases followed by restriction fragment length polymorphism (RFLP) analysis for determining the number and size of the DNA fragments are used to establish the similarity between two phytoplasmas. RFLP analysis was applied for the detection and identification of papaya yellow leaf crinkle (PYC) and mosaic (PM) diseases caused by phytoplasmas. The RFLP profiles of these phytoplasmas were identical. But the phytoplasma causing dieback (PDB) symptoms exhibited distinctly different RFLP pattern (Gibb et al. 1996). The Argentinian alfalfa witches' broom (ArAWB) phytoplasma could be detected by RFLP analysis of partial 16S rRNA gene to two ArAWB isolates. When digested with 16 restriction enzymes the profiles revealed distinct differences between the ArAWB isolates and when disgested with six enzymes, the profiles revealed distinct differences between the ArAWB and the reference strain (AshY1^{T}). Restriction patterns unique for the group and an exclusive *Hinf* 1 restriction site were found in the ArAWB phytoplasma DNA (Conci et al. 2005). Tomato big bud (TBB) phytoplasma belonging to group 16SrII occurring in Brazil was detected by RFLP analysis using endonucleases *Hha*I and *Rsa*I (Mello et al. 2006).

2.2.3.3 Polymerase Chain Reaction-Based Assays

Polymerase chain reaction (PCR)-based assays have been primarily responsible for providing more reliable and sensitive diagnosis of various diseases caused by phytoplasmas in several crop plants all over the world. PCR formats have been successfully applied for the detection of phytoplasmas in both symptomatic and asymptomatic plants which may have to be eliminated to restrict the subsequent spread of the disease (s) especially those affecting perennial plants. General and specific primers located in the 16S rDNA intergenic spacer (IGS) and the 23S rDNA regions of the phytoplasma genomes have been employed frequently. A cloned fragment of a plasmid from the phytoplasma is sequenced to identify oligonucleotide primers for PCR. Amplified DNA fragments of the predicted size are then obtained from the DNA extracted from plants or insects infected by the target phytoplasma, whereas no amplification occurs in healthy plant or insect DNA. The PCR-based methods have been shown to be more sensitive (>500 times) than hybridization-based assays as in the case of aster yellows phytoplasma (Goodwin et al. 1994).

High quality phytoplasma DNA unassociated with PCR inhibitors of plant or insect origin is the basic requirement for performing the PCR assay as in the case of detection of other microbial plant pathogens. For the extraction of DNA of phytoplasmas infecting woody and herbaceous plants, a simple and efficient procedure was developed. Commercially available microspin-column matrices were employed in place of phenol, chloroform or alcohol conventionally used for precipitating nucleic acids. Effective extraction of pathogen DNA could be achieved in less than one hour from different kinds of plant tissues such as whole leaves, petioles, midribs, roots and dormant buds. The phytoplasmas causing pear decline, Western-X disease, peach yellow leafroll, peach rosette, apple proliferation, Australian grapevine yellows and *Vaccinium* wiches' broom diseases were detected by PCR assays (Green et al. 1994). In order to overcome the difficulty of purifying the phytoplasmas present exclusively in the phloem tissues and low abundance in infected plants, the suppression subtractive hybridization (SSH) method was modified and applied to selectively amplify DNA of the stolbur phytoplasma infecting periwinkle plant. Plasmid libraries were constructed and the origins of the inserts were verified by hybridization and PCR screenings. After a single round SSH, there was still a significant level of contamination with plant DNA (around 50%). Inclusion of a second round of subtraction (double SSH) resulted in an enhancement of purity of phytoplasma DNA to 97%. This modified SSH procedure may be useful for obtaining high quality DNA from other phytoplasmas also (Cimmerman et al. 2006).

The procedure of preparing crude extract preparation used earlier was adapted to extract DNA from tissues of apple trees infected by '*Candidatus Phytoplasma mali*'. End-point and real-time PCR detection of '*Ca. P. mali*' were employed to compare this extraction procedure with the method used earlier, for the efficiency of extraction of purified DNA. The crude extract proved fully adequate for phytoplasam detection in samples from 86 in vitro and 35 in vivo apple shoots or plants and ten periwinkle plants. Different TaqMan MGB- or SYBR Green- based real-time PCR protocols could be effectively performed by using the DNA extracted from the infected apple trees by the improved method. This new procedure was useful by providing suitable DNA samples for accurate detection of '*Ca. P. mali*' in different types of apple and periwinkle samples (Aldaghi et al. 2009). As the PCR-based assays are more sensitive, precise and rapid, detection of a large number of phytoplasmal diseases has been possible and some of them are listed in Table 2.10.

Table 2.10 Detection of phytoplasmas in infected plants by PCR-based assays

Phytoplasma	Primers used	References
American aster yellows	16S rRNA gene sequences	Ahrens and Seemüller 1992
Apple proliferation	16S rRNA gene sequences	Giunchedi et al. 1993
	Sequences of N6 rDNA (specific)	Malisano et al. 1996
	Primer pair fO1/rO1	Lorenz et al. 1995
Carrot aster yellows	Universal phytoplasma-specific primer pair P16SF/P235R	Guo et al. 2003
Coconut Kalimantan wilt	Primer pairs P1/P7 and R16F2n/R16R2	Warokka et al. 2006
Cocounut Kerala wilt	Primer pair P4/P7	Edwin and Mohankumar 2007
Coffee crispiness	Primer pairs P1/P7 and FU5/rU3	Galvis et al. 2008
Elm yellows	Sequences of 16S rDNA	Boudon-Padieu et al. 2004
Green ash yellows	Universal primer pair P16SF/P235R	Guo et al. 2003
Jujube witches' broom	Primer pair P1/P7	Han 2005
Maize bushy stunt	Universal primer pair P16SF/P235R	Guo et al. 2003
Oat stunt (aster yellows)	Primer pair P1/P7	Urabana Vičienė et al. 2006
Papaya dieback	Sequences of 16S rRNA and 16S–23S IGS region	Siddique et al. 1998
Parsley yellows	Universal primer pair P1/P6	Khadhair et al. 1998
Paulownia witches' broom	Primer pair rp3/rp4	Nakamura et al. 1996
Pear decline	16S rRNA gene sequences	Firrao et al. 1994
Pinus phytoplasma	Sequences of 16S rRNA gene	Schneider et al. 2005
Plum leptonecrosis	Sequences of N6 rDNA fragments	Malisano et al. 1996
Potato stolbur	R16mF2/mR2	Paltrinieri and Bertaccini 2007
Rose witches' broom	*tuf* gene sequences	Gao et al. 2008
Rubus stunt	Sequences of ITS region	Davies 1998
Stone fruit yellows	Primers amplifying 237-bp fragment	Jarausch et al. 1998
Sweet potato little leaf	16S rRNA gene sequences	Gibb et al. 1995
Tomato big bud	R16mF2/mR2	Paltrinieri and Bertaccini 2007
Withania little leaf	Sequences of 16S rDNA	Khan et al. 2009

Primers for the detection of phytoplasmas causing apple proliferation (AP) and pear decline (PD) diseases were designed based on the sequences of ribosomal and nonribosomal DNA. A ribosomal primer pair fU5/rU3 was employed to initiate amplification of the target DNA from all 42 samples from PD-infected pear trees and 36 samples from AP-infected apple trees. A pair of group-specific primers fO1/rO1 derived from the 16S rRNA gene was used for the detection of closely related phytoplasmas associated with AP, PD and European stone fruit yellows. Detection of PD phytoplasma using these primers was more sensitive and specific than detection by DAPI fluorescence test. Restriction enzyme analysis of the PCR amplicons obtained with primer pairs fO1/rO1 and fPD/rO1 was found to be useful for differentiating AP and PD phytoplasmas. However, the pathogen-specific primers were not able to detect all strains of AP and PD phytoplasmas (Lorenz et al. 1995).

'*Candidatus Phytoplasma mali*' causative agent of the apple proliferation (AP) disease prevalent in several European countries was detected by applying PCR and RFLP analyses of 16S rRNA gene and non-ribosomal DNA fragment. Among the 31 symptomatic apple trees tested, the presence of three AP subtypes (AT-1, AT-2 and AP-15) was recognized. AT-1 subtype was found to be predominant in the north-eastern Italy. Subsequent nucleotide sequence analysis of PCR-amplified 1.8 kb (P1/P7) fragment containing the 16S rDNA, the 16S-23S intergenic ribosomal region and the 5′-end of the 23S rDNA, indicated that at least two phytoplasmal genetic lineages within the AT-1 subtype were distinguishable. Further, in silico single nucleotide polymorphism (SNP) analysis based on the 16S rDNA sequence could be employed to differentiate AT-1 subtype from AT-2 and AP-15 subtype. Molecular markers on 16S rDNA identified in this study, have the potential for use in the epidemiological investigations (Casati et al. 2010).

A PCR-based assay was developed for the detection of European stone fruit yellows phytoplasma involved the amplification of a 237-bp DNA fragment from total DNA extracts derived from over 300 stone fruit samples. A high correlation (97%) was observed between the results obtained with specific and universal primers employed for PCR amplification (Jarausch et al. 1998). By using primers based on the 16S rRNA gene and 16S–23S intergenic spacer (IGS) region, the presence of papaya dieback (PDB) phytoplasma was detected in different plant tissues such as leaves, stems and roots, but not in mature leaves of infected papaya plants. When the PDB-affected leaf, stem and fruit samples were examined by electron microscopy, PDB phytoplasma cells could not be observed in sieve elements of tissues of plants at different stages of symptom expression. However, PCR assays indicated the presence of the phytoplasma DNA in these tissues. Variations in the results of PCR assay and electron microscopy may possibly be due to low concentrations and irregular distribution of PDB phytoplasma in infected plants (Siddique et al. 1998).

The PCR assay has been shown to be an effective tool to establish with certainty the identity of different isolates of the phytoplasma under investigation, in the same location or distant locations. Application of universal primers to sequences in the 16S rRNA gene and group-specific primers to sequences in the ITS region of the phytoplasma inducing stunting symptoms in *Rubus* plants, revealed that the infected plants contained two distinct isolates. One isolate associated with Rubus stunt disease was similar to the phytoplasmas included in the group V (elm yellows), whereas the second isolate exhibited similarity to the members of group III (X-disease) (Davies 1998). The cotton phyllody phytoplasma was detected and differentiated from faba bean phyllody phytoplasma, by subjecting PCR amplicons to RFLP analysis (Marcone et al. 1999).

Parsley (*Pteroselinum hortense*) yellows phytoplasma was detected and identified to be a member of aster yellows (AY) phytoplasma group by amplifying the phytoplasma DNA extracted from infected leaves using a 16S rDNA universal primer pair P1/P6. The expected PCR product of 1.5-kb was amplified by the primer pair and the identity was confirmed by amplification with the specific primer pair R16R1/F1 which was designed based on the sequences of AY phytoplasma 16S rDNA. The expected DNA fragment of 1.1-kb was amplified by the specific primer set in the direct and nested PCR assays (Khadhair et al. 1998). Based on rDNA sequence information from an Australian isolate of European stone fruit yellows (ESFY) phytoplasma, primers capable of functioning at high annealing temperatures were designed. The specificity of detection was enhanced in addition to lowering of the risk of false positives. The new primers PA2F/R employed in direct PCR assay, could detect in tissue cultures of apple, pear, and apricot infected with AP, PD, and ESFY phytoplasmas. For detection of phytoplasmas in field samples, a nested PCR format was required because of low concentration and uneven distribution of phytoplasmas. As the inhibitors present in the fruit plant extract can inhibit PCR amplification, a dilution of 1:2,000 was recommended. By using appropriate restriction enzyme to cleave at specific sites, it was possible to distinguish the phytoplasmas. When *Tai*I enzyme was used, ESFY could be distinguished from AP/PD phytoplasmas, whereas *Tsp*509 was useful to differentiate AP/ESFY and PD phytoplasmas. *Taq*I was used to distinguish AP from ESFY and PD phytoplasmas. Furthermore, phytoplasma strains could be identified by direct PCR amplification by specific primers followed by RFLP analysis. This procedure was shown to be effective in detecting the phytoplasmas in micropropagated fruit tree planting materials (Henrich et al. 2001).

Clover phyllody (CPh) phytoplasma has *nusA* gene coding for NusA protein which is a transcription factor involved in elongation, transcription termination and antitermination. In order to assess the possibility of using *nusA* as a tool for detection and differentiation of phytoplasma lineages within the group 16SrI, the *nusA* gene sequences of '*Candidatus Phytoplasma asteris*' related strains were amplified in the PCR followed by RFLP analysis of PCR products. The results of RFLP analysis of amplicons indicated that *nusA* gene sequences may assist in the detection of '*Ca. P. asteris*'-related strains. RFLP analysis of PCR products using *MseI* and *Isp*5091 restriction enzymes differentiated subgroups 16SrI-A, 16SrI-B and 16SrI-C. Since NusA is a ubiquitous and universally conserved protein among bacteria, *nusA* gene has the potential to be useful in aiding the identification of phytoplasmas (Shao et al. 2006).

Detection of phytoplasmas in infected herbaceous and woody plants was tested by employing five simplified DNA preparation procedures for PCR amplification. Thin free-hand cross-sections made from infected herbaceous plants stored in acetone were treated by (i) grinding in NaOH, (ii) sonicating in water, (iii) microwaving in water, (iv) boiling in NaOH and (v) placing directly in PCR tube. PCR amplification was performed with a universal phytoplasma-specific primer pair in a reaction buffer containing 0.5% (v/v) Triton X-100, 1.5 mM magnesium chloride and 10 mM Tris-HCl. All the five procedures provided phytoplasma template DNA suitable for PCR amplification from periwinkle, carrot and maize, whereas grinding, microwaving and boiling procedures allowed positive amplification from woody green ash plants. The phytoplasmas detected were Eastern X-disease, aster yellows, ash yellows, elm yellows in periwinkle plants, aster yellows in carrots, maize bushy stunt in maize and ash yellows in green ash plants. The methods of extraction of phytoplasma DNAs developed in this investigation are much simpler and rapid and the DNA can be used directly as a template for the single-step PCR or two-step nested-PCR amplification (Guo et al. 2003).

Reddening of corn, a disease of unknown etiology, was investigated by applying molecular tool. Direct PCR assay employing universal phytoplasma primer pair P1/P7 resulted in the amplification of the expected fragment of 1,880-bp from 12 of 16 samples collected from symptomatic corn plants, while none of the four samples from asymptomatic plants gave amplification. Nested PCRs with F1/B6 and R16(1) F1/R1 primer pairs led to amplification of the expected fragments ca. 1,700- and 1,100-bp respectively. The phytoplasma detected in infected corn plants was identified as a member of the subgroup 16SrXII-A (stolbur group) based on RFLP analysis of PCR-amplified 16S rDNA and *tuf* gene sequences, selective amplification of phytoplasma DNA using primer pair G35p1m and similarity of 16–23S rDNA gene sequences. This investigation reporting the infection of stolbur phytoplasma in corn appears to be the first record of a phytoplasma belonging to this subgroup in corn (Duduk and Bertaccini 2006).

The incidence of oat stunt disease phytoplasma inducing stunting of plants and formation of sterile, deformed yellow spikes in infected plants was observed in Luthuania. By employing phytoplasma universal primer pairs P1/P7 and R16F2n/R16F2 in the PCR assay, amplification of phytoplasma-specific 1.8 kb and 1.2 kb rDNA products was accomplished. The RFLP analysis of the PCR-amplified product from 16S rDNA indicated tht oat stunt phytoplasma belonged to the group 16SrI (aster yellows group) and subgroup 16SrI-B. This report appears to be the first to record the incidence of AY group phytoplasma in oats (Urabana Vičienė et al. 2004).

Lethal yellowing (LY) disease infecting palms was diagnosed by employing universal rRNA primer pair P1/P7, nested LY-group specific rRNA primer pair 503f/LY 16Sr or LY phytoplasma-specific nonribosomal primer pair LYF1/R1. The phytoplasma could be detected in the embryos from fruits of diseased Atlantic coconut palms. The distribution of LY phytoplasma in sectioned tissues from PCR-positive embryos was traced by applying in situ PCR and digoxigenin-11-deoxy-UTP (DIG) labeling of amplified products. Colorimetric assay was used to detect DIG-labeled DNA products (Cordova et al. 2003). The etiology of Kalimantan wilt disease of coconut occurring in Indonesia was investigated. The phytoplasmal etiology was tested by extracting the DNA from the infected coconut using two procedures viz., a modified CTAB method involving grinding of coconut trunk tissues in pre-warmed CTAB, instead of liquid nitrogen and a small scale DNA extraction method. Both methods were found to be effective in providing suitable DNA preparation for PCR assay. The nested PCR format employing the universal primer pair P1/P7 in the first round of amplification, followed by use of primer pair R16F2n/R16R2 to amplify the PCR amplicon of the first round PCR, was effective in detecting the phytoplasma in samples from both symptomatic (95.1%) and asymptomatic (67.3%) plant tissues. The phytoplasma DNA was amplified from 95 of 116 samples (81.9%), indicating the presence of the causative agent in the coconut plants tested (Warokka et al. 2006).

Elm yellows (EY) disease occurring in North America and several countries in Europe exhibit differences in severity of symptoms. A survey was undertaken to assess the extent of disease incidence in the elm conservatories in France and Western Europe. EY phytoplasma was detected in symptomatic trees by using the

PCR assay for amplification of pathogen DNA fragment, followed by RFLP analysis of conserved regions of 16S rDNA of the phytoplasma or of the EY group-specific nonribosomal DNA fragment FD9. The pathogen could not be detected in asymptomatic elm plants. Three types of EY group phytoplasmas could be distinguished on the basis of *Tru*91 restriction pattern of the FD9 fragment. One type resembled the American EY1, while the second type resembled the European ULW and the third pattern was slightly different from either of these two types. All elm phytoplasmas were clearly different from phytoplasmas associated with grapevine yellows and alder yellows phytoplasmas (Boudon-Padieu et al. 2004).

Pinus silvestris (Scots pine) and *P. halepensis* (Aleppo pine) trees exhibiting abnormal shoot branching and dwarfed needles were examined for the presence of phytoplasma. The presence of phytoplasma bodies could not be observed in the sections stained with DAPI or electron microscopic observations of ultrathin sections, possibly because of low concentrations and irregular distribution of the phytoplasma. PCR amplification, on the other hand, using universal phytoplasma primer pairs, provided positive amplification of the phytoplasma DNA present in the samples collected from symptomatic and nonsympotomatic plant parts of both *P. sylvestris* and *P. halepensis*. PCR amplification produced a 880-bp fragment from the 16S rRNA gene of the phytoplasama present in the infected samples. In addition, nonsymptomatic trees surrounding the infected trees also proved to be infected by the phytoplasma pathogen (Fig. 2.11). The 16S rRNA gene sequences of the phytoplasma present in both species of *Pinus* were nearly identical. But the phytoplasma in *Pinus* spp. was found to be distantly related to other phytoplasmas. The closest relatives were members of the palm lethal yellowing and rice yellow dwarf groups and '*Candidatus Phytoplasma cataneae*' which share between 94.5% and 96.5% 16S rRNA gene sequence homology. Hence, the phytoplasma infecting *Pinus* spp. was considered to represent a coherent, but discrete taxon and named as '*Candidatus Phytoplasma pini*' (Schneider et al. 2005).

Carrot crops showing symptoms suggestive of infection by Mollicutes (phytoplasmas and spiroplasmas) were observed in Washington State. To establish the nature of the

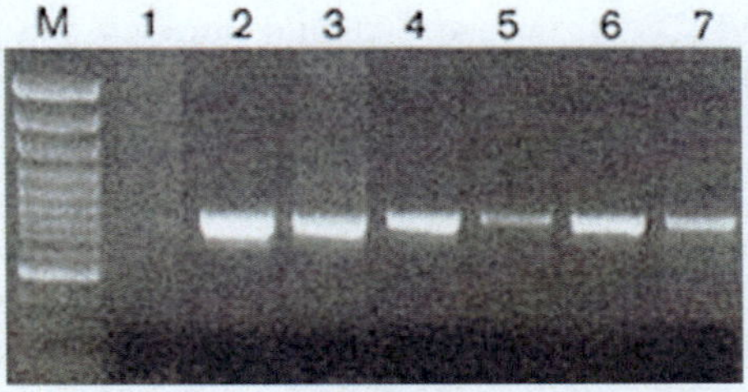

Fig. 2.11 Detection of pine phytoplasma in periwinkle used as diagnostic host plant employing PCR assay. M: 100 bp DNA ladder; Lane 1: healthy periwinkle; Lane 2: apple proliferation as positive control; Lane 3: symptomatic shoot; Lane 4: nonsymptomatic shoot of symptomatic tree; Lane 5: root from symptomatic tree; Lane 6: shoot from nonsymptomatic tree; Lane 7: root from nonsymptomatic tree (Courtesy of Schneider et al. 2005; Society for General Microbiology, Reading, UK)

cause(s) of the diseases, primers specific for phytoplasmas and spiroplasmas were employed in a PCR-based assay to amplify a specific product of 16S rDNA. The PCR amplicon was subjected to RFLP analysis. About 81% of affected plants with dark purple or yellow purple leaf symptoms tested positive for *Spiroplasma citri*. Thus the carrot purple leaf disease was proved to be due to *S. citri*. Other carrot plants exhibiting mild purple discoloration of leaves showed the infection by clover proliferation group 16SrVI, subgroup 16SrVI-A phytoplasmas and aster yellows group 16SrI, subgroup 16SrI-A phytoplasma alone or in combination with *S. citri*. Further, the incidence of beet leafhopper-transmitted virescence agent (BLTVA) yellows of phytoplasma origin was also demonstrated by employing the PCR-based assay (Lee et al. 2006).

Grapevine yellows (GY) are associated with three molecularly distinguishable phytoplasmas. Flavescence dorée (FD) disease is due to the phytoplasma belonging to the subgroup 16SrV-C (elm yellows group), while Bois noir (BN) disease is associated with the phytoplasma belonging to the subgroup 16SrXII-A [stolbur (STOL)] phytoplasma. A mixed infection of European stone fruit yellows (apple proliferation group, 16SrX-B subgroup) is also observed. Molecular characterization of RFLP analysis of *rpS3* and *SecY* genes of Flavescence dorée phytoplasma in Serbia was useful to distinguish this strain from a strain occurring in Italy (Duduk et al. 2004). In a later investigation, typical symptoms of GY were observed in grapevine red cultivar 'Merlot'. By using generic primers for phytoplasmas, positive amplification of expected pathogen DNA fragment was accomplished. In the second round amplification using specific primers for FD and BN, the presence of BN (stolbur) was detected in three samples and other 17 samples as well as the healthy grapevine control samples tested negative. In order to confirm infection of grapevines by stolbur phytoplasma, a multiplex nested PCR format was applied for direct detection of FD and stolbur phytoplasmas. This protocol provided reliable results based on which testing of mother plants during growing season was taken up. In Bulgaria, infected canes were eradicated, if necessary, as a preventive measure (Avramov et al. 2008).

Oats, barley and *Triticosecale* (*Triticum* L. × *Secale* L.) plants showed symptoms similar to those caused by phytoplasmas infecting other host plant species. In order to establish the nature of the cause(s) of the disease of these cereals, a phytoplasma-characteristic fragment of 16S rDNA was amplified in a nested PCR format. The universal primer pair P1/P7 was employed for the first round amplification, followed by amplification of the PCR amplicon of the first round by the second primer pair R16F2n/R16R2 (F2n/R2). The PCR products were resolved by electrophoresis on 1% agar gel stained with ethidium bormide and the DNA bands were visualized by using a UV trans-illuminator. The nested PCR format confirmed phytoplasma infection in symptomatic plants. RFLP analysis of amplified 16S rDNA fragment indicated that the diseased oats were infected by 16SrI group (AY group) phytoplasma strains belonging to the subgroup 16SrI-A, whereas barley and *Triticosecale* were infected by strains belonging to the subgroup 16SrI-B. The phytoplasma strains infecting oat, barley and *Triticosecale* were designated oat yellow (Oat Y), barley deformation (BaDef) and *Triticosecale* stunt (TrSt) phytoplasma respectively (Urabana Vičienė et al. 2004).

The possibility of involvement of a phytoplasma in soybean green stem syndrome was explored. Nested PCR assay using a combination of two phytoplasma-specific universal primer pairs (P1/P7-R16F2n/R16R2 and R16mF2/R16mR1-R16F2n/R16R2) amplified 16S rDNA producing the expected product of 1.2 kb from the DNA extracted from 48 of 121 soybean leaf samples tested. RFLP analysis of the nested PCR products indicated that only two of these samples contained phytoplasmas. In these two samples, phytoplasma belonging to the aster yellows (AY) phytoplasma group 16SrI, subgroup 16SrI-A and a newly established subgroup 16SrI-O were identified. Phylogenetic analysis using partial 16S rDNA sequences (1.2 kb) from representative phytoplasma strains clustered soybean phytoplasmas into two distinct phytlogenetic lineages that are consistent with the two groups defined by RFLP analysis. However, no correlation was found between the presence of phytoplasmas and the occurrence of green-stem syndrome in soybean (Lee et al. 2002).

Specific amplification of the 16S-23S rRNA gene of alfalfa witches' broom phytoplasma occurring in Sultanate of Oman was accomplished by using the phytoplasma-specific universal primer pairs. RFLP profiles of amplicon obtained with P1/P7 primer pair identified this phytoplasma as a member of peanut witches' broom group (16SrII or faba bean phyllody). The enzyme profiles indicated that this phytoplasma was different from all others included in the subgroup 16SrII, except Australian tomato big bud and hence it was classified in subgroup 16SrII-D. The PCR product of P1/P7 primer pair amplification of alfalfa phytoplasma (AlfWB) DNA was sequenced and it exhibited 99% similarity with papaya yellow leaf crinkle (papaya YC) phytoplasma from New Zealand. The results indicated that AlfWB phytoplasma was a new phytoplasma species with closest relationship to papaya YC phytoplasma from New Zealand and Chinese pigeonpea witches' broom phytoplasma from Taiwan (Khan et al. 2002).

Black pepper (*Piper nigrum*) phyllody phytoplasma was detected by applying the nested PCR format. A 1.20-kb DNA fragment coding the phytoplasma 16S rDNA was consistently amplified. Cloning and sequencing of this fragment showed that it contained 1,230 nucleotides. The sequences of the gene were found to be highly similar to that of the members of aster yellows group (16SrI) phytoplasma. The black pepper phyllody phytoplasma was identified as a member of aster yellows (AY) group based on sequence homology and phylogenetic studies (Bhat et al. 2006).

Carrots are known to be infected by aster yellows (AY) and beet leafhopper-transmitted virescence-associated (BLTVA) phytoplasmas. Another disease inducing symptoms different from these two diseases was examined to ascertain the cause(s) of this disease by performing PCR using phytoplasma universal primer pair P1/P7. No amplification of DNA from the infected plants occurred with this primer pair, suggesting the involvement of some other cause. As a next step, PCR assays were carried out using primers specific to phytoplasmas and spiroplasmas. RFLP analysis of PCR-amplified 16S rDNA sequences revealed that about 81% of affected carrot plants exhibiting dark purple or yellow purple leaf symptoms tested positive for *Spiroplasma citri* (Fig. 2.12). Clover proliferation phytoplasma (16SrVI, subgroup 16SrVI-A) was detected in 18% of plants showing purple

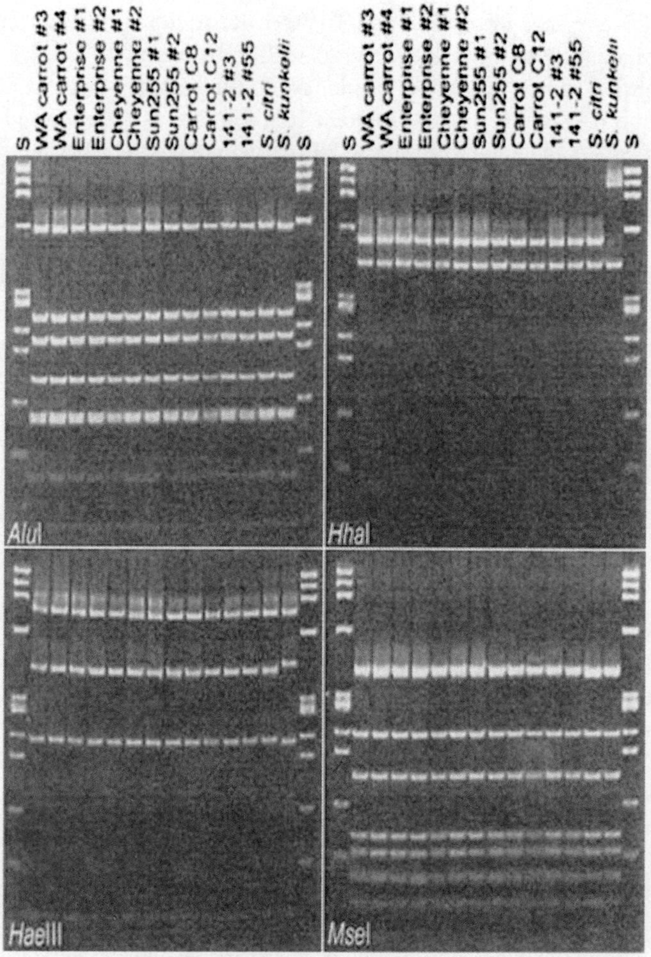

Fig. 2.12 Detection of *Spiroplasma citri* in carrot plants by RFLP analyses of 16S rDNA amplified by PCR with pathogen 16S rDNA-based primers. PCR products are digested with restriction enzymes and separated by polyacrylamide gel elelctrophoresis. Lane S: ΦX174RF1 DNA *Hae*III digest; fragment sizes (bp) from top to bottom: 1,353, 1,078, 872, 603, 310, 281, 271, 234, 194 and 72 (Courtesy of Lee et al. 2006; The American Phytopathological Society, St. Paul, MN)

discoloration of leaf margins, whereas 11% of infected plants contained aster yellows group (16SrI), subgroup 16SrI-A. Some of the carrot platns were co-infected with *S. citri* and either AY or Clover prolifieration group phytoplasma (Lee et al. 2006) [Appendix 13].

In another investigation, carrot plants exhibiting reddening of leaves were examined for possible infection by phytoplasmas. Direct PCR assay with universal phytoplasma primer pair P1/P7 resulted in production of amplicons with expected size of ca. 1,800-bp from 36 of 40 symptomatic carrot samples tested. No amplification occurred from negative control samples. RFLP analysis was carried out after digestion of amplicons with restriction enzymes *Hha*I, *Taq*I and *Tsp*5091. Restriction profiles from 31 carrot samples were identical to the reference strain CHRY, belonging to aster yellows (AY) subgroup 16SrI-A, while restriction profiles from four samples were identical to the reference strain EAY, belonging to subgroup 16SrI-B. One sample showed restriction profile identical to the reference strain STOL belonging to subgroup 16SrXII-A (Duduk et al. 2008).

In a further study, carrot plants infected by phytoplasmas were tested by applying PCR assay followed by RFLP analyses and/or sequencing of phytoplasma rDNA and ribosomal protein genes *l22* and *ss3*, *tuf*, putative aa kinase plus ribosomal recycling factor genes and DNA helicase gene. Phytoplasma belonging to 16SrI-A and 16SrI-B ribosomal protein subgroups respectively were identified by RFLP analyses in 13 of 15 symptomatic plants tested. Asymptomatic plants tested negative for phytoplasmas. Sequence analyses of the phytoplasma genes confirmed the identity of the phytoplasmas in carrots. Current classification of aster yellows phytoplasmas relies on PCR amplification of 16S rDNA followed by RFLP analyses and/or sequencing. This approach is not useful for differentiating the strains of this group. Hence, four additional genes mentioned above, were used for detection and differentiation of aster yellows phytoplasma strains infecting carrot crops. Presence of interoperon sequence heterogeneity was detected and phytoplasma strains were identified as belonging to 16SrI group, but not assigned to 16S rRNA or ribosomal protein subgroup. This appears to be the first attempt to use several molecular markers for the detection and differentiation of phytoplasmas (Duduk et al. 2009).

An endemic disease of coffee known as coffee crispiness occurring in Colombia for over 60 years was investigated to ascertain the nature of the causative agent. Electron microscopy and DAPI fluorescence test indicated that a phytoplasma was the possible presumptive disease-causing agent. A nested PCR assay was performed employing the universal primer pair P1/P7 followed by FU5/rU3 primer pair. A 16S rDNA product of 941-bp was amplified exclusively from the DNA of symptomatic plants. Nested PCR assay could detect phytoplasma causing coffee crispiness disease (CCD) at concentrations as low as 51 cells/g of plant tissues. Based on the results obtained, the CCD phytoplasma was considered to belong to 16SrIII (X-disease) group (Galvis et al. 2007).

Polymerase chain reaction (PCR) and restriction fragment length polymorphism (RFLP) analyses have been employed to detect and identify the phytoplasmas associated with different diseases affecting various crops. DNA isolated from canola (*Brassica napus* and *B. rapa*) plants showing yellows symptoms and symptomatic dandelion (*Taraxacum officinale*) was used to amplify 16S rDNA fragments by PCR employing two pairs of universal primers P1/P6 and R16F2n/R2. RFLP analysis of the amplified DNA fragments using endonucleases *Alu*I, *Hha*I, *Hpa*II, *Mse*I, *Rsa*I and *Sau*3AI revealed the presence of two distinct types of phytoplasmas in canola plants infected by yellows disease. Phytoplasmas were detected in symptomatic dandelion plants and in canola and alfalfa fields where severe alfalfa witches' broom occurred. One of these two phytoplasmas was identified as a member of subgroup 16SrI-A, while another was considered as a member of a distinct group in the aster yellows group on the basis of unique RFLP patterns (Wang and Hiruki 2001).

Pumpkin yellow leaf curl (PYLC) disease occurring in Australia was found to be due to '*Candidatus Phytoplasma australiense*', earlier reported as the causal agent of papaya dieback (PDB), Australian grapevine yellows and strawberry green petal diseases. The primer pair P1/P7 was employed for the amplification of 16S rRNA gene of the phytoplasma from pumpkin plants exhibiting yellow leaf curl disease. The amplicons were digested with restriction enzymes *Alu*I and *Rsa*I, while the amplified product obtained using *tuf* gene primers were digested with *Hpa*II and *Hind*III restriction enzymes. The amplified products had the same banding patterns after digestion with *Alu*I and *Rsa*I. The RFLP banding patterns of the diseased pumpkin samples were indistinguishable from that of '*Candidatus Phytoplasma australiense*'. In addition, the diseased pumpkin samples gave the same banding pattern as the PDB reference phytoplasma, when the *tuf* gene PCR products were analyzed. As pumpkin was found to be a host for '*Ca. Phytoplasma australiense*', it can be a potential source of infection for papaya, grapevine and strawberry crops (Streten et al. 2005).

Coconut tissue samples were collected from 50 sites with coconuts showing varying levels of mortality due to lethal yellowing (LY) disease. A 1.8 kb rDNA fragment was amplified from all samples by PCR assay employing the primer pair P1/P7. In the nested PCR format, the amplification of the products of P1/P7 primer pair was carried out using the 16SrIV group-specific primer pair LY16Sf/LY16S-23Sr. A product of the expected size of 1.7 kb was generated from all samples. Evaluation of samples by PCR with pathogen-specific nonribosomal primer pair LYR1/LYF1 amplified a 1.0 kb product providing the specific detection of LY phytoplasma in the coconut tissues from all sites in Jamaica (Myrie et al. 2007). LY-like disease known as Cape Saint Paul wilt disease (CSPWD) was investigated to ascertain the possibility of transmission of CSPWD through coconut fruit. Hence, PCR amplification was performed using tissues from different plant parts. Employing phytoplasma universal primer pair P1/P7 in standard PCR or followed with a nested PCR using CSPWD-specific primer pair G13f/AwkaSR, phytoplasma infection was detected in the trunks, peduncles, spikelets, male and female flowers of four infected West African Tall (WAT) coconut palms. The presence of the phytoplasma was detected by nested PCR format in nine of 52 embryo DNA samples and confirmed by cloning and sequencing of the amplified product. One of eight fruits from asymptomatic palms also was nested PCR-positive. However, the presence of the phytoplasma could not be detected in any one of the seedlings or plantlets grown from these fruits. Although phytoplasma DNA could be detected in the embryos, no evidence was obtained to confirm the transmission of CSPWD phytoplasma through coconut fruits (Nipah et al. 2007).

A new set of phytoplasma primers from nonribosomal *secA* gene which encodes SecA, the ATP-dependent force generator in the bacterial precursor protein translocation cascade system, was designed. The *secA* gene sequences amplified by semi-nested PCR offer an additional approach to phytoplasma diagnostics and strain identification. All phytoplasmas (34) including aster yellows, clover phyllody, Australian tomato big bud, coconut lethal yellowing, elm witches' broom, pepper stolbur etc., from a wide range of taxonomic groups gave distinct PCR products of 482- or 488-bp. This was possible, despite the fact that the *secA* gene is a single-copy gene in the phytoplasma genome, while the rRNA operon is present in two copies. The *secA* PCR product was readily distinguishable from other bands that were occasionally amplified by the primers. Phylogenetic analysis of *secA* gene sequences showed similar clustering of phytoplasmas when compared with clusters resolved by similar sequence analysis of a 16-23S ISR-23S rRNA gene contig or of the 16S rRNA gene alone (Hodgetts et al. 2008).

Jujube witches' broom (JWB) disease affects jujube tree cultivation seriously in Asian countries. For rapid detection and differentiation of JWB phytoplasma, oligonucleotide primer pair 16Sr(V)F/R for PCR amplification was designed based on the sequences of 16S rRNA of the pathogen. The PCR employing phytoplasma universal primer pair P1/P7 consistently amplified DNA in all tested phytoplasma isolates, but not from the DNA from the healthy plants. In the nested PCR assay, the primer pair 16Sr(V)F/R amplified DNA in all tested JWB and related phytoplasmas including ligustrum witches' broom phytoplasma of the 16S rRNA group V, but no DNA amplification occurred from the DNA of other phytoplasma strains such as groups 16SrI (aster yellows) and 16SrXII (stolbur group) which include mulberry dwarf and chrysanthemum witches' broom phytoplasmas. Nested PCR assay using universal primer pair P1/P7 and 16SrV group-specific pair 16Sr(V)F/R could detect group V phytoplasmas rapidly and specifically (Han 2005).

Australian lucerne yellows (ALuY) disease causes significant losses in lucerne seed crops in Australia. Electron microscopic observations revealed the presence of pleomorphic bodies in the phloem cells of infected plants indicating ALuY disease might be due to a phytoplasma. Oligonucleotide specific to phytoplasma 16S–23S rRNA intergenic spacer region (ISR) were used for amplification in PCR assays on DNA extracted from plants with and without symptoms. The RFLP enzyme profiles obtained for PCR products amplified from ten yellows-affected lucerne samples were found to be identical. The phytoplasma detected in ALuY-diseased lucerne was distinct from the widespread tomato big bud (TBB) phytoplasma, based on the RFLP patterns obtained following digestion with restriction enzymes *Mse*I, *Alu*I, *Rsa*I and *Hpa*II. The ALuY phytoplasma was most closely related to faba bean phyllody (FBP) phytoplasma or phytoplasma group 16SrII (Pilkington et al. 2003).

Ashleaf maples showed symptoms of possible infection by a phytoplasma. The disease was transmitted to *Catharanthus roseus* used commonly as a bioindicator for the detection of phytoplasmas. Nested PCR amplification using universal primer pairs fA/rA and R16F2n/R16R2 and aster yellows-specific primers R16(I)F1/R16(I)R1 revealed that a strain of aster yellows (AY) phytoplasma was the causative agent of ashleaf maples disease (Kamińska and Śliwa 2005). Phytoplasma causing phyllody disease of the annual ornamental plant *Celosia argentea* was detected and identified by amplifying the 16S rRNA gene sequences of the pathogen using universal primer pair P1/P7, R16F2n/R16R2 and template DNA extracted from infected plants. The 1.2-kb 16S rDNA product of nested PCR, primed by primer pair R16F2n/R16R2 was subjected to single enzyme digestions with eight restriction endonucleases. RFLP analysis revealed that the phytoplasma infecting *C. argentea* belonged to the 16SrI aster yellows phytoplasma group, subgroup I-M (Samuitiene and Navalinskiene 2006). *Gaillardia aristata* yellows (GaiY) phytoplasama was detected and identified by employing PCR assay. Phytoplasmal 16S rRNA gene sequences were amplified using universal primer pairs P1/P7 and R16F2n/R16R2. The RFLP analysis of the 1.2-kb rDNA product subjected to single enzyme digestions with nine restriction enzymes showed that GaiY phytoplasma belonged to group 16SrI (AY phytoplasma group), subgroup I-C (clover phyllody, CPh subgroup). Further, natural infection of *Gaillardia* by phytoplasma strains belonging to subgroups 16SrIII-B (clover yellow edge subgroup) and 16SrI-A (AY subgroup) was also detected by the PCR format (Valiūnas et al. 2008).

Phytoplasma disease-like symptoms were observed in *Rhododendron* hybridum plants in Czech Republic. The nature of the causative agent was investigated by employing transmission electron microscrope (TEM), PCR amplification and RFLP analysis of the amplicons. TEM observations revealed the presence of pleomorphic, cell wall-less bodies with a size range of 50–600 nm in diameter. DNA extracts from infected tissues were used as template for direct PCR to specifically amplify a portion of 16S rRNA gene, using the universal phytoplasma primer pair R16F2/R16R2. The PCR amplification products were resolved by staining in SYBR Green I and visualization of DNA bands. After digestion of amplicon with restriction enzymes *Alu*I, *Bsu*RI, *Hha*I, *Hpa*II and *Rsa*I were resolved on an 8% polyacrylamide gel followed by staining with ethidium bromide. Comparison of the amplified sequences and phylogenetic analysis confirmed that the phytoplasma infecting *Rhododendron* hybridum plants belonged to the clover proliferation group (16SrVI), subgroup 16SrVI-A and it was identified as '*Candidatus Phytoplasma trifolii*' (Přibylova et al. 2009).

Application of electron microscopy and nested PCR assay was effective for the detection of the phytoplasma infecting red and white currant plants exhibiting full blossom symptoms. Twenty out of 41 plants analyzed tested PCR-positive. RFLP analysis and sequencing of the 16S rRNA gene of the phytoplasma isolates indicated that the phytoplasma isolates belonged to the apple proliferation and aster yellows groups. In addition, infection by European stone fruit yellows and pear decline phytoplasmas was also detected (Navrátil et al. 2001). Australian lucerne yellow (ALuY) disease causes significant losses of lucerne seed crops. Electron microscopic observations revealed the presence of pleomorphic bodies in the phloem cells of infected plants indicating ALuY disease might be due to a phytoplasma. Oligonucleotides specific to phytoplasma 16S–23S rRNA intergenic spacer region (ISR) were used for amplification in PCR assays on DNA extracted from plants with and without symptoms. The RFLP enzyme profiles obtained for PCR products amplified from ten yellows-affected lucerne samples were found to be identical. The phytoplasma detected in ALuY-diseased lucerne plants was distinct from the widespread tomato big bud (TBB) phytoplasma, based on the RFLP patterns obtained following digestion with restriction enzymes *Mse*I, *Alu*I, *Rsa*I and *Hpa*II. The AluY phytoplasama was most closely related to faba bean phyllody (FBP) phytoplasma or phytoplasma group 16SrII (Pilkington et al. 2003). Ashleaf maples showed symptoms of possible infection by a phytoplasma. The disease was transmitted by grafting to *Catharanthus roseus* used commonly as a bioindicator for the detection of phytoplasmas. Nested PCR amplification using universal primer pairs fA/rA and R16F2n/R16R2 and aster yellows-specific primers R16(I) F1/R16(I)R1 revealed that a strain of aster yellows (AY) phytoplasma was the causative agent of ashleaf maples disease (Kamińska and Śliwa 2005).

Citrus huanglongbing (HLB) or greening disease has been shown to be due to a phytoplasama based on electron microscope observations and DAPI fluorescence test. PCR assay usisg phytoplasma-specific primer sets fU5/rU3 nested with primer set P1/P7 detected the phytoplasma in 110 (78%) positive samples. A 1,785-bp amplicon was obtained with primer set P1/P7. Sequence analysis revealed a 100% identity of the sequence in the region of 16S rDNA and 16S–23S rRNA intergenic transcribed spacer (ITS) to three strains of '*Candidatus Phytoplasma asteris*' (onion yellows). Of the 141 samples, 89 (63.1%) samples were positive for '*Ca. Liberibacter asiaticus*'. Observations under electron microscope showed the presence of both walled and wall-less bodies in the phloem sieve-tubes of HLB-affected citrus plants. The results indicated that both '*Ca. Liberibacter asiaticus*' and a phytoplasma related to '*Ca. P. asteris*' were detectable in citrus exhibiting HLB symptoms in PR China (Chen et al. 2009).

Nested PCR format offers higher level of sensitivity of detection of phytoplasmas compared to standard PCR assay (Llop Bonaterra et al. 2000). However, the introduction of a second round of amplification that substantially enhances the risks of contamination in routine analysis, is required to confirm the molecular identification of phytoplasmas. Co-operational PCR (Co-PCR) carried out in a single reaction, minimizes the risks of contamination (associated with nested PCR format) and has similar sensitive levels as nested PCR and real-time PCR assays. Furthermore, it is possible to couple Co-PCR procedure with dot blot hybridization making it possible to characterize the nucleotide sequence.

A protocol based on Co-operational (Co)-PCR was developed for the detection and characterization of fruit tree phytoplasmas. A triprimer reaction coupled with hybridization using general and specific probes were employed for the detection of '*Candidatus Phytoplasma mali*', '*Ca. Phytoplasma prunorum*' and '*Ca. Phytoplasma pyri*'. The sensitivity of Co-PCR coupled with dot blot hybridization was compared with single round, nested PCR and real-time PCR formats (Table 2.11). Phytoplasma DNA was detected by standard PCR only in undiluted purified sample. On the other hand, Co-PCR, nested PCR and real-time PCR formats were able to detect the phytoplasmas at a dilution of 1:1,000, providing a sensitivity of 1,000 times more than single round PCR assay. No amplification was evident in healthy plant and water control samples. Specific hybridization and color precipitation was observed for all phytoplasma isolates, when the general probe was used. Use of the 16Sr X group probe allowed specific detection of two isolates belonging to this group of isolates. No hybridization could be seen when isolates from other groups or DNA from asymptomatic plant material were tested (Fig. 2.13). The Co-PCR protocol was validated by testing field samples collected from *Malus*, *Prunus* and *Pyrus* spp. and *Olea europea* and compared with seven phytoplasmas maintained in the indicator plant *Catharanthus roseus*. The Co-PCR format constitutes a high throughput

Table 2.11 Comparison of sensitivities of different PCR assays tested using tenfold dilutions of purified DNA from three phytoplasma isolates (Bertolini et al. 2007)

PCR assay	Isolate number 1 (16SrX-A)	8 (16SrX-B)	28 (16SrX-C)
Conventional PCR (P1/P7)	Undiluted	Undiluted	Undiluted
PCR(P1/P7) with Co-PCR conditions	1/10[a]	1/10	1/10
Nested PCR (R16F2n/R16R2)	$1/10^3$	$1/10^3$	$1/10^3$
Real-time PCR [P1/R16 (X)F/r]	$1/10^3$	$1/10^3$	$1/10^3$
Co-PCR (P7/R16F2n/R16R2)	$1/10^3$	$1/10^3$	$1/10^3$

[a] The highest dilution at which positive amplification occurred

detection procedure for simultaneous detection and characterization of members of the 16Sr X phytoplasma group that attracts quarantine regulations. The relative simplicity and high sensitivity of the Co-PCR assay makes it a preferable method for large scale use in routine detection of phytoplasmas, especially when real-time PCR equipment is not available (Bertolini et al. 2007).

Terminal-restriction fragment length polymorphism (T-RFLP) analysis is a direct DNA-profiling method that usually targets rRNA. This genetic fingerprinting method uses a fluorescently labeled oligonucleotide primer for PCR amplification and the digestion of the PCR products with one or more restriction enzymes.

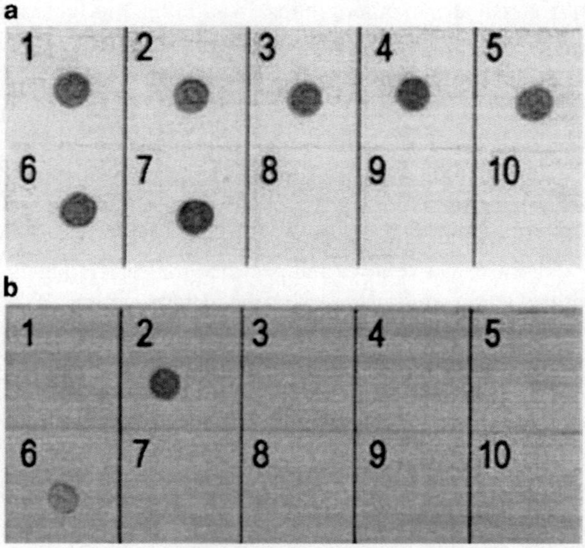

Fig. 2.13 Detection of fruit tree phytoplasmas by co-operational (Co)-PCR assay coupled with dot blot hybridization. Colorimetric detection of Co-PCR products: (**a**): hybridization with general probe; (**b**) hybridization with the 16Sr X group probe. 16Sr phytoplasma groups are indicated by numbers. 1: 16SrI-A; 2: 16SrX-A; 3: 16SrIII-A; 4: 16SrVII-A; 5: 16SrV-A; 6: 16SrX-C; 7: 16SrXII-A; 8 and 9: symptomless controls from *Prunus salicina*; 10: PCR cocktail control (Courtesy of Bertolini et al. 2007; British Society for Plant Pathology, Hertfordshire, UK)

This generates labeled terminal restriction fragments (TRFs) of various lengths depending on the DNA sequence of the phytoplasma present and the enzyme used to cut the sequence. The results of T-RFLP are obtained through TRF separation by high resolution gel electrophoresis on automated DNA sequencer. The laser scanning system of the DNA sequencer detects the labeled primer and from this signal the sequencer can record corresponding fragment sizes and relative abundances. Resulting data is very easy to analyze, being presented as figures for statistical analysis and graphically for rapid visual interpretation. T-RFLP procedure, because of its simplicity, has the potential for rapid, and reliable diagnosis of phytoplasmal diseases.

The T-RFLP technique has shown that by employing primers based on 23S ribosomal gene, it can resolve almost all of the 16S groups of phytoplasmas present in 37 samples tested. In addition, through the design of primers that amplify chloroplast DNA, an internal control can be built into the system to guard against any potential problems of PCR inhibitors. Primers were so selected to amplify part of the 23S rRNA gene and to provide improved resolution between the taxonomic groups compared to conventional RFLP analysis of the 16S rRNA. By using the restriction enzymes *Bsh*12361 and *Mse*I on the PCR products and fragment analysis in the range of 68–640 bp, the effectiveness of T-RFLP protocol was evaluated for identification and classification of 37 isolates from ten of the 16Sr groups. Distinct and unambiguous T-RFLP profiles were produced for nine of the ten taxonomic groups, such that almost all isolates within a group shared the same profile and they could be distinguished from isolates in other groups. In addition, this technique also identified the presence of mixtures of phytoplasmas from different groups in samples. Another advantage of this technique is the use of primers amplifying a terminal restriction fragment (TRF) product from host plant DNA to be employed as a built-in internal control to detect PCR inhibition, if any. The possibility of simultaneous detection and taxonomic grouping of phytoplasmas present in samples using a single PCR has also been indicated. Some of the phytoplasmas detected and classified by using T-RFLP protocol include chrysanthemum yellows, clove phyllody, aster yellows, peach X disease, ash yellows and pear decline etc. (Hodgetts et al. 2007).

Highly sensitive diagnostic systems for the determination of PCR amplified products which obviate the need for electrophoresis and associated methods have become essential for routine large-scale application. The immunoenzymatic detection of PCR products for identification of phytoplasmas causing apricot chlorotic leafroll (ACLR), plum leptonecrosis (PLN) and pear decline (PD) was successfully performed. The PCR-ELISA format was found to be highly reliable and sensitive, since the infected tissue extracts showed very high absorption, visible even to the nacked eye. The PCR-ELISA procedure exhibited a greater sensitivity compared to the methods requiring electrophoresis after amplification with universal primers fP1/rP7 or with specific primers. Further, some blackberry samples with rubus stunt symptoms were not amplified with B1/rULWS pair, while the fAT/rPRUS pair in some cases amplified the samples containing PD. Such cross-reactions were not encountered in the case of PCR-ELISA procedure. The great ease with which the results can be read, makes this technique preferable for large-scale application (Pollini et al. 1997) (Table 2.12).

Table 2.12 Comparative sensitivities of PCR–ELISA and gel electrophoresis resolution for phytoplasma detection (Pollini et al. 1997)

		Dilution limits	
Phytoplasma	PCR–ELISA[a]	Amplification with P1/P7	Amplification with specific primers
OY	1/160	1/10	1/20
RS	1/160	1/10	1/20
PD	1/160	1/20	1/20
PLN	1/80	1/10	1/20
ACLR	1/80	1/10	1/20

[a]Absorption values of at least three times greater than the values for healthy control are taken as positive reaction

2.2.3.4 Real-Time Polymerase Chain Reaction

Conventional PCR assay may not be suitable for detection of the phytoplasma, when the incidence of disease reaches very high levels within short periods. Hence, a more sensitive and rapid method like real-time PCR assay is required to handle large number of samples. Real-time PCR assay detects amplicons during the exponential phase of the PCR using fluorescence-labeled primers or probes, allowing extremely sensitive, specific and rapid detection of the target phytoplasma DNA. Other advantages of the real-time PCR are the nonrequirement of post-amplification steps like electrophoresis, reduction of contamination risks and absence of false positive results.

A quantitative real-time PCR (Q-PCR) assay and bioimaging method were employed to quantify and localize the branch-inducing phytoplasma in *Euphorbia pulcherrima* and peach yellow leafroll (PYLR) strains in *Catharanthus roseus.* Q-PCR assay showed that infection levels in infected plants was in the range of 5.3×10^2–3.7×10^5 cells/µg of plant DNA in *E. pulcherrima* and 4.8×10^6 cells/µg of plant DNA in *C. roseus.* The Q-PCR method provided sensitive and quantifiable data for phytoplasma cells. Infected plants contained the highest concentration of phytoplasma in the source regions indicating that phytoplasmas multiply fastest here. The bioimaging method using vital dyes specific for DNA and membrane potential, detected phytoplasmas in sieve tubes. The dyes allowed physiologically active phytoplasma cells to be discriminated from total population in a given phloem region (Christensen et al. 2004).

Real-time PCR assays for the specific diagnosis of Flavescence dorée (FD), Bois noir (BN) and apple proliferation (AP) phytoplasmas and a universal pair for the detection of phytoplasmas belonging to groups 16SrV, 16SrX and 16SrXII were developed. The primers designed were able to amplify the expected fragments from the DNA from phytoplasmas in field-collected samples from grapevines infected with FD and BN diseases. The group-specific assays provided highly efficient detection comparable to nested PCR (Galetto et al. 2005). In a later investigation, grapevine yellows (GY) phytoplasma was detected by employing a real-time PCR assay, using TaqMan minor groove binder probes. Two amplicons for group-specific detection of FD and BN phytoplasmas plus a universal phytoplasma amplicon were also included. FD and BN amplicons were designed to amplify species-specific genomic DNA fragments and the universal amplicon to amplify the 16S rDNA region. The reactions were specific and no cross-reactivity with other phytoplasma strains, plant or insect vector DNA was evident. The real-time PCR had greater sensitivity, since phytoplasmas could be detected in several PCR-negative and in all PCR-positive samples. The real-time protocol developed in this investigation was shown to be reliable, specific and sensitive and it could be readily applied for throughput diagnosis of grapevine yellows disease (Hren et al. 2007). In another study, TaqMan probes and primers were designed on the 16S rRNA gene sequences of FD, BN and AY phytoplasma genomes. Another TaqMan assay targeting a grapevine gene encoding chloroplast chaperonin 21 was employed to ascertain the DNA quality and to note the absence of PCR inhibition. The efficiency of the two TaqMan assays was compared with nested PCR format. The sensitivity, specificity and speed of detection of FD, BN and AY phytoplasmas in the samples from grapevine, periwinkle and other plant species were similar. However, the real-time PCR assay required less labor and had reduced risk of sample contamination. This diagnostic assay has the potential for analyzing large number of samples, as may be required for certification and control programs (Angelini et al. 2007).

A real-time PCR conjugated with the SYBR® Green I dye was developed for the sensitive and simultaneous quantitative detection of all phytoplasmas '*Ca. Phytoplasma pyri*', '*Ca. P. prunorum*' and '*Ca. P. mali*' from 16SrX group affecting pear, apple and stone fruit species. The selected primers amplified specifically a target of 217-bp fragment from 16Sr gene region of the 16SrX group and not from any other tested phytoplasma groups. The sensitivity of this technique was similar to nested PCR (ten copies of the amplified target per microlitre). The concentration of the phytoplasmas estimated by the real-time PCR protocol in infected pear, plum and apricot trees ranged from 9.7×10^3–3.0×10^5 phytoplasmas per g of tissue. This procedure has the potential for simultaneous detection, in a single reaction, of all the three quarantine-regulated phytoplasmas affecting fruit tree hosts (Torres et al. 2005).

Table 2.13 Relative period of time required after inoculation for detection of '*Candidatus Phytoplasma mali*' by different detection methods (Aldaghi et al. 2007)

	Period of time				
Detection methods	15 days (%)	40 days (%)	70 days (%)	4 months[a] (%)	7 months[a] (%)
Indexing					
Periwinkle	0	26	71	100	100
Apple	0	0	0	10	75
Conventional PCR[b]					
Periwinkle	NT	50	96	100	100
Apple	0	35	80	90	100
Real-time PCR[c]					
Periwinkle	NT	71	100	100	100
Apple	5	70	100	100	100

NT Not tested

[a] One leaf midrib was used for extraction of DNA

[b] Amplified with universal primer pair fU5/rU3

[c] Protocol developed in this investigation

Advantages of employing real-time PCR assay such as accuracy, dynamic range, high throughput capacity and nonrequirement of post-PCR manipulations were exploited for the quantitative detection of apple proliferation (AP) phytoplasma transmitted to periwinkle plants. A new primer pair specific to AP, the qAP-16S-F/R and the AP-MGB probe were employed in a newly-optimized real-time PCR format. This probe successfully detected only AP isolates, but not closely related European stone fruit yellows (ESFY) and pear decline (PD) phytoplasmas. The method was adapted to allow quantification of the phytoplasma in periwinkle leaves. For quantification, a calibration curve was built on serial dilutions of a plasmid containing the amplified fragment (16Sr rRNA gene of the phytoplasma). The detection limit of the assay was one copy of the cloned phytoplasma DNA in the reaction. By using the calibration curve, the concentrations in the tested samples were calculated at 2×10^5–10^6 individuals per mg of fresh midribs samples (Aldaghi et al. 2007).

A new approach for the detection of the apple proliferation (AP) phytoplasma in plant materials using a multiplex real-time PCR assay for simultaneous amplification of a fragment of AP phytoplasma 16S rRNA gene and the host *Malus domestica* chloroplast gene coding for tRNA leucine which was used for the first time as the internal analytical control to validate the false-negative results caused by PCR inhibition. Pathogen detection was based on the highly conserved 16S rRNA gene to ensure amplification of different AP phytoplasma strains. The primer/probe set designed in this investigation allowed specific detection of all AP strains examined, since the primers and the probe matched the sequences in the variable region of the 16S rRNA gene starting at position 82 and ending at position 157. There was no variability within this region among sequences of the four AP strains from Spain, France, Italy and Germany. While the conventional PCR assay failed to detect the phytoplasmas, because of low yield of isolated DNA, the real-time PCR assay provided accurate, reproducible positive results rapidly. Hence, the real-time PCR assay could be particularly useful at low infection levels, facilitating early detection of AP phytoplasma in infected apple plants (Baric and Dalla-Via 2004).

TaqMan real-time PCR and four PCR-based assays were evaluated for their potential for research and routine applications for the detection of apple proliferation (AP) phytoplasma. Of the 162 DNA samples of isolates of AP phytoplasma tested by real-time PCR, PCR-ELISA and PCR with fU5/rU4, fO1/rO1 and AP5/AP4 primary pairs, positive results were obtained respectively for 119, 117, 115, 115 and 114 samples. The highest test sensitivity ranged from 100% for real-time PCR to 95.8% for PCR with AP5/AP4 primer pair. The test specificity for real-time PCR was the highest among the diagnostic tests employed, in addition to being not susceptible to PCR inhibition. Further, TaqMan real-time PCR format had the simplest and fastest testing process involving a minimum of handling steps, requiring shortest test duration of 3:15 h, as against 8:15 h needed for PCR–ELISA test. The main disadvantage of real-time PCR assay is the high purchase price of the specialized equipment and relatively high material costs. But the high material cost may be compensated by the lower personnel cost to some extent. PCR with primer pair AP5/AP4 was found to be the least costly test assay. As real-time PCR assay has a high potential for automation, it may be more suitable for large-scale routine diagnosis of AP disease (Baric et al. 2006).

A new TaqMan minor groove binding (MGB) probe and new PCR conditions were formulated for rapid, specific and sensitive detection of '*Candidatus Phytoplasma mali*'. The new probe could distinguish a single mismatch between '*Ca. P. mali*' and '*Ca. P. prunorum*'. This protocol was able to detect one plasmid copy in water and 100 copies in healthy plant DNA extracts. Comparison of sensitivities of standard PCR, real-time PCR and the real-time PCR protocol developed in this investigation showed that the new protocol could detect *Ca. P. mali* in periwinkle plants at 3 months after inoculation, whereas standard PCR and real-time formats required 4 months for successful detection of the phytoplasma in inoculated periwinkle plants (Table 2.13) (Aldaghi et al. 2007).

A quantitative, real-time TaqMan PCR assay was developed for the detection and quantification of a group 16SrVI phytoplasma in extracts of DNA from tomato and potatoes infected by Columbia Basin potato purple top (CBPPT) phytoplasma. Primers and probes were designed from the 16S rRNA gene of CBPPT phytoplasma which is closely related to the beet leafhopper-transmitted virescence agent (BLTVA). The real-time PCR assay detected readily the group 16SrVI phytoplasma in DNA from infected tomatoes, potatoes and the aster yellows (group 16SrI) and potato witches' broom group (group 16SrIX) phytoplasma in periwinkle plants. The dilution end point for DNA from infected tomatoes was 5–50 pg for real-time PCR, while nested PCR assay detected phytoplasma in 500 pg, but not 50 pg of DNA from infected tomato plants. A dilution series of 5 pg–50 ng of DNA from six potato plants showing symptoms of purple top disease indicated that there was considerable variation in the pathogen concentration as reflected by phytoplasma DNA contents in these plants (Fig. 2.14). Aster yellows and pigeon pea witches'

Table 2.14 Correlation between percentages of infestation determined by SSEB-ELISA method and seedling infection under greenhouse conditions (Frommel and Pazos 1994)

	Percentage	
Seedlot	Seedling infestation in greenhouse conditions	Seedling infestation as as determined by SSEB–ELISA
1	12.00	17.93
2	14.00	16.30
3	1.00	3.26
4	8.75	10.30
5	2.50	5.00
6	14.00	18.29
7	6.50	8.54
8	8.75	10.97
9	6.50	6.10
10	12.00	7.60
11	9.00	12.19
12	6.50	8.50

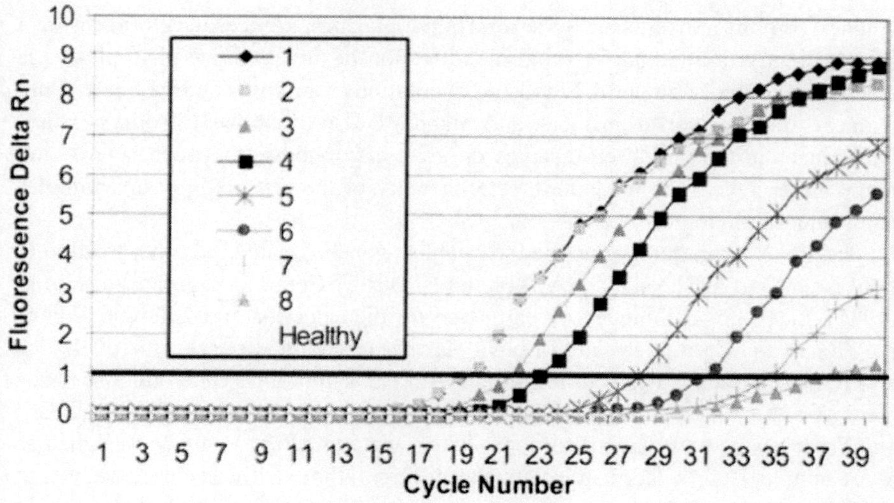

Fig. 2.14 Detection of potato purple top phytoplasma in the potato plants by real-time polymerase chain reaction. Fluorescence patterns for foliage samples from eight potato plants showing disease symptoms and a healthy Shepody potato plant are indicated (Courtesy of Crosslin et al. 2006; The American Phytopathological Society, St. Paul, MN)

broom phytoplasmas were also detected by the real-time PCR assay in infected periwinkle plants. As the real-time PCR assay is less labor-intensive than nested PCR assay, it may be more suitable for rapid and quantitative detection of phytoplasmas (Crosslin et al. 2006).

As the visible symptoms of corn stunt disease caused by *Spiroplasma kunkelii* appear only closer to flowering time, it is essential to detect the spiroplasma infection in the early stages of crop growth to reduce disease spread and consequent losses. The gene encoding a novel adhesion-like protein has been identified in the pathogen genome. Adhesion-like protein has a role in the adhesion of spiroplasma cells to gut cells of insect vectors during early stages of infection (Berg et al. 2001; Davis et al. 2005). A field-deployable real-time PCR protocol was developed for rapid, specific and sensitive detection of *S. kunkelii* using the primers designed based on the sequences of adhesion-like gene and fluorigenic probe. The real-time PCR format could detect *S. kunkelii* DNA at a concentration of as low as 5 fg in the corn plant and vector tissue extracts. The sensitivity of real-time PCR assay was 100 fold greater than the standard PCR format. No detectable fluorescence from the DNA of *S. citri*, healthy corn plant or AY-phytoplasma-infected plant tissues could be observed, indicating the specificity of detection of the phytoplasma in the infected tissues (Wei et al. 2006).

2.2.3.5 Heteroduplex Mobility Analysis

Heteroduplex mobility analysis (HMA) is based on the delay in the rate of migration of a DNA heteroduplex in comparison with a DNA homoduplex to identify mismatches or deletions in DNA sequences. Sympotms like chlorosis, stunting and witches' broom are induced by several phytoplasmas. Hence, development of a simple diagnostic method for rapid identification of the causative phytoplasma is essential, especially for certification and quarantine programs. The HMA procedure requires only universal primers and standard DNA extracted from samples. Differences in the migration distance on gel electrophoresis between hetero- and homoduplex reveal the nucleotide heterogeneity of the test phytoplasma samples and standard phytoplasma DNA.

The HMA technique was applied to establish genetic relatedness among different field isolates of aster yellows (AY) phytoplasmas by Ceramic-Zagorac and Hiruki (1996). Later, this technique was employed for the detection and differentiation of the French and German isolates of phytoplasmas infecting *Populus nigra* cv. Italica and tomato inducing witches' broom and stolbur symptoms. Heteroduplex bands were absent, when typical aster yellows phytoplasma DNA was used as a standard and French isolate of witches' broom phytoplasma infecting *P. nigra* cv. Italica as a test sample. But the German isolate was different, though it was closely related to the French isolate. The phytoplasma causing stolbur and big bud symptoms in tomato was differentiated by HMA procedure. A high degree of nucleotide sequence heterogeneity existed between the DNA strands of reference stolbur and aster yellows phytoplasmas. Hence, they were placed in two different groups (Cousin et al. 1998). The HMA in conjunction with PCR showed a shift in the mobility for a heteroduplex formed in combination with the chestnut little leaf (CLL) and jujube witches' broom (JWB) phytoplasmas. No change in the mobility for the heteroduplex was observed for the combination with CLL and each of paulownia witches' broom and mulberry dwarf phytoplasma DNAs. The combination of HMA and PCR was demonstrated to be a useful technique for the detection and differentiation of phytoplasmas (Han et al. 1998).

2.2.3.6 DNA Array Technology

Nested PCR followed by RFLP analysis after digestion with a range of restriction enzymes has been applied for the detection and classification of phytoplasmas, but the procedures are laborious and time-consuming. Later, real-time PCR methods have been employed for the detection of phytoplasmas rapidly and reliably. Although real-time PCR assays are less laborious than nested PCR procedure and equally sensitive, they do not provide a basis for classification of the phytoplasmas detected in different host plant species. DNA array technology, an essentially reverse blotting method, has been demonstrated to be useful for detection and identification of plant pathogenic fungi, bacteria and viruses. In the recent years, DNA array technology has been applied for the detection and classification of phytoplasmas infecting some crops.

Application of ligase detection reaction (LDR) associated with hybridization on universal arrays has been shown to be a rapid and reliable tool for detection and identification of phytoplasmas of quarantine importance. Phytoplasmas involved in the etiology of grapevine Flavescence dorée (FD) were established as the cause of the disease by using LDR procedure. Probes surrounding the polymorphic sites specific for each group of phytoplasmas were designed to distinguish between FD phytoplasmas (16SrV-C and 16SrV-D) from that one associated with Bois noir disease (BN, 16SrXII-A). The discriminating and common probes were designed to carry the SNP in their 3 end and an optimal length of the probe was fixed to obtain a melting temperature (Tm) of 60°C. LDR experiments were performed on selected DNA samples from grapevine plants showing typical symptoms of GY. Two samples reacted with stolbur-specific probes (zip1 and zip 3), but not with those (zip 11 and d13) designed for groupV hybridization. Moreover, all the grapevine samples reacted with phytoplasma universal probe (zip 15), while no signal was recorded for DNA amplicon of 16Sr XII-A infecting grapevine amplified with 16SrF1/R1(V) primer pair. The results indicated that LDR technique may be useful for detection of phytoplasmas as an alternative and rapid technique in infected plants/planting materials that may have to be eliminated as a preventive disease management measure (Frosini et al. 2002).

The procedure using consensus PCR followed by oligonucleotide microarray hybridization was developed for identification of phytoplasmas belonging to 16Sr groups. The consensus primers were employed to obtain a relatively small PCR product for increased sensitivity. The most suitable region for design of capture oligonucleotides (COs) for identification of phytoplasma groups is the 5 end of the 16S rRNA gene, since this region contains several polymorphic regions. Specific COs (21–33 nucleotides long) were designed to detect and identify individual phytoplasma groups and two COs were designed for hybridization to all known phytoplasmas. The use of an array grid allowed 16 sample analyses simultaneously on one array slide. The array could efficiently identify samples from 16SrI, 16SrII, 16SrIII, 16SrV, 16SrVI, 16SrIX, 16SrX and 16SrXII ribosomal groups. The ribosomal group-specific COs were able to detect phytoplasmas belonging to target groups with high specificity. Specific signals were generally between 20,000 and 50,000 relative units (RUs), whereas nontarget signals were between 0 and 1,000 RU. The array can be used for rapid identification of phytoplasmas, in most cases, to group level for a relatively large number of samples, since 16 hybridization reactions can be processed on each array slide and several slides can be processed at a time (Nicolaisen and Bertaccini 2007) [Appendix 15].

2.3 Detection of Bacterial Pathogens in Seeds and Planting Materials

Bacterial plant pathogens are known to be carried on or in the true seeds and vegetatively propagated plant materials such as tubers, corms, cuttings and setts that form the primary sources of infection for the crops grown in the next season. The infected seeds and planting materials may or may not exhibit recognizable symptoms of infection and they may be carried inadvertently to long distances by travelers or when importation is made for the use of growers or as exchange of genetic materials by researchers. It is necessary to detect the bacterial pathogens in these plant materials to prevent the introduction of new bacterial pathogen(s) into the new areas or to restrict the spread of the existing bacterial pathogens in a geographical location (Narayanasamy 2008).

2.3.1 Detection of Bacterial Pathogens in Seeds

Infection of seeds by bacterial pathogens may not be recognized by visual examination in most cases. It becomes necessary to employ special techniques for the detection of bacterial infection in seeds.

2.3.1.1 Biological Methods

The bacterial pathogens present in the seeds may be isolated in appropriate medium that favors their rapid development. A semi-selective medium (XTS agar) was used for the isolation of *Xanthomonas campestris* pv. *translucens* (*Xct*), causing black chaff disease of wheat, from seeds of wheat (*Triticum aestivum*). The XTS agar medium containing Difco nutrient agar, glucose, gentamycin and cephalexin inhibited more than 91% of the seed-associated saprophytic bacteria and allowed *Xct* present in seed washings to develop. This simple procedure may be applied to assess the levels of wheat seed contamination by *Xct.* The results of laboratory seed assays reflecting the colony forming units (cfu) were favorably correlated with development of black chaff disease in the field (Schaad and Forster 1985). Washed and surface-sterilized rice seeds or their extracts at an appropriate dilution were plated in the specific medium. Fluorescent colonies developed from washed rice seeds placed on King's medium B, after 10 days, indicating the infection by *Pseudomonas fuscovaginae* (Agarwal et al. 1989).

A leaf inoculation method was developed for the detection in rice seeds of *X. oryzae* pv. *oryzicola,* causing bacterial leaf streak disease. Seed washings are inoculated on rice leaf segments placed on 1% agar amended with benzimidazole (75–100 ppm) to maintain freshness of inoculated leaf segments. The agar plates are then incubated in a moist chamber. Typical leaf streak lesions followed by bacterial ooze may be observed in the inoculated leaf segments, indicating the seed infection by the bacterial pathogen (Xie and Mew 1998).

The possibility of employing bacteriophages for the detection and identification of seedborne bacteria was explored by some researchers. The bacterial pathogen *X. campestris* pv. *oryzae* (*Xco*) was detected and differentiated by employing OP_1 phage. Based on the sensitivity of isolates of *Xco* to the phage, they were grouped into 15 lysotypes (Ou 1985). Two races of *X. axonopodis* pv. *malvacearum* showed distinct differences in their sensitivity to the phage, providing a reliable basis for their detection and differentiation (Freigoun et al. 1994). Requirement of species-specific phages, development of resistance in bacterial pathogen to phages and production of false negatives seem to limit the applicability of phages for detection and discrimination of species/pathovars of bacterial pathogens.

2.3.1.2 Immunoassays

Specific antigenic determinants present in or on bacterial pathogens capable of reacting with antibody probes have been identified in some of the pathogenic bacterial species. Bacterial pathogens are more amenable for detection by immunoassays, when compared to fungal pathogens present in seeds. Both polyclonal and monoclonal antibodies (PAbs and MAbs) specific to bacterial species/ pathovars have been employed for the detection of seedborne bacterial pathogens. The MAbs are more specific and sensitive in resolving the presence and identity of bacterial pathogens in plants, seeds or planting materials that may or may not exhibit symptoms of infection.

Agglutination-/Diffusion-Based Tests

Agglutination or precipitation in test tubes or microscope glass slides produces visible precipitates, indicating a positive relationship between the bacterial antigen and the antibody. *Pseudomonas syringae* pv. *phaseolicola*, *Xanthomonas campestris* pv. *phaseoli* were detected in bean and pea by agglutination tests (Trujillo and Saettler 1979; van Vuurde and van den Bovenkamp 1981; Ball and Reeves 1992). The virobacterial agglutination (VBA) test was performed by sensitizing *Staphylococcus aureus* cells with specific MAbs. *P. syringae* pv. *phaseolicola* and *P. syringae* pv. *pisi* could be detected in bean and pea seeds respectively (Lyons and Taylor 1990). Ouchterlony double diffusion test was shown to be effective for detection and differentiation of pathovars of *P. syringae*. *P. syringae* pv. *pisi* present in pea seeds was specifically detected by using antibodies generated against glutaraldehyde-fixed pathogen cells (Mazarei and Kerr 1990).

Immunofluorescence Microscopy

The efficacy of PAbs and MAbs generated against *X. campestris* pv. *campestris* (*Xcc*) was assessed for the detection of this pathogen causing black rot of crucifers in the cabbage seeds by using immunofluorescence (IF) microscopy technique. The PAb 94, MAbs 20H6, 2F4, 18G12 and a mixture of MAbs 20H6, 18G12, 2F4 and 16B5 could be employed for efficient detection of *Xcc* in seed extracts. However, some virulent *Xcc* strains derived from seedlots did not react with MAbs 10C5, 2F4, 18G12, 17C12 and 16B5. No relationship was noted between IF reactions of *Xcc* strains using MAbs and reactions of *Xcc* strains in pathogenicity tests (Franken 1992). An indirect immunofluorescence colony staining method was developed for the detection of important seedborne bacterial pathogens of tomato. The method involves the use of specific antiserum for initial binding of target bacteria and visualization of positive colonies with commercially available secondary antiserum conjugated with FITC and observed under a fluorescence microscope. The assay is easy to perform and results can be easily assessed by visual scoring or image analyzer. If MAbs or recombinant antibodies (commercially available) are used, the sensitivity of the assay can be enhanced remarkably (Veena and van Vuurde 2002).

Enzyme-Linked Immunosorbent Assay

Detection of bacterial pathogens has been accomplished more frequently by employing enzyme-linked immunosorbent assay (ELISA) and its variants than any other immunodiagnostic test. *Pseudomonas syringae* pv. *pisi* (*Pss*) was efficiently detected in pea seeds by using specific MAbs in indirect and competitive ELISA formats. In addition, strains of *Pss* were differentiated by these assays (Candlish et al. 1988). The presence of *Erwinia stewartii*, causative agent of Stewart's bacterial wilt disease in corn, was detected by employing double antibody-sandwich (DAS)-ELISA and four different procedures in corn seed tissues. DAS-ELISA procedure was found to be the most appropriate method, using PAbs for capture and MAbs for detection. This assay detected *S. stewartii* antigen in seeds from plants inoculated with an antibiotic-tolerant strain of *E. stewartii.* The presence of viable *E. stewartii* in seeds from inoculated plants was confirmed by culturing the pathogen. A high positive correlation between recovery of bacteria from seeds and ELISA responses was observed (Lamka et al. 1991).

Seed treatment with chemicals is a common practice adopted for elimination of seedborne fungal/ bacterial pathogens. In order to assess the effect of treatment of corn seeds with fungicides, insecticides or both on the sensitivity of ELISA test in detecting *E. stewartii* was determined. Seed treatment with Actellic, Apron, Captan, Cruiser, Gaucho, Maxim, Poricho, Thiram and Vitavax either singly or in combination did not affect significantly the sensitivity of ELISA-based seed health test to detect E. *stewartii.* Seed treatments did not affect the number of positive and negative results of ELISA-based seed health test (Pataky et al. 2004).

Application of double antibody sandwich (DAS)-ELISA procedure allowed the detection of *X. campestris* pathovars, *undulosa*, *cerealis* and *phleipratensis* in pure cultures and wheat seeds contaminated with *Xcu*. Pathovars *undulosa*, *cerealis* and *translucens* are pathogenic to wheat, whereas pv. *phleipratensis* infects *Phleum pratensis*. The lower limit of detection was 5×10^3 cells/ml (approximately 1,000 cells/well). A semi-selective enrichment broth (SSEB) improved the recovery of *Xcu* in cultures mixed up with contaminating bacteria commonly found on wheat seeds. The enriched samples gave two-to threefold increases in A_{405nm} absorption in ELISA test, when viable cells of *Xcu* were present. By enrichment, *X. campestris* pvs. *undulosa*, *cerealis*, *translucens* and *phleipratensis* were detected in samples that originally had less than 5×10^2 cfu/ml. Semi-selective enrichment combined with ELISA (SSEB-ELISA) allowed for determination of the percentage of infestation (PSI). The PSI of naturally infested seedlots was obtained by growing seed samples in the greenhouse under conditions optimal for disease development. Percentages of seed infestation determined by SSEB-ELISA procedure were highly correlated with PSI ($r = 0.87$, $p \leq 0.05$). The SSEB–ELISA procedure provides a highly sensitive and reliable detection and enumeration method for seeds infested with <100 cfu/ml and the results can be obtained within 24 h, providing a useful tool for fast initial screening of wheat seedlots in certification programs (Frommel and Pazos 1994). Detection of *X. translucens* causing barley bacterial blight disease in seeds by conventional dilution plate and ELISA tests was standardized. ELISA test was found to be fast and reliable. However, this test identified up to the genus *Xanthomonas*, but not *X. translucens* specifically. As all *Xanthomonas* spp. are considered to be pathogenic, ELISA test was found to be specific enough for seed testing, using isolates from dilution plate assays (Olsen et al. 2002).

Xanthomonas campestris pv. *oryzae* (*Xco*) was detected in seeds and leaves of 60 rice accessions. The MAbs generated specifically against *Xco* reacted with all 178 tested strains of the pathogen obtained from diverse geopgraphical locations. These strains could be classified into four groups based on their reactions with four MAbs (Zhu et al. 1998). A later study by Huang et al. (1993) showed that 63 strains of *Xco* could be classified into nine reaction types consisting of four serovars and seven subserovars. In another investigation, *X. oryzae* pv. *oryzae* (*Xoo*) present in contaminated rice seed was detected and identified by using pathovar-specific MAbs (Gnanamanickam et al. 1994). Infection of cabbage seedlings by *X. campestris* pv. *campestris* (*Xcc*) on the seed beds cannot be visually recognized. However, by employing ELISA test, latent infection of cabbage seedlings by *Xcc* could be detected before the appearance of any visible symptoms of black rot or blight induced by different strains of *Xcc* (Shigaki et al. 2000). *Xanthomonas campestris* pv. *carrotae*, causing carrot bacterial leaf blight disease was detected by employing DAS-ELISA test. Color development, indicating positive reactions, was observed, when the seed-wash suspenstions prepared from seedlots had contamination levels >10^5 cfu/g of seeds. The absorbance values were not significantly different from control samples (uncontaminated seeds), if the seed-wash suspensions from seedlots had contamination levels <10^4 cfu/g of seeds. However, DAS–ELISA test was found to be much less sensitive than the seed-wash dilution plating assay (Meng et al. 2004).

Table 2.15 Detection of *Clavibacter michiganensis* subsp. *michiganensis* (*Cmm*) in tenfold dilutions of naturally infected tomato seed extract (de León et al. 2007)

	Cmm concentration (cfu/ml)				
Detection technique	10^4	10^3	10^2	10	1
YPGA medium[a]	+	±	–	–	–
mSCM medium[a]	+	+	+	+	–
IMS-plating	+	+	+	+	±
Immunofluorescence (IF)	+	±	–	–	–
DAS–ELISA	±	–	–	–	–
Direct PCR	–	–	–	–	–

[a]YPGA Yeast peptone glucose agar medium; mSCM semi-selective agar medium

Immunomagnetic separation (IMS) is a process by which immunomagnetic beads (IMBs) are coated with specific antibodies allowing selective trapping of target bacterial cells from the sample. Then they can be resuspended in smaller volumes of solution and plated on non-selective medium. This procedure combines the advantages of serologic recognition, target bacteria concentration and growth of characteristics of colonies on culture medium, factilitating the isolation required for positive diagnosis. IMS technique was employed for selective isolation of *Clavibacter michiganensis* subsp. *michiganensis* (*Cmm*), causing the tomato bacterial canker disease, from tomato seeds to improve the specificity and sensitivity with respect to standard isolation of this pathogen. Two commercial PAbs for *Cmm* were evaluated for their efficacy in binding magnetic beads. The PAb-PRI, recommended by the European Plant Protection Organization (EPPO) was found to be better than PAb-BR. Different concentrations of immunomagnetic beads (IMBs) were tested. IMS with 10^6 IMBs/ml coated with polyclonal antiserum at 1/3,200 dilution recovered more than 50% of target cells from the initial inocula of 10^3–10 cfu/ml. The threshold detection was lower than 10 cfu/ml, even in seed extracts containing seed debris and high populations of non-target bacteria. The IMS technique permitted *Cmm* isolation from naturally infected seeds with higher sensitivity and it was faster than direct isolation on semi-selective medium used earlier.The IMS method is suitable for routine testing and an advantageous alternative tool for detection of *Cmm* in tomato seedlots (de León et al. 2006) [Appendix 16].

In a later investigation, the effectiveness of IMS-plating procedure was evaluated for the detection of *Clavibacter michiganensis* subsp. *michiganensis* (*Cmm*) in tomato seeds in comparison to DAS-ELISA, immunofluorescence (IF) and direct PCR assays. Different seedlots and different strains of *Cmm* were also tested. IMS-plating method allowed the detection of less than 10 cfu/ml of the pathogen in all assayed samples. ELISA and IF tests could detect *Cmm* at concentrations of 10^4 and 10^3 cfu/ml respectively. In the case of direct PCR assay, DNA extraction was required to obtain results in seed extracts containing 10^3 cfu/ml or more. IMS-plating method provided the best results regarding sensitiviy and specificity for *Cmm* detection allowing the recovery of viable bacteria from seed extracts. When combined with standard protocols, IMS-plating method could improve their effectiveness appreciably (Table 2.15) (de León et al. 2007).

2.3.1.3 Detection by Nucleic Acid-Based Assays

Generally the nucleic acid-based techniques have greater levels of specificity and sensitivity than immunoassays and they have been shown to be capable of detecting some bacterial pathogens whose concentrations are below the detection limits of immunoassays. However, these methods are relatively more expensive and involve handling of hazardous chemicals. With the development of nonradioactive probes labeled with digoxigenin, use of nucleic acid-based techniques has become a preferable approach. By employing specific probes and primers based on pathogen DNA sequences, it is now possible to detect, identify and quantify the seedborne bacterial pathogens reliably and rapidly.

Nucleic Acid Hybridization Technique

Detection by nucleic acid hybridization methods has been successfully achieved in the case of *Pseudomonas syringae* pv. *phaseoli* in beans (Schaad et al. 1989; Prossen et al. 1991), *Xanthomonas campestris* pv. *phaseoli* in beans (Gilbertson et al. 1989), *X. oryzae* pv. *oryzae* and *X. oryzicola* in rice seeds (Cottyn et al. 1994) and *P. glumae* in rice seeds (Tsushima et al. 1994). Presence of a large number of rRNA molecules (> 10,000 copies/cell) has been recognized in actively growing bacterial cells. These rRNA molecules vary widely in the nucleotide sequences and these variations may be used as a basis for the identification of bacterial species/genera. A sequence of rRNA gene of *X. oryzae* pv. *oryzae* that specifically hybridized only with this pathovar was identified and employed for its detection (De Parasis and Roth 1990). Likewise, a probe specific for *P. glumae* causing bacterial grain rot of rice was designed and hybridization between the DNA of all strains of *P. glumae* and the probe occurred, but not with *X. oryzae* pv. *oryzae* and several other bacterial species tested, indicating the specificity of the assay (Tsushima et al. 1994).

Restriction Fragment Length Polymorphism Analysis

Restriction fragment length polymorphism (RFLP) analysis, using specific restriction enzymes that cleave the pathogen DNA at specific sites, may be useful for detection and differentiation of bacterial pathogens. Digestion with *Pst*1 may produce RFLP patterns specific for a group of strains within a pathovar as in *X. oryzae* pv. *oryzae* (*Xoo*). Based on RFLP patterns, strains of *Xoo* occurring in the Philippines could be identified and differentiated. The strains of *Xoo* present in the USA appeared to be not closely related to the Asian strains (Leach and White 1991). Strains in India mostly belonged to the pathotype 1b, as indicated by RFLP analysis (Yashitola et al. 1997).

Polymerase Chain Reaction-Based Assays

Polymerase chain reaction (PCR) assay, its variants and PCR in combination with other diagnostic methods, have been demonstrated to be very effective in detecting and quantifying seedborne bacterial pathogens. *P. syringae* pv. *phaseolicola* (*Psp*), causing halo blight disease was detected in bean seeds by employing the primers from DNA sequences of the phaseolotoxin gene of the pathogen. The presence of contaminating saprophytic bacteria did not affect the results of the PCR assay (Prossen et al. 1991). A rapid PCR-based protocol was developed for concurrent detection of the causal agents of bean common blight *X. campestris* pv. *phaseoli* (*Xcp*) and bean halo blight, *P. syringae* pv. *phaseolicola* (*Psp*) in bean seed. The G + C-rich oligonucleotide primers were designed from the phaseolotoxin gene cluster of *Psp*. The HB14 primers specifically directed the amplification of a 1.4 kb DNA fragment from 19 isolates of *Psp*, but not from 62 other bacterial strains including *Xcp* and species. The HB 14 primers for *Psp* and X4 primers specific for

Xcp were able to amplify discrete DNA fragments from bean seeds contaminated with these two pathogens. When the primers were used in combination, they successfully detected individual and mixed infections of bean common and halo blights and yielded distinct DNA fragments from batches of seeds containing as few as one infected seed in 10,000 seeds (Audy et al. 1996).

Xanthomonas campestris pv. *campestris* (*Xcc*), causative agent of black rot of crucifers is a worldwide problem of economic importance, requiring a rapid and sensitive diagnostic method to detect the pathogen in the seed. Initial inoculum carried by an infected seed is a critical factor determining the severity of the disease. A PCR-based technique was developed based on the primers designed to specifically amplify a 619-bp fragment of the *hrpF* gene from *X. campestris*. Amplification products were not detected from other *Xanthomonas* spp. or from other pathogenic or epiphytic bacteria occurring on cruciferous plants. Primers targeting a 360-bp section of the ITS region from *Brassica* spp. were used as internal control to avoid false-negative results. The assay readily detected *X. campestris* pv. *campestris* infections in plants and it was more sensitive and specific than traditional plating methods and a commercially available ELISA test. A seed-washing protocol developed in this investigation allowed detection of a single artificially infected seed among 10,000 healthy seeds using the multiplex PCR format (Berg et al. 2005).

Bacterial leaf blight of carrot caused by *X. campestris* pv. *carotae* is an important seedborne disease with a worldwide distribution. The 3S primer pair specifically directed the amplification of the ~350-bp target fragment from boiled bacterial cell extracts. Similar amplification of ~350-bp target fragment occurred from all *X. campestris* pv. *carotae*-like strains from carrot seeds and plants. The PCR-based seed assay detected *X. campestris* pv. *carotae* from seedlots with combination rates ranging from 2×10^2 to 2.3×10^8 cfu/g of seed. The PCR assay provided results similar to a seed-wash dilution plating assay and proved to be more sensitive than ELISA test (Meng et al. 2004) [Appendix 17]. Citrus variegated chlorosis caused by *Xylella fastidiosa*(*Xf*), a xylem-limited bacteria, could be detected by PCR assay. The pathogen present was detected in all main fruit vascular bundles as well as in seed and dissected seed parts, although no deformity in seed was recognizable. *Xf* in seed was identified by cloning and sequencing the specific amplification product obtained by standard PCR assay employing specific primers. Transmission of *Xf* through infected seeds to seedlings was reported for the first time (Li et al. 2003).

The BIO-PCR technique combines biological and enzymatic amplification of PCR targets and provides improved sensitive detection and estimations of viable cells of the bacterial pathogens present in seeds. The PCR-assay that amplifes a segment of the phaseolotoxin (*tox*) gene cluster, although specific for *P. syringae* pv. *phaseoli* (*Psp*), causative agent of bean halo blight disease, is time-consuming. Further, it needs cost-adding Southern blots to detect the small numbers of pathogen cfu found in some seed extracts. To obviate these problems, an agar-plating step prior to PCR analysis was introduced in the BIO-PCR format which provides the added benefit of biological amplification of the PCR targets in seed extracts, resulting in reduction of chances of detecting dead cells or free DNA. Aliquots of a seed extract are plated onto King's medium B and after 45–48 h, pooled washings from the plates are used for direct nested PCR. Positive detection results were reproducible. The detection threshold of BIO-PCR was 3×10^3 cfu/sample of 3,000 seeds washed in 1,500 ml of buffer. Elimination of false positives and negatives, because of the presence of dead cells and potential PCR inhibitors in seed, increased sensitivity and nonrequirement of DNA extraction prior to amplification are the distinct advantages of BIO-PCR over direct PCR assay (Schaad et al. 1995a). A modified BIO-PCR assay, allowing *P. syringae* pv. *phaseolicola* present in the bean seed extract to multiply in a semisolid medium for 18 h prior to PCR amplification, provided a more sensitive detection of *P. syringae* pv. *phaseolicola* (Mosqueda-Cano and Herrera-Estrella 1997). A disease-free nuclear stock of mother plants of gysophila was established by applying the BIO-PCR technique to eliminate plants infected by *Erwinia herbicola* pv. *gysophilae* (Manulis et al. 1998). BIO-PCR procedure alone could detect *Clavibacter michiganensis* subsp. *michiganensis* (*Cmm*) in tomato seeds that were naturally infected by or artificially contaminated with the pathogen. The limit of detection of *Cmm* by BIO-PCR protocol was one infected seed in lots of 10,000 seeds (Hadas et al. 2005).

In order to improve the sensitivity of detection of bacterial pathogens, an immunologic step prior to PCR amplification was incorporated. Immunogenetic separation (IMS) was performed to concentrate the population of *Acidovorax avenae* subsp. *citrulli* (*Aac*) present in watermelon seeds. An increase in sensitivity (by 100 fold) of detection was achieved by the combination of IMS and PCR over direct PCR assay. Super-paramagnetic beads pre-coated with sheep antirabbit antibodies were coated with purified IgG fractions of anti-*Aac* as per manufacturer's instructions. The detection limit of IMS-PCR format was 10 cfu/ml. It was possible to detect as low as 0.1% seed infection by this combination of techniques. The IMS-PCR procedure was not affected by the PCR inhibitors present in the seeds, as in the case of direct PCR assay. Further, IMS-PCR could detect *Aac* in 100% of seedlots with 10% infestation, while seedlots with 5 % and 0.1% infestation showed positive reaction in 80% of tests. Seed treatment with fungicide did not affect the result of IMS-PCR assay (Walcott and Gitaitis 2000). The possibility of watermelon seeds getting infected from the inoculum from blossom was tested by IMS-PCR technique. *A. avenae* subsp. *citrulli* was detected in 44% seedlots assays, despite lack of fruit symptoms. The blossoms may prove to be potential site of ingress for fruit and seed infestation by this pathogen (Walcott et al. 2003). The IMS-PCR procedure was found to be effective for the detection of *Pantoea amanitis*, causing center rot disease in naturally infected onion seeds. Species-specific primers and PAbs were employed in this assay which had a detection limit of 10–10^3 cfu/ml of seed-wash (Walcott et al. 2002).

Immunomagnetic separation (IMS)-PCR and BIO-PCR assays have been shown to significantly enhance the sensitivity of detection of bacterial pathogens present in seed-wash/extract and other plant tissues. In order to reduce labor required for BIO-PCR assay, a highthroughput 96-well membrane BIO-PCR technique was developed for the ultrasensitive detection of *P. syringae* pv. *phaseolicola* (*Psp*) in seed washings using available classical PCR primers and probe. The sequences of a tox-argK chromosomal cluster of the *Psp*-specific phaseolotoxin gene were used for designing the primers and probe which were confirmed to be specific to *Psp*. Samples were filtered under vacuum in 96-well membrane plates. After incubating on soft agar medium for 48–52 h, each well is washed with sterile water and used immediately for nested (two-step) PCR or real-time PCR format. Classical PCR was unable to detect *Psp* at mean concentrations of 40 cfu/ml. BIO-PCR detected *Psp* in five out of six samples at 40 mean cfu/ml, but none at mean concentrations of 4.2 and 0.4 mean cfu/ml. In contrast, membrane BIO-PCR detected the pathogen in all six samples tested containing as few as 0.4 mean cfu/ml. When large number of samples has to be handled, the 96-well plate allows two replications of 40–45 samples including controls (Schaad et al. 2007).

Immuno-capture (IC)-PCR assay combines the ability of specific antibodies to capture the bacterial cells and the sensitivity and specificity of the PCR primers for amplification of pathogen-specific DNA fragment. IC-PCR procedure was developed for the sensitive detection of *Acidovorax avenae* subsp. *citrulli* (*Aac*), causative agent of bacterial fruit blotch (BFB) of watermelon. All strains of *Aac* tested, produced 360-bp fragment using IC-PCR and direct PCR formats. No amplification occurred from the DNA of ten different bacterial species and strains tested. The detection limits of IC-PCR and standard PCR assays were about 50–100 cfu/ml and 10^4 cfu/ml respectively. The IC-PCR assay detected *Aac* reliably in one cantaloupe, two honeydew melon and two watermelon seed varieties, indicating that IC-PCR assay is an accurate, sensitive, rapid and cost-effective technique (Xiao et al. 2007) [Appendix 18].

Real-Time Polymerase Chain Reaction

Real-time polymerase chain reaction (RT-PCR) assays have been shown to be more rapid, reliable, sensitive and specific than standard PCR format for detection and quantification of bacterial pathogens in plants and seeds. Panicle blight disease of rice caused by *Burkholderia glumae* causes serious losses in southern United States. A real-time PCR procedure was developed for the detection of *B. glumae* in seeds by employing specific primers based on the sequences of 16S–23S rDNA ITS region of several representative isolates of the pathogen from Arkansas and Japan. The real-time PCR procedure was effective in detecting and identifying *B. glumae* in rice seedlots as well as in whole plants. This real-time PCR protocol was found to be highly sensitive, detecting as few as 1–10 cells in water without the need for DNA extraction and gel electrophoresis. In addition to providing qualitative and quantitative estimation of pathogen population, the assay time was reduced from 6 h to just 2 h. The development of the real-time PCR assay for detection of *B. glumae* in rice seeds appears to be a crucial step in implementing a clean seed program and preventing crop losses through improved seed sanitation (Sayler et al. 2006) [Appendix 19].

Pantoea stewartii, causative agent of Stewart's bacterial wilt disease of maize is transmitted through corn seed at a very low frequency. A TaqMan®-based real-time PCR assay targeting the *cpsD* gene was developed for reliable and specific detection of *P. stewartii* in corn seeds and leaves. The primers and probe selected were able to detect all the 14 reference strains of *P. stewartii*. The TaqMan® PCR assay had a detection limit of 1 pg of purified DNA and 10^4 cfu/ml (10 cells/ reaction) in pure culture. Direct processing of seeds by the assay could detect 50 cells/reaction. As the real-time PCR procedure is sensitive, specific and rapid, it may be suitable for certifying maize seeds meant for export (Tambong et al. 2008).

A combination of magnetic capture hybridization (MCH) and multiplex real-time PCR procedure was developed for the simultaneous detection of *Acdiovorax avenae* subsp. *citrulli* (*Aac*), causing bacterial fruit blotch and the fungal pathogen *Didymella bryoniae*, causative agent of gummy blight disease, in the cucurbit seed samples. Sequestered template DNAs were subsequently amplified by multiplex real-time PCR, using pathogen-specific TaqMan PCR assays. Detection thresholds of *Aac* at 10 cfu/ml and of *D. bryoniae* at 10^5 conidia/ml in mixtures of pure cultures of the two pathogens were achieved, representing a tenfold increase in sensitivity over the direct real time PCR assays for these two pathogens separately. The comparative frequencies of detection of *Aac* for direct real-time PCR assay and MCH real-time PCR formats were 25% (1/4 samples) and 100% (4/4 samples) respecitively at the same DNA concentration. However, MCH did not improve detection sensitivity for *D. bryoniae* relative to direct real-time PCR using conidial suspensions or seed washes from *D. bryoniae*-infested cucurbit seed. MCH real-time PCR procedure could be effectively applied for detection of both pathogens in watermelon and melon seed samples (n = 5,000 seeds/sample) in which 0.02% of the seeds were infested with *A. avenae* subsp. *citrulli* and 0.02% of the seeds were infested with *D. bryoniae* (Ha et al. 2009).

Clavibacter michiganensis subsp. *michiganensis* (*Cmm*), causing bacterial canker disease in tomato is of quarantine importance to many countries. *Cmm* spreads primarily through infected seeds and seedlings. It is essential, therefore, to detect *Cmm* in symptomless seeds to contain the disease spread. Primers and probe were designed based on the sequences of the ITS region with an expected length of 270-bp. With TaqMan probe 20, strong signals were recorded only from *Cmm* strains, but not from other bacterial species including different subspecies of *C. michiganensis*. The tomato seeds were tested directly without extracting DNA and adding reagents that arrested endogenous PCR inhibitors, resulting in reduction of time, test costs and chances of cross-contamination across samples. The detection threshold was 10 cfu when pure culture of *Cmm* was used. In tomato seeds, the pathogen could be detected at contamination levels higher than 1% within a period of 2 h (Zhao et al. 2007).

Tomato bacterial canker disease agent *C. michiganensis* subsp. *michiganensis* (*Cmm*) has to be detected and quantified to assess the population of viable cells that will be responsible for the incidence of the disease. In order to discriminate the viable pathogen cells from dead ones, a quantitative real-time PCR procedure using the DNA binding dye ethidium monoazide (EMA) was developed. EMA can selectively penetrate dead cells and bind to DNA as a bifunctional cross-linking agent. It can inhibit amplification of the target DNA from dead cells in PCR assay. Viable cells with intact membrane are resistant to the entry of the dye. In the real-time PCR assay the viable *Cmm* cells could be discriminated from dead cells with 1 μg/ml of EMA in the sample, but only when the bacterial concentration was 10^7 cfu/ml or lower. The primers and TaqMan probe designed by aligning the ITS sequences of the five

C. michiganensis subspecies were highly specific for *Cmm* at the subspecies level. The detection limit of the direct real-time PCR was 10^3 cfu/ml in seed samples with an apparent sensitivity of 2 cfu of target cells in PCR solution. The real-time PCR protocol developed in this investigation has the potential for quantifying viable pathogens which forms the basic requirement for possible prediction of disease incidence, if the infected seedlots are sown in the following season (Luo et al. 2008).

2.3.2 Detection of Bacterial Pathogens in Propagative Planting Materials

Infected propagative planting materials such as tubers, corms, runners, cuttings, setts etc., form the primary sources of infection of microbial plant pathogens, necessitating early reliable detection of pathogens for the eliminatination of the infected planting materials. This approach assumes greater importance, since use of disease-free planting materials is the basic step for profitable crop production.

2.3.2.1 Detection by Biological Methods

Rapid and accurate indexing procedures to detect latent infections are essential to certify the plant materials for freedom from infection by microbial pathogens. Classical agar-plating technique is employed as a preliminary step to isolate the microorganism(s) consistently associated with the disease under investigation. It may be necessary to develop a suitable medium that differentially encourages the development of the pathogen concerned. For example, the cefazolin trehalose agar (CTA) medium was more effective for the isolation of *Xanthomonas campestris* pv. *manihotis* (*Xcm*) infecting cassava from contaminated planting materials than a starch-based semiselective medium used earlier. The CTA medium may be of use for efficient isolation of *Xcm* from infected planting materials which may not exhibit symptoms of infection, thus facilitating effective elimination of latently infected planting materials (Fessehaie et al. 1999). Likewise, MMG semiselective medium was used for the isolation of *X. campestris* pv. *vitans* (*Xcv*), causing bacterial spot of lettuce, allowing reliable identification of *Xcv* (Toussaint et al. 2001). After bringing the bacterial pathogens into pure culture, they are identified based on cultural, morphological, biochemical, immunological and genomic properties.

2.3.2.2 Detection by Immunoassays

Enzyme-Linked Immunosorbent Assay

Visual examination of seed potato tubers may not be sufficient for obtaining reliable results, while selecting tubers free from infection by bacterial pathogens like *Erwinia* spp. that may remain latent. To prevent the escape of infected tubers, an ELISA protocol was formulated after anaerobic amplification in DPEM medium of *E. chrysanthemi*, causative agent of soft rot disease. This step enhanced the sensitivity of the test and the level of detection improved from 10^5 to $<10^3$ bacteria/ml. The latent infection by *E. chrysanthemi* could be detected in up to 10% of 133 seedlots that had been earlier certified by visual examination as disease-free. Hence, this ELISA format was recommended for large scale adoption for certification programs and for forecasting disease outbreaks in Switzerland (Cazelles et al. 1995). The presence of *E. chrysanthemi* in potato tubers was detected by using a specific MAb (6A6) in the triple antibody ELISA test. The limit of detection of this assay was 10^7 cfu/ml (Singh et al. 2000).

Clavibacter michiganensis subsp. *sepedonicus* (*Cms*), causing bacterial ring rot disease was detected in field-collected potato tubers by ELISA test which was found to be equally sensitive as PCR assay. The effectiveness of detection of *Cms* depended on several factors such as inoculum dose ($>10^9$ cfu), cultivar and interval after planting (Slack et al. 1996). The extent of repeatability (within analyst variation) and reproducibility (among analyst variation) were analyzed to assess the reliability of ELISA tests for detection of *Cms* in potato tissues. The standard deviation (SD) for repeatability of ELISA test was small, but increased at higher absorbance values, whereas the SD for reproducibility was greater and also increased at higher absorbance values indicating the need for adjusting the bacterial concentration at optimal levels (De Boer and Hall 2000).

Ralstonia solanacearum (*Rs*), causing bacterial wilts of plants and brown rot of potato tubers, is under quarantine regulations. Latent infection of potato tubers being visually unrecognizable is of much concern for the personnel of quarantine and certification programs. ELISA formats were found to be suitable for the detection of *Rs* in potato tubers. The sensitivity of ELISA was significantly increased by including an enrichment step prior to ELISA test performed on nitrocellulose membrane for detection of *R. solanacearum* (Priou et al. 1999; Elphinstone et al. 2000). A sensitive and specific ELISA format was developed for the detection of *Rs* in symptomless potato tubers. This protocol involves an efficient enrichment step prior to an indirect double antibody sandwich (DAS)-ELISA format, using specific MAb 8B-IVIA. This MAb reacted with 168 typical *Rs* strains and did not react with other 174 other pathogenic or unidentified bacterial species isolated from potato. The modified Wilbrink broth was used for incubation of the isolates of *Rs* for 72 h at 29°C. The threshold of detection by this ELISA format was 1–10 cfu/ml of individual potato extract containing the pathogen. The results of detection of *Rs* by ELISA test were corroborated by those of conventional methods. This diagnostic test has the potential for use in routine testing in certification programs and epidemiological studies (Caruso et al. 2002).

Immunofluorescence Test

Immunofluorescence (IF) tests have been shown to be effective in detecting the bacterial pathogens in vegetatively propagated planting materials. *C. michiganensis* subsp. *sepedonicus* (*Cms*) was detected in potato tissues by IF test with varying sensitivities (De Boer et al. 1994). A protocol combining the use of rhodomine-labeled oligonucleotide probe and indirect IF test based on specific MAb with a FITC-labeld conjugate was useful in the efficient detection and accurate identification of *Cms* by microscopic examination (Li et al. 1997).

2.3.2.3 Detection by Nucleic Acid-Based Techniques

Nucleic Acid Hybridization Technique

Specific DNA probes were employed for the detection of *Erwinia carotovora* subsp. *atroseptica* (*Eca*) (Ward and De Boer 1984, 1990) and *C. michiganensis* subsp. *sepedonicus* (*Cms*) (Verreault et al. 1998) in potato tubers. *Cms*, the causative agent of potato ring rot disease, was detected by labeled probes obtained from unique three single-copy DNA fragments designated Cms 50, Cms 72 and Cms 85 isolated from CS3 strain of *Cms* by subtractive hybridization. These probes specifically hybridized to all North American strains of *Cms* tested by Southern hybridization. The detection limit of a PCR assay using a primer pair that amplified Cms 85 was found to be 100 cfu/ml of pathogen cells in potato core fluid (Mills et al. 1997). Fluorescent in situ hybridization (FISH) technique was applied for the detection of *Ralstonia solanacearum* race 3 biovar 2, causing brown rot disease of potato tubers, which exhibited strong fluorescent signal. Pectin methyl esterase (Pme) is required for the development of *Rs*, after initiation of infection. Primers designed based on the sequences of the *pme* gene coding for Pme amplified an expected product of 800-bp involved in the biosynthesis of Pme by *Rs*. By using this sequence as a probe, PMR-coding plasmids of *Rs* genome library were identified. This study indicated that Pme was required for the growth of *Rs*, but not for its pathogenicity (Wullings et al. 1998). *Clavibacter xyli* subsp. *xyli* (*Cxx*), causing sugarcane ratoon stunting disease, was detected in sugarcane stalks by tissue hybridization assay using the 560-bp PCR product amplified from the intergenic region of 16S–23S rDNA of *Cxx* with two universal primers (Pan et al. 1998).

Polymerase Chain Reaction-Based Assays

Clavibacter michiganensis subsp. *sepedonicus* (*Cms*), infecting potato tubers, is frequently contaminated with other pathogenic and saprophytic bacterial species. It will be difficult to detect *Cms* by hybridization technique under such conditions. A pair of primers Sp1f and Sp5r was employed for specific PCR amplification of a 215-bp fragment of *Cms*. This primer pair did not amplify DNA from phenotypically and serologically related bacteria isolated from tubers or stem of potato plants. When PCR, ELISA and IF tests were evaluated for their sensitivities of detection of *Cms*, PCR was found to be more sensitive than other two methods. Tubers infected by ring rot pathogen that tested negative for IF and ELISA tests, were found to be PCR-positive, indicating the higher level of sensitivity of PCR assay (Li and De Boer 1995). Use of digoxigenin (DIG)-labeled PCR and employing the primer pair MSIF1/CMSIR1, resulted in high sensitivity on nylon membrane and the signal was detected by a colorimetric assay using alkaline phosphatase. This assay could detect *Cms* in asymptomatic field-grown potato tubers (Lee et al. 2001). A rapid and efficient method of extracting DNA from *C. michiganensis* subsp. *michiganensis* (*Cmm*) and *Ralstonia solanacearum* (*Rs*), pathogens of quarantine importance, was developed using a mixture of lysozyme and proteases combined with minimized TRIS/HCl/BSA buffer volume. The limit of detection of these pathogens by this PCR format, was between 10^4 and 10^5 cfu/ml of potato tuber extracts and the results could be obtained rapidly and reliably (Niepold 1999).

A new potato disease named Zebra Chip (ZC) due to the presence of intermittent dark and light symptom in affected tubers was investigated to establish the nature of the causative agent. Positive transmission through grafting and psyllids and detection of bacteria-like organisms (BLOs) by electron microscopy in the phloem tissues of potato and tomato plants inoculated by grafting and psyllid feeding indicated the possible involvement of the BLO in ZC disease. PCR-amplified 16S rDNA sequences from samples representing different geographic locations including USA, Mexico and Gautemala were almost identical to the 16S rDNA of '*Candidatus Liberibacter solanacerum*' which was reported earlier from solanaceous plants in New Zealand and the USA. For the detection of '*Ca. L. solanacearum*', a rapid PCR protocol was developed and the results could be obtained in less than 90 min. The data confirmed the association of '*Ca*. L. *solanacearum* with ZC disease affecting potato tubers (Secor et al. 2009)

Primers capable of specifically amplifying DNA fragments of *Erwinia carotovora* subsp. *atroseptica* (*Eca*) were employed for the detection of the pathogen in potato tuber and stem tissues. The PCR assay was shown to be more sensitive than ELISA test using MAbs specific to *Eca* (De Boer and Ward 1995). Techniques such as PCR assay providing higher levels of sensitivity and specificity are required for eliminating tubers latently infected by bacterial pathogens, since they are likely to carry and introduce the inoculum into new locations in the next season (Sledz et al. 1999). A one-step PCR-based procedure was developed for the rapid and reliable detection of all five species of *Erwinia* including subsp. *carotovora* and *atroseptica* and all biovars of *E. chrysanthemi* in micropropagated plants. By employing primers for amplifying conserved region of the 16S rRNA gene, detection of the target pathogens could be accomplished. Further, inclusion of an enrichment step prior to PCR amplification resulted in enhancement of the sensitivity of detection by about 200 folds (Toth et al. 1999).

A quantitative competitive PCR protocol was developed for the detection of *E. carotovora* subsp. *atroseptica* (*Eca*). This assay incorporated a competitor PCR template cloned into *Escherichia coli* in the vector pGEM-T. Predetermined numbers of *E. coli* were added to potato peel extract, either pre-inoculated with *Eca* or naturally contaminated tubers. *Eca* concentration was estimated by comparing the ratio of products generated from *Eca* target DNA and competitor template DNA following PCR. The assay results were not affected by the presence of other erwinias and saprophytes. All serogroups of *Eca* were detected, leading to formulation of a key relating product ratios to the likelihood of incidence of blackleg disease. The competitive PCR assay has the potential for use in quarantine and certification programs (Hyman et al. 2000).

Potato scab pathogens *Strepromyces scabies*, *S. turgidiscabies* and *S. aureofaciens* were detected by employing species-specific primers. The forward (leading strand) primer was compatible with the highly variable γ-region of the rRNA gene. The reverse primer was compatible with the variable ε- region (*S. scabies* and *S. turgidiscabies*) or β-region (*S. aureofaciens*). A total of 1,245 scab lesions were tested from potato cvs. Matilda and Sabina grown in two geographical regions of Finland. After incubating harvested or stored tubers at room temperature (18–21°C) under humid conditions for a few days, bacterial growth from scab lesions was collected for DNA extraction and PCR amplification. *S. scabies* and *S.turgidiscabies* were detected in the same potato fields, tubers and scab lesions. The relative incidence of *S. scabies* was high in freshly harvested tubers, but was much lower than that of *S. turgidiscabies* after storage. Both pathogens were transmitted through seed tubers in cultivars 'Matilda' and 'Sabina' after 24 weeks of storage at 4°C. The strain 317 earlier identified as nonpathogenic, was found to be *S. aureofaciens* pathogenic to potato. *S. aureofaciens* was also detected by the PCR assay using the species-specific primer pair (Lehtonen et al. 2004).

A multiplex PCR–ELISA procedure was developed for the detection of *Clavibacter michiganensis* subsp. *sepedonicus* (*Cms*), using primers based on the sequences of three single-copy unique DNA sequences. The three sequences were simultaneously amplified from the genomes of all 42 strains of *Cms* that were tested including variant mucoid forms, but not from strains representing five related subspecies and *Rathayibacter rathayi* and *Rhodococcus faciens*. The lowest limit of detection by gel electrophoresis was estimated to be approximately 300 cfu/ml, when pathogen cells were spiked into potato core fluid. The sensitivity increased approximately tenfold using PCR-ELISA format. When an engineered sea anemone DNA fragment was included, it was amplified by the Cms 72 primer set and it provided simultaneous signal, indicating proper functioning of the system. Further, addition of hydrolyzed casein to the reaction mix reduced markely or eliminated the PCR inhibition by plant cell components or contaminants. The multiplex PCR–ELISA technique was shown to be verifiable for analysis of tuber and stem samples based on the amplification of multiple sites in its genome. The specificity and sensitivity of this assay were of high order, surpassing that of gel electrophoresis of PCR products and serological approaches. The major advantages of the PCR–ELISA multiplex detection format are the elimination of erroneous positive reactions by amplification of unique Cms DNA sequences, verification of the presence or absence of the pathogen at three loci, inclusion of an internal control fragment to eliminate false negatives and prevention of inhibition of PCR amplification using hydrolyzed casein (Mills and Russell 2003).

Amplification of unique DNA fragment of target pathogen is monitored by agarose gel electrophoresis of ethidium bromide-stained amplified DNA fragments. As an alternative to this approach, a colorimetric detection of PCR products was developed enabling visualization of the DNA products, based on initial amplification of target DNA followed by enzyme-linked oligonucleosorbent assay (ELOSA). In ELOSA, the PCR product is labeled with a digoxigenin (DIG) reporter group by incorporating DIG-labeled dUTP's in the PCR. ELOSA relies on annealing of separate biotinylated and digoxigenated probes to the amplified nucleic acid. These complexes are then captured on streptavidin-coated microtiter plates and detected using an anti-digoxigenin antibody conjugate. When the substrate for the enzyme attached to the anti-DIG is added, a color develops, the intensity of which can be quantified spectrophotometrically. ELOSA is more expensive than gel electrophoresis, but it is faster and easier to quantify the target DNA in many samples, using a microtiter plate reader. As there is no need to handle ethidium bromide, ELOSA is safer to apply for the detection of bacterial pathogens.

The ELOSA technique was applied for the detection of *C. michiganensis* subsp. *sepedonicus* (*Cms*) in infected potato tubers. PCR amplification was accomplished by employing primer sets Cms 50 and Cms 72 designed through subtractive hybridization. For all strains of *Cms*, the ELOSA test appeared to be equal or greater in sensitivity, when compared to gel electrophoresis for evaluating PCR products using either primer. In bacterial suspensions, fewer than three cells/10 μl of reaction mix were detected, after PCR amplification and hybridization with specific DIG-labeled DNA probes in ELOSA test. In naturally infected potato tuber samples representing three cultivars, the diagnostic sensitivity of PCR–ELOSA was 96%, while the specificity exceeded 99%. The sensitivity of ELOSA test was not influenced by colony morphology, primer sets or potato cultivar for the detection of *Cms* in tubers obtained from production areas. As a high-throughput technique, PCR–ELOSA protocol, utilizing two unique *Cms* DNA sequence regions, provides higher level of reliability than that of classical PCR and ELISA tests (Baer et al. 2001).

Real-Time Polymerase Chain Reaction

The standard PCR and its variants require longer time, because of the post-PCR amplification steps. Further, they are not useful for quantitative estimation of pathogen population in the plant samples. Real-time polymerase chain reaction (RT-PCR) technology provides more sensitive and reliable results for detection and identification of microbial pathogens present in the asexually propagated plant materials. A combination of real-time PCR and BIO-PCR procedures led to a highly sensitive detection of strains of race 3 biovar (bv) 2 of *Ralstonia solanacearum* (*Rs*) in asymptomatic tubers. Primers and probes of real-time PCR detected all 17 strains of bv 2 including 12 from potato and five from geranium plants. Other strains of *Rs* and 13 other bacterial species associated with potato were not detected by this procedure, indicating the high level of specificity of detection of target pathogen only by this technique. In contrast, the standard real-time PCR format could not detect all strains of *Rs*, revealing the suitability of application of the combination of real-time PCR and BIO-PCR assays, when accuracy and reliability of results are required (Ozakman and Schaad 2004).

Potato bacterial ring rot (BRR) disease caused by *Clavibacter michiganensis* subsp. *sepedonicus* (*Cms*) is included in the category under 'zero tolerance' regulations in the certified seed industry. Hence, detection of symptomless infections with a high degree of accuracy and sensitivity becomes essential to reduce the inoculum in certified seed potato stocks. A real-time PCR assay was developed using the cellulase A (CelA) gene sequence for designing primers. The CelA primers were specific to *Cms* grown in vitro and did not detect any other coryneform bacteria or bacteria pathogenic to potato. The assay detected 69 strains of *Cms*. The CelA real-time PCR assay was more sensitive than immunofluorescence assay (IFA). The primer pair Cms 50/72a detected *Cms* in infected potato tuber cores blended with healthy tuber cores in simulated seedlot contamination experiments. The CelA primers were more sensitive in detecting symptomless infections in seed tubers naturally infected by *Cms*. When tested prior to planting of potato tubers, this format exhibited greater level of sensitivity compared to IFA and ELISA test. The CelA real-time PCR format proved to be a very robust detection aid for BRR disease pathogen (Gudmestad et al. 2009).

A combination of BIO-PCR and an automated real-time fluorescence assay was employed for the detection of *C. michiganensis* subsp. *sepedonicus* (*Cms*) in symptomless potato tubers. PCR primers and a fluorescent probe for use in Perkin Elmer 7700 automated real-time PCR detection system (TaqMan®) were designed from a *Cms*-specific generic DNA fragment. Three techniques, (i) plating aliquots into agar media, (ii) standard PCR and (iii) BIO-PCR were evaluated for their efficacy of *Cms* detection. These three procedures detected *Cms* in 4, 8 and 26 tuber samples respectively. The BIO-PCR combined with TaqMan® automation-based detection was found to be a rapid and reliable method, when large number of potato tuber extracts had to be tested, especially when the samples contained low pathogen population (Schaad et al. 1999).

C. michiganensis subsp. *sepedonicus* (*Cms*), causative organism of potato bacterial ring rot disease, was detected by employing TaqMan PCR into which an internal reaction control component was integrated. The reaction control cloned into the plasmid pCmsC4, consisted of a sequence unrelated to *Cms* flanked by the primer sequences used in the TaqMan PCR, thus eliminating the need for multiplexing. The limit and specificity of detection were not affected by the presence of the internal control. Addition of SYBR Green allowed melt analysis of PCR products. The 242-bp reaction control amplicon, could be distinguished easily from the 152-bp primary diagnostic target amplicon. The internal control was effective in detecting inhibition or reaction failure. Use of the internal control facilitated the application of TaqMan real-time PCR for routine testing of potato tubers for the presence of *Cms* (Smith et al. 2008).

Phytotoxin thaxtomin is a pathogenicity determinant for *Streptomyces* spp. causing potato sacb diseases. Genes encoding thaxtomin synthetase (*txtAB*) are present in pathogenicity island characteristic of genetically diverse plant pathogenic *Streptomyces* spp. SYBR Green quantitative real-time PCR assay was developed using primers designed to anneal to the *txtAB* operon of *Streptomyces* spp. to detect and quantify populations of these pathogens in potato tubers and soil. This assay was specific for pathogenic strains of *Streptomyces* spp. with a detection limit of 10 fg of the target DNA or one genome equivalent. The cycle threshold (Ct) values were linearly correlated with the concentration of target DNA ($R^2 = 0.99$) and they were not affected by the presence of plant DNA extracts. The amount of *Streptomyces* DNA in total DNA extracts from 1 g of asymptomatic and symptomatic tuber tissues was quantified using this protocol. The pathogen concentration ranged from 10 to 10^6 pg. This real-time PCR assay using the primers designed from the sequences of *txtAB* operon was rapid, accurate and cost-effective. Further, quantification of pathogenic strains of *Streptomyces* spp. present in potato tubers was also effectively accomplished by this protocol (Qu et al. 2008).

Grapevine bacterial blight disease caused by *Xylophilus ampelinus* (*Xa*), is primarily disseminated through infected propagative materials. Hence, elimination of such infected planting materials based on reliable and rapid diagnostic techniques can be expected to effectively restrict the spread of this disease. A real-time protocol based on a 5′-nuclease and minor groove binding (MGB) probe was developed to overcome the limitations of the methods available earlier. This protocol was rapid and sensitive, providing reliable detection of *Xa* in plant tissues and facilitating precise identification of the isolated bacteria. The limit of detection of this protocol was approximately 100 cells of *Xa* in plant tissue extract, proving to be more sensitive than the nested PCR format by tenfold. The real-time PCR technique could detect all isolates of *Xa* obtained from different geographic locations. The target sequence appeared to be conserved and specific in *Xa* isolates, since no specific signal was observed in samples containing other bacterial species infecting grapevines. The high sensitivity of real-time PCR assay allowed detection of *Xa* in low number in tissues, thus offering the possibility of detecting latent infection in planting materials which carry the infection to long distances as well as to the subsequent generations of planting materials (Dreo et al. 2007).

DNA Array Technology

An array of species-specific oligonucleotide probes representing different bacterial pathogens infecting potato is built on a solid support and then probed with labeled PCR products amplified from potato sample. Conserved primers are employed to amplify common bacterial genome fragments from extracts of potato tubers containing bacterial pathogens. The presence of DNA sequences unique to pathogenic species may be inferred by hybridization to species-specific oligonucleotide probes within the array. Thus, simultaneous detection of several pathogens in a sample is possible by applying DNA array technique.

Oligonucleotides in the 16S–23S intergenic spacer (IGS) region of the ribosomal gene clusters that are specific to each of *Clavibacter michiganensis* subsp. *sepedonicus* (*Cms*), *Erwinia carotovora* subsp. *atroseptica* (*Eca*), *E. carotovora* subsp. *carotovora* (*Ecc*) and *E. chrysanthemi* (*Ec*) were designed and formatted into an array by pinspotting on nylon membranes. Conserved ribosomal primers were employed for the amplification of specific DNA fragment from different bacterial species (in culture) and labeling simultaneously with digoxigenin-dUTP. It was possible to recognize distinct hybridization patterns, following hybridization of amplicons to the array. The digoxigenin label was detected serologically.

Hybridization patterns were recorded as gray values for each hybridized spots. A consistent pattern of hybridization was observed for multiple strains of each species or subspecies isolated from diverse geographical locations. The bacterial species could be detected and identified accurately present in mixed cultures as well as in inoculated potato tuber tissues (Fessehaie et al. 2003).

Nucleic Acid Sequence-Based Amplification Assay

The PCR-based techniques have been used for the detection and quantification of various bacterial pathogen populations. However, PCR amplification methods are not well-suited for investigations such as population dynamics and assessment of effect of control strategies applied against crop pathogens such as chemical application and eradication measures that require determination of viable cells. PCR is based on amplification of DNA which is relatively stable. Hence, the DNA from lysed target cells persisting for long time will also be included in the quantitative estimation by PCR-based assays. To assess the quantity of viable, living pathogen cells, a method that can detect and quantify RNA of pathogen cells will be useful, since the RNA is degraded rapidly upon cell death. A nucleic acid sequence-based amplification (NASBA) technique was developed to amplify a specific sequence of human immunodeficiency virus (HIV) (Kievits et al. 1991) and adapted for the detection of human bacterial pathogens (Van der Vliet et al. 1994). In this technique the specific 16S rRNA sequences of the target pathogen are amplified from viable pathogen cells and hence, detection of RNA coincides with cell viability. Consequently, NASBA procedure enables specific detection of living cells of the target pathogens.

The NASBA technique was developed for the direct detection of *Ralstonia solanacearum* (*Rs*). NASBA exclusively detected RNA, but not the DNA, from samples of nucleic acids from *Rs* treated with RNase. The ability of NASBA to assess cell viability was demonstrated in two sets of experiments. Viable and chlorine-killed cells of *Rs* were added to a potato tuber extract and tested by NASBA and PCR assays. In NASBA, only extracts spiked with viable cells, resulted in specific signal after Northern blot analysis, whereas in PCR assay, targeting 16S rDNA sequences, extracts with both viable and killed cells of *Rs* resulted in specific signals. In another experiment, survival of *Rs* on metal strips was determined by employing NASBA, PCR and dilution plating assays. A positive correlation was established between NASBA and dilution plating methods that detect culturable living cells of *Rs*. In contrast, PCR amplification led to positive reactions also in samples containing dead cells of the pathogen. Loss of NASBA signal coincided with a loss of culturability of *Rs* present in the sample, indicating the capacity of NASBA to discriminate the living from dead cells of *Rs*. The limit of detection of *Rs* by NASBA method was 10^4 cfu/ml of potato tuber extract, equivalent to 100 cfu/reaction. This corresponds with less than one bacterial cell, assuming that a metabolically active cell contains ca. 10^5 copies of RNA (Bentsink et al. 2002).

As the large scale applicability of NASBA technique was found to be limited because of the lack of robustness and the possibility of cross-contamination while performing Norhern hybridization, an improved AmpliDet RNA procedure was developed based on a combination of NASBA amplification and homogeneous real-time detection of amplicons with molecular beacons. Real-time detection of amplicons reduced labor input and contamination risks drastically. This technique was applied for the detection of *R. solanacearum* in potato tuber samples, in sealed tubes to reduce the contamination risks. AmpliDet RNA assay provided reliable detection of specific 16S rRNA sequences of *Rs* in total RNA extracts from potato tuber samples in 90 min at a level of 10 cells/ml of *R. solanacearum* race 3, biovar 2. The pathogen could be detected reliably in 18 naturally infected potato tuber samples containing varying concentrations of cells of *R. solanacearum* (Van der Wolf et al. 2004).

2.3.3 Detection of Bacterial Pathogens in Postharvest Produce

Bacterial pathogens infect more frequently plant organs/tissues such as roots, stem, leaves and floral structures. But only a few bacterial pathogens are associated with postharvest produce such as fruits and vegetables. Very often saprophytic bacteria may be found along with other pathogens causing general rotting to different levels during transit or storage depending on the environmental condtions. Conventional methods of culturing the bacterial pathogens infecting fruits and vegetables are adopted and purified cultures of pathogens are used for identitication as per the procedures described earlier. Molecular techniques have been employed for detection and identification of some bacterial pathogens isolated from the infected produce.

Enzyme-linked immunosorbent assay (ELISA) test was used to detect *Xanthomonas axonopodis* pv. *citri* (*Xac*), causing canker spots on citrus fruits, The market value of the infected fruits is reduced drastically. According to European Union (EU) legislation, the European countries have to analyze samples of fruits with suspicious symptoms, when they are imported from countries where citrus bacterial canker (CBC) is prevalent (Anon 2005). In order to increase the bacterial density, a modified PSB (MPSB) medium amended with Fe-EDTA was used for enrichment by incubating fruit rinse water for 24 h prior to ELISA test. When the efficacy of ELISA test and bacteriophage test (BPT) was evaluated, ELISA test was found to be more sensitive by detecting *Xac* in all citrus fruits, whereas BPT detected *Xac* only in 44% of the fruits tested. Further, ELISA required only a short time for detecting *Xac* on citrus fruits compared to BPT (Jin et al. 2001).

Nucleic acid-based techniques have been shown to be very sensitive and rapid in detecting the bacterial pathogens present in postharvest produce. Two oligonucleotide primers designed from the sequences of a unique DNA fragment of the plasmid pEA29 were used for the detection by PCR of *Erwinia amylovora* (*Ea*) in leaves, axillary bud and mature fruit calyx from apple trees infected by fire blight disease (McManus and Jones 1996). PCR assay was employed to monitor the populations of *Ea* in the calyxes of apple fruits under storage conditions (Hale and Taylor 1999). Citrus variegated chlorosis (CVC) disease of sweet orange caused by *Xylella fastidiosa* (*Xf*) was detected by employing PCR assay. The presence of *Xf* was detected in all main fruit vascular bundles as well as in seeds and samples of peduncle, peel, endocarp, septum, locular membrane and central axis of infected fruits of three sweet orange cultivars 'Pera', 'Natal' and 'Valencia' (Li et al. 2003).

Strains of citrus bacterial canker (CBC) pathogen *Xanthomonas axonopodis* pv. *citri* (*Xac*) were rapidly detected and differentiated by employing the real-time PCR format using a portable PCR equipment. The assay provided results within a period of 4 h and had an apparent sensitivity of less than 10 cfu of *Xac* cells from single lesions (Mavrodieva et al. 2004). A detection system for *Xac* needs to be optimized for analysis of fruits with suspected lesions and it has to be fast and sensitive to the maximum extent. An integrated approach that includes isolation, pathogenicity assays and molecular techniques was adopted for the diagnosis of *Xac*. Bacterial isolation, three conventional PCR protocols and real-time PCR with SYBR Green or TaqMan probe were compared. Canker lesions had *Xac* in 11 of 15 fruit samples, as revealed by PCR assay. In 16 out of 130 lesions analyzed, *Xac* was isolated and pathogenicity on grapefruit leaves was confirmed. By using real-time PCR SYBR Green or TaqMan probe, *Xac* was detected in 58 and 80 lesions respectively. The pathogen was detected by conventional PCR in 39–52 lesions depending on the protocol applied (Fig. 2.15). When detection methods were applied to naturally infected samples, real-time PCR format was the most efficient technique for the detection of *Xac*, especially if TaqMan probe was used. However, conventional PCR assay could be used as a screening test for reliable detection of *Xac* in commercial citrus fruits (Golmohammadi et al. 2007).

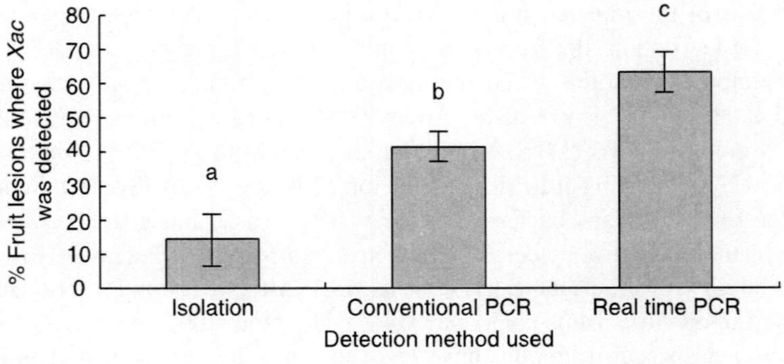

Fig. 2.15 Comparative efficacy of detection of *Xanthomonas axonopodis* pv. *citri* (*Xac*) by isolation, conventional PCR and real-time PCR methods in citrus fruit lesions. Histograms indicate the average of 15 samples with the standard error bars (Courtesy of Golmohammadi et al. 2007; The Society for Applied Microbiology/Wiley-Blackwell, Oxford, UK)

2.4 Detection of Phytoplasmal Pathogens in Planting Materials

The nonculturable nature of the phytoplasmas exclude the possibility of studying their morphological characteristics in pure cultures. On the other hand, *Spiroplasma* species have been characterized based on the cultural characterisitics. Various techniques based on the immunological and genomic properties have been applied for the detection and differentiation of phytoplasmas present in plants. The phytoplasmas spread primarily through planting materials obtained from infected mother plants or asymptomatic seedlings which have been inoculated by the vector insects under natural conditions. Phytoplasmas induce partial or total sterility in infected plants which may not produce viable seeds. In such cases, transmission of phytoplasmas through seeds is unlikely. The plausible way of checking the spread of the phytoplasmal diseases is to rigorously test the mother plants prior to preparing asexually propagated planting materials from them by applying diagnostic techniques that can provide reliable results rapidly. The usefulness and comparative efficiency of various diagnostic methods applicable for detection of phytoplasmas infecting plants have been discussed earlier in this chapter. Certification programs are in operation in several countries to certify the planting materials for freedom from economically important diseases including those caused by phytoplasmas (Narayanasamy 2001).

Potatoes are infected by many fungal, bacterial, phytoplasmal, viral and viroid diseases. Infection of potato seed tubers by phytoplasmas has been examined less extensively compared to diseases caused by other kinds of pathogens. In the investigation carried out in Italy, eight batches of seed potatoes collected from different locations were planted in the greenhouse and sprouts growing from the selected tubers were tested for the presence of phytoplasmas by employing universal primers P1/P7 in direct PCR followed by nested PCR with primer pairs R16F2/R2 and R16(1)F1/R1. The primer pair R16mF2/mR2 was also used in direct PCR assay. RFLP analyses were performed with *Tru*I and *Taq*I for phytoplasma identification. Small shoots (2–15 cm long) collected from either single potato or from batches of three potatoes each were used for DNA extraction and tested for the presence of phytoplasmas. The assays showed that phytoplasmas belonging to diverse ribosomal groups were present in the potato tubers. Based on RFLP analyses on 16S rDNA gene, phytoplasma belonging to ribosomal subgroups 16SrI-B (related to '*Candidatus Phytoplasma asteris*'), 16SrI-C (related to clover phyllody), 16SrII-D (related to tomato big bud from Australia), 16SrX-A (related to '*Ca*. P. mali') and 16SrXII-A (related to stolbur) were identified. The protocol employed in this investigation provided reliable results rapidly (Paltrinieri and Bertaccini 2007). The standard PCR and its variants have been employed extensively to detect the phytoplasmas in plant hosts from which propagative materials are prepared for planting in the subsequent seasons. If the freedom of planting materials from phytoplasmas is not ensured by employing reliable diagnostic techniques, cultivation of crops, using infected planting materials repeatedly, will soon become uneconomical in a few generations.

Appendix 1: Media for Isolation of Bacterial Plant Pathogens

A. General Media	
Brinkerhoff's medium (Brinkerhoff 1960)	
Dextrose	20 g
K_2HPO_4	50 g
Calcium carbonate	10 g
Agar	15 g
Water	1 l
Dye's medium (Dye 1962)	
Glucose	10 g
K_2HPO_4	2 g
Ammonium phosphate	1 g
$MgSO_4$	0.2 g
NaCl or KCl	0.2 g
Water	1 l
King's B agar medium (Matsuyama et al. 1998)	
Peptone	20 g
K_2HPO_4	1.5 g
$MgSO_4.7H_2O$	1.5 g
Agar	15 g
Glycerol solution (1%)	1,000 ml
PH	7.2
Medium 523 (Kado 1971)	
Sucrose	10 g
Casein acid hydrolysate	8 g
Yeast extract	4 g
KH_2PO_4	50 g
Magnesium sulfate	0.3 g
Agar	15 g
Water	1 l
Nutrient broth	
Bactopeptone	5 g
Beef extract	3 g
Water	1 l
Potato–peptone–glucose–agar (PPGA) medium	
Potato extract	500 ml
Bactopeptone	5 g
Glucose	5 g
NaCl	3 g
Na_2HPO_4	3 g
K_2HPO_4	0.5 g
Agar	18 g
Water	500 ml
Tetrazolium medium (TTC) (Kelman 1954)	
Dextrose	10 g
Peptone	10 g

(continued)

Appendix 1 (continued)

Cis amino acids	1 g
Agar	18 g
Water	1 l
Wakimoto's medium (Wakimoto 1960)	
Potato	200 g
Surcrose	15 g
Peptone	5 g
$Na_2HPO_4.12\ H_2O$	2 g
Calcium nitrate	0.5 g
Water	1 l
B. Semi-selective/selective media	
Cefazolin trehalose agar (CTA) medium (Fessehaie et al. 1999)	
K_2HPO_4	3 g
Na_2HPO_4	1 g
$MgSO_4.7H_2O$	0.3 g
NH4Cl	1 g
D (+)-trehalose	9 g
D (+)-glucose	1 g
Yeast extract	1 g
Cefazolin	0.025 g
Lincomycin	0.0012 g
Phosphomycin	0.0025 g
Cycloheximide	0.25 g
Agar	14 g
Distilled water	1 l
MMG semiselective medium (Toussaint et al. 2001)	
Maltose	10 g
Tryptone	5 g
K_2HPO_4	2.75 g
Trace elements	0.02–1 mg
Methyl green (1% aqueous solution)	2 ml
Amoxicillin	32 mg
Cephalothin	32 mg
Cycloheximide	50 mg
Agar	15 g
Distilled water	1 l
MSCFF medium (Maringoni et al. 2006)	
Peptone	5 g
Meat extract	3 g
Sucrose	5 g
Agar	15 g
Skim milk powder	5 g
Congo red	0.05 g
Chlorothalonil	0.01 g
Thiophanate methyl	0.01 g

(continued)

Appendix 1 (continued)

Nalidixic acid	0.01 g
Nitrofurantoin	0.01 g
Oxacillin	0.001 g
Sodium azide	0.001 g
Distilled water	1 l
PCCG medium (Hara et al. 1995; Ito et al. 1998)	
Potato semisynthetic agar (PSA) medium	
Potato decoction from	300 g
Peptone	5 g
Sucrose	15 g
$Ca(NO_3)_2$	0.5 g
$Na_2HPO_4.12H_2O$	2 g
Agar	18 g
PCCG medium	
PSA	1,000 ml
Gella gum	18 g
Crystal violet	5 mg
Polymyxin B	4×10^5 units
Chloramphenicol	7.5 mg
Cycloheximide	50 mg
Tetrazolium chloride	2.5 mg
T-5 Semi-selective agar medium (Gaitaitis et al. 1997)	
NaCl	5 g
Ammonium phosphate	1 g
K_2HPO_4	1 g
$MgSO_4.7H_2O$	0.2 g
D-tartaric acid	3 g
Phenol red	0.01 g
Agar	20 g
Water	1 liter
After autoclaving add	
Bacitracin	10 mg
Vaniomycin	6 mg
Cycloheximide	75 mg
Novobiocin	45 mg
Penicillin G	5 mg
Adjust pH to 7.4	
XTS selective medium (Schaad and Forster 1985)	
Nutrient agar (Difco)	23 g
Glucose	5 g
Add the following antibiotics after autoclaving:	
Cycloheximide (stock solution of 100 mg/ml in 75% ethanol)	200 mg

(continued)

Appendix 1 (continued)

Gentamycin (stock solution of 10mg/ml in 75% ethanol)	8 mg
Cephalexin (stock solution of 10 mg/l in 75% ethanol)	10 mg

Appendix 2: Detection of *Acidovorax avenae* subsp. *citrulli* (*Aac*) by Pulsed-Field Gel Electrophoresis (PFGE) (Walcott et al. 2000; Burdman et al. 2005)

1. Grow the target bacteria (*Aac*) on nutrient agar (NA) at 28°C for 48 h; suspend the cells in a buffer containing 100 mM EDTA and 100 mM NaCl, pH 8 and adjust the bacterial cells to an OD_{600} of 0.3 (approximately 5×10^8 cfu/ml at 600 nm using a colorimeter).
2. Centrifuge the suspension at 10,780 × g for 2 min; resuspend the pellet in 300 µl of the buffer used earlier (step 1) and add 60 µl of 10 mg/ml lysozyme solution.
3. Prepare agarose plugs as follows: mix 500 µl of molten low-melting point agarose (10 mM Tris, pH 8, 10 mM $MgCl_2$, 10 mM EDTA, pH 8 and 2% (w/v) Sea Plaque GTG agarose (FMC BioProducts, Rockland, USA) with cell suspensions and pipette the mixture into disposable plastic plug-molds (Bio-Rad Laboratories, USA).
4. Solidify the agar plugs at 4°C for 20 min; cut the agar plugs into 5 × 5 × 1 mm pieces; rinse four times with 1 ml of sterile 1 × TAE (30-min rinses); replace 1 × TAE (40 mM Tris, 40 mM acetic acid and 1 mM EDTA) with 200 µl of 1 × restriction enzyme (*Spe*I, New England Biolabs, USA), buffer B (Promega Corp, USA) and incubate the plugs at 25°C for 15 min.
5. Replace restriction buffer with 200 µl of fresh restriction buffer containing 1 µl of *Spe*I and incubate at 37°C overnight.
6. Replace the restriction buffer with 500 µl of lysing solution without proteinase K; incubate at 55°C for 2 h; replace the lysing solution with fresh lysing solution without proteinase K and incubate the agarose plugs at 25°C for an additional 2 h.
7. Load the agarose plugs into wells of a 1% (w/v) pulsed-field certified agarose (BioRad Laboratories) gel (15 × 15 cm 1% Seakem Gold agarose); seal the molten low-melting-point agarose and use concatemeric lambda DNA (New England Biolabs) as a molecular marker.
8. Incubate the loaded gel in 0.5 × Tris-borate-EDTA (TBE) buffer (45 mM Tris; 45 mM boric acid and 1 mM EDTA, pH 8) for 30 min prior to electrophoresis and perform electrophoresis on a contour-clamped homogeneous electric field gel electrophoresis unit (CHEF-DRIII, Bio-Rad Laboratories) for 22 h at 6V/cm in 0.5 × TBE buffer at 14°C with initial and final switch times of 5 and 45 s respectively.
9. Stain the gels with a solution of 0.5 µg of ethidium bromide per ml for 30 min and capture the images under ultraviolet (at 295 nm) transillumination using a Kodak DC-40 digital camera.

10. Perform two independent PFGEs for each isolate for comparison of results for similarity/variation.

Appendix 3: Detection of Bacterial Pathogens by Slide Agglutination Test (Lyons and Taylor 1990)

Preparation of Somatic Antigen

1. Harvest the growth of target bacteria from 20 petriplates into 100 ml distilled water; centrifuge at 12,000 × g for 30 min; resuspend the bacterial cells in 50 ml cold acetone added with constant stirring to a total volume of 200 ml; centrifuge at 250 × g for 5 min; resuspend the bacterial cells in 10 ml cold acetone; lyophilize and store at −20°C.
2. Transfer approximately 2 g lyophilized bacterial cells to 100 ml cold distilled water; stir well; add slowly 100 ml 1 N trichloroacetic acid; stir the mixture for 18 h in the cold and centrifuge at 12,000 × g for 30 min at 4°C.
3. Collect the supernatant; neutralize with saturated sodium hydroxide; add four volumes of cold 95% ethanol (−20°C) and mix thoroughly.
4. Centrifuge at 200 × g for 10 min; remove the white floccular precipitate after discarding the supernatant; dissolve the precipitate in 5 ml distilled water; add 20 ml cold 95% ethanol; centrifuge at 200 × g for 5 min; resuspend the precipitate in 5 ml cold 95% ethanol; lyophilize and store dried powder at −20°C.

Production of Antiserum

1. Immunize New Zealand white rabbits with 0.5 ml of thick suspension of bacterial cells (10^8–10^9/ml) in saline or 10 mg of dried somatic antigen in 0.5 ml sterile distilled water emulsified with 0.5 ml Freund's incomplete adjuvant intramuscularly, after bleeding for obtaining normal control serum; provide a second injection after 7 days and a third one, if necessary; collect the antiserum after bleeding; mix the antiserum with equal part of glycerol and store at 2–4°C.

Production of Staphylococcus aureus Reagent

1. Grow the authentic Cowan 1 strain of *S. aureus* on nutrient agar at 37°C for 24–48 h; prepare a cloudy suspension of bacterial cells (10^3–10^7) in glycerol broth containing 1.6 g nutrient broth (Difco), 30 ml glycerol and 170 ml distilled water; transfer 1–2 ml aliquots to sterile storage containers; add an equal volume of sterile 3 mm hollow glass beads (Creative Beadcraft Ltd, Amersham, UK) and place the containers directly at −80°C without pre-cooling.

2. Thaw one container; transfer the contents to 10 ml nutrient broth; distribute the culture uniformly by shaking well to 40 nutrient agar plates and incubate at 37°C for 72 h.
3. Harvest the bacteria from the plates by flooding with phosphate-buffered saline (PBS) containing 0.02% sodium azide and gently scrapping surface of plates with a glass streaker.
4. Centrifuge for 30 min at 300 × g; resuspend the deposit, after discarding the supernatant, in 20 ml 1.5% formaldehyde solution; mix well for 90 min; heat at 80°C for 30 min and cool rapidly to room temperature.
5. Wash twice by centrifugation; resuspend in PBS containing 0.05% sodium azide and mix with PBS containing 0.05% sodium azide at the rate of one volume of packed cells to nine volumes of buffer.
6. Sonicate the suspenstion gently for 30 s to eliminate clumps and store at 2–4°C until required for use.

Conjugation of Working Reagent

1. Use 170 μl antiserum (1:1 mixture with glycerol), 4 ml PBSA, 830 μl *S. aureus* reagent and 100 μl filtered saturated alcoholic basic fuchsin 100 μl and store this preparation at 2–4°C.

Slide Agglutination Test

1. Mix the conjugated working agent and the target organism either from pure culture or infected plant tissue extracts for several seconds on a glass microscope slide (Multitest slides) using a sterile wood toothpick; use 7 μl conjugated working reagent + sufficient bacteria to produce a thin suspension or5 μl plant extract + 5 μl conjugated reagent for mixing on the slide.
2. Maintain positive and negative controls with homologous bacteria and test reagent alone respectively.
3. Observe for the formation of granular agglutination indicating the positive reaction.

Appendix 4: Detection of *Xanthomonas campestris* pv. *vesicatoria* by Enzyme-Linked Immunosorbent Assay (ELISA) Formats (Tsuchiya and d'Ursel 2004)

Preparation of Bacterial Cells (Antigen)

1. Grow the target bacterial species on suitable (YPA) medium containing 7 g yeast extract, 16 g agar, 1 l distilled water, pH 7.2 by incubating at 28°C for 2 days;

collect the bacteria with 10 ml of 0.01 M phosphate-buffered saline (PBS), pH 7.2 and wash the cells with PBS three times by centrifugations at 10,000 × g for 10 min.

2. Weigh the wet cell precipitates; adjust the stock concentration of 10 mg/ml and prepare the standard cell suspension spectrophotometrically to about 10^8 cfu/ml.

Preparation of Plant Samples

1. Collect the leaves with symptoms at random in the fields; cut them separately; treat them as described below, maintaining healthy controls.
2. Collect leaves from inoculated plants; wash them by successively soaking them in water, 30 s in 70% EtOH and then 5 min in 0.5% NaClo and rinse thoroughly with sterile distilled water.
3. Excise two disks for each sample with a 10 mm diameter cork borer; homogenize in 1 ml of 0.1 M phosphater buffer, pH 7.0 with a tissue grinder; transfer to 1 ml microcentrifuge tubes; centrifuge for 1 s at 5,000 rpm; store the supernatant at 4°C and adjust the fractions for coating with 1 M sodium carbonate buffer, pH 9.6 just before use.

ABC-Enzyme-Linked Immunosorbent Assay

1. For noncompetitive immunoassay, coat 96-well plates first with samples (200 µl/well) suspended in 0.1 M carbonate buffer at 37°C for 4 h and wash three times with PBS buffer and also in subsequent steps as described below:
2. Block unspecific reactive sites with a solution of 1% skim milk powder for 2 h at room temperature (RT); add monoclonal antibodies (MAbs) (100 µl/well); incubate at RT and wash.
3. Characterize the MAbs by following one of the two methods described below:

For avidin/biotin peroxidase-conjugate (ABC) method, use an ABC Kit (Vector Laboratories, USA) with 2,2 -azino-di-3-ethyl-benzothiazoline-b-sulfonic acid (ABTS) (Boehringer, Germany) as substrate for peroxidase (A_{max} + 405 nm)

For the second method, incubate with goat anti-mouse immunoglobulin M (IgM) peroxidase conjugate (Chemi-ConInternational, USA) for 30 min; add a solution of 0.04% *O*-phenylenediamine dihydrochloride (OPD) (A_{max} = 492 nm) and 0.12% hydrogen peroxidase in 0.5 M phosphate–citrate buffer pH 5, as substrate.

4. Competitive immunoassay allows reaction between MAbs and the samples diluted in PBS for 4 h in microcentrifuge tubes at RT; dispense in duplicates to the antigen precoated 96-will plate; determine the amount of MAbs retained on the wells as per the second method described above (step 3).

Appendix 5: Detection of *Erwinia chrysanthemi* Using Fimbria-Specific Antibody in ELISA Technique (Singh et al. 2000)

Production of Antiserum

1. Immunize New Zealand white rabbit for producing polyclonal antibodies (PAbs) with glutaraldehyde-fixed cells of target bacterial species intramuscularly; provide five injections at 2-week intervals; collect the blood by cardiac puncture and separate the antiserum.
2. Immunize the mice for producing monoclonal antibodies (MAbs) by intraperitoneal injection of glutaraldehyde-fixed bacterial cells.
3. Fuse the splenocytes from immunized mice with FOX-NY cells using 50% polyethylene glycol (PG) and dimethyl sulphoxide as fusogens.
4. Use Dulbeccos modified Eagles Medium supplemented with hypoxanthine, aminopterin and thymidine to select suitable hybridomas for screening for specificity of reaction.

Screening Hybridomas for Specific MAbs

1. Wash 24-h old pathogen cells from pure cultures; suspend them in 0.01 M phosphate buffered saline (PBS), pH 7.2 and adjust the cell population to an OD of 1.0 at 620 nm.
2. Transfer 100 µl bacterial cells diluted to 1:1 in 0.1 M carbonate buffer, pH 9.6 to each of the microplate well and incubate overnight at 4°C.
3. Block the monspecific sites with 5% skim milk powder for 20 min; wash and add hybridoma fluid containing MAbs.
4. Carry out all incubations (one h at 37°C) and washing steps as in standard ELISA protocol.
5. Detect the reactive MAbs using goat ant-mouse antibodies conjugated with alkaline phosphatase and *p*-nitrophenol phosphate at 0.5 mg/ml in 1 M diethanolamine buffer, pH 9.8 and incubate for 45–60 min.
6. Determine the color intensity at 405 nm using an ELISA reader.

Detection of Target Pathogen in Plant Tissues

1. Prepare the 1% plant extracts using the sample buffer consisting of 2 g KH_2PO_4, 11.5 g Na_2HPO_4, 0.14 g disodium EDTA, 0.02 g thimerosal and 0.2 g lysozyme in 100 ml water.

2. Coat the microtiter wells with PAbs (1:5,000 dilution of raw serum in 0.1 M carbonate buffer, pH 9.6).
3. Transfer the plant extracts to each well to capture the antigen by PAbs present in the wells.
4. Follow the steps as per standard ELISA procedure.

Appendix 6: Detection of *Erwinia chrysanthemi* by Immunogold Labeling and Electron Microscopy Technique (Singh et al. 2000)

Immunogold Labeling

1. Collect the target bacterial cells from a 24-h old culture; suspend in PBS at a concentration of 1.0 OD at 620 nm; mix with an equal volume of hybridoma 6A6 culture fluid and incubate at 4°C overnight.
2. Wash the cells three times in PBS; suspend them in 200 µl of PBS and add 15 µl of gold beads conjugated with goat anti-mouse immunoglobulin G and immunoglobulin M (IgG + IgM, Bio/Can Scientific, Canada) and incubate the mixture for 1 h at room temperature with agitation.
3. Wash the cells with PBS three times and resuspend in 200 µl PBS.

Electron Microscopy

1. Place the bacterial cell suspension on copper-coated electron microscope grids at 25 µl/ grid and air dry.
2. Stain with 1% phosphotungstic acid and view under the electron microscope.

Appendix 7: Detection of *Clavibacter michiganensis* subsp. *michiganensis* by Flow Cytometry (Chitaara et al. 2006)

Labeling of Target Bacterial Cells

1. Grow the target pathogen on 1% glucose-nutrient agar medium (Oxoid) for 24 h at 25°C; harvest the bacterial cells; resuspend in 0.2 M sodium phosphate buffer (SP1), pH 7.4; adjust the OD at 620 nm to approximately to 0.35 to give 10^6–10^7 cfu/ml concentration.

2. Prepare nonviable bacterial cells by exposing the bacterial cells to 80°C for 30 min.
3. Mix the viable (non-treated) and non-viable (heat-treated) cells in different ratios: 100/0, 80/20, 50/50, 20/80, 0/100 respectively.
4. Incubate the bacterial cells in SP1 for 1 h at 28°C in the presence of 10 μM cFDA (carboxy-fluorescein diacetate), 10 μM Calcein AM (Calcein acetoxy methyl ester) separately or in combination with 10 μM PI (propidium iodide) for double labeling; incubate the samples stained with PI alone for 20 min at room temperature.
5. Centrifuge the suspension at 11,000g; wash; resuspend in SP1, pH 7.4 and place them on ice till required.

Flow cytometric Analysis

1. Analyze individual cells using a FACS Calibur flow cytometer (Becton-Dickinson Benelux NV, Belgium) equipped with an air-cooled argon ion laser (excitation wave length 488 nm) operated at 15 mW.
2. Use a logarithmic amplification of the incoming signal and a sample analysis time of approximately 2 min.
3. Quantify based on the flow rate that is determined to be 4.8 μl/min.
4. Use non-treated (viable) cells as negative controls for Calcein AM and cFDA and heat-treated (non-viable) unstained cells for PI.
5. Sort out the bacterial cells labeled with Calcein AM, cFDA or PI based on the green and red fluorescence of cells at 530 and >670 nm respectively.
6. Note the high green fluorescence signals from Calcein AM- and cFDA-stained cells (viable) and high red fluorescence signal from PI-stained cells (non-viable).

Appendix 8: Detection of *Erwinia amylovora* by Polymerase Chain Reaction (PCR) (Stöger et al. 2006)

Extraction of DNA

1. Cut wood slices from naturally infected or inoculated plant samples; place 0.5–0.7 cm slices/leaf discs into 1.5-ml collection tube; add extraction solution of REDExtract-N-Amp™ Plant PCR Kit and incubate at 95°C for 10 min; follow all steps as per the manufacturer's recommendations.
2. Dilute the extract with 100 μl of dilution buffer; mix by vortexing and use 4 μl of the diluted extract for PCR.
3. Adopt the modified procedure for enhancement of sensitivity; place the wood slices (instead of leaf discs) into 1.5-ml collection tube; add 150 μl extraction solution supplemented with 0.1% Triton X-100 (v/v) and 0.05% Nonidet NP-40 and incubate at 95°C for 30 min.

4. Transfer 50 μl of the extract to a new tube; dilute it with 50 μl dilution buffer and store at 2–8°C till use.
5. Prepare a 1:30 dilution (v/v) of extract with a 1: 1 mixture of extraction: dilution solution and use 4 μl of diluted extract for PCR.

Polymerase Chain Reaction

1. Perform all steps as per manufacturer's instructions by providing the following PCR conditions: 95°C for 5 min; 35 cycles each of 95°C for 15 s, 53°C for 30 s and 72°C for 45 s and final step at 72°C for 10 min.
2. Analyze amplified PCR products on 1.5% agarose gels stained with ethidium bromide and photographed.

Appendix 9: Detection of *Xanthomonas campestris* pv. *vesicatoria (Xcv)* by PCR Assay (Park et al. 2009)

PCR Amplification

1. Use the primers XCVF and XCVR designed using the sequences of *rhs* family gene of *Xcv* to amplify a product of 517-bp.
2. Perform the PCR amplification using a PTC-225™ thermocycler (MJ Research, USA) in a final volume of 50 μl containing 10 mM Tris-HCl, 50 mM KCl, 1.5 mM $MgCl_2$, 0.01 % gelatin, 0.2 mM of each dNTP, 10 pM of each primer and 2 Units of *Taq* polymerase (Promega™, USA) and approximately 50 ng of genomic DNA of test microorganism.
3. Provide the following conditions: 25 cycles, each consisting of 60 s at 94°C, 30 s at 58°C and 60 s at 72°C with initial denaturation of 5 min at 94°C and final extension of 10 min at 72°C.
4. Resolve an aliquot of 8 μl of PCR product on a 1.0% agarose gel, stained with ethidium bromide and visualized on a UV transilluminator.

Dot Blot Analysis

1. Spot aliquots of 100 ng of each sample of DNA isolated from test pathogen (*Xcv*) onto Hybond-N+ nylon membrane (Amersham Pharmacia Biotech UK).
2. Label the PCR product of *Xcv* as probe with ^{32}P dCTP using random primed method as per manufacturer's recommendations (Ladderman™ Labeling Kit, Takara, Japan).

3. Perform prehybridization and hybridization in hybridization buffer consisting of 0.75 M NaCl, 75 mM sodium citrate, 0.5% SDS, 0.1% BSA, 0.1% FiColl, 0.1% polyvinyl pyrrolidone and 50 μg/ml denatured salmon sperm DNA at 65°C for 18 h.
4. Wash the filter twice for 10 min each in 2 × SSC containing 0.1% SDS at room temperature and twice for 15 min each, in 0.1 × SSC containing 0.1% SDS at 65°C.
5. Carry out autoradiography at - 70°C with CUR1 X-ray film (AGFA, Belgium).

Appendix 10: Detection of *Xanthomonas axonopodis* pv. *citri* (*Xac*) by Immuno-Capture (IC) and Nested (N)-PCR Assay (Hartung et al. 1996)

Magnetic Immunocapture of Target Pathogen Cells

1. Quantify the selected MAb (A1 for *Xac*) spectrophotometrically at 280 nm and incubate with paramagnetic Dynabeads M-280 precoated with sheep anti-mouse IgG (Dynal) overnight at 4°C with shaking at 125 rpm.
2. Coat 40 μl (0.4 mg) of beads with 4 or 0.04 μg of MAb (10 or 0.1 μg of MAb/10^6 beads) depending on the experiment; separate the bead-MAb complex from the solution by applying a magnet to the wall of the tubes; eliminated unbound MAb by four washes for 30 min/wash with two volumes of phosphate buffered saline (PBS) with 0.1 bovine serum albumin (BSA) and incubate the complex with 500 μl of serially diluted target bacterial suspensions (from 10 to 10^8 cells/ml of Tris-HCl buffer, pH 7.2) for at least 1 h at 4°C with gentle shaking.
3. Separate the immunocaptured bacterial cells magnetically from the solution; wash three times with PBS buffer; suspend the bacteria in 10 μl sterile water; boil the beads for 5 min and centrifuge at 13,000 × g briefly.

PCR Reactions

1. Use primers based on the sequence of 572-nt of *Xac* insert in the plasmid pFL1 (Cruchem, USA).
2. Use reaction buffer with 3 mM $MgCl_2$, 0.5 μM each primer, 125 μM nucleotides and 1.25 units *Taq* polymerase per 25 μl of reaction.
3. Provide the following conditions for the first round of PCR: 35 cycles of 95°C for 30 s, 58°C for 60 s and 72°C for 60 s.
4. For nested (N)-PCR, use 1μl from the first reaction product as template with primers 94-3 bio (5 -biotin) and 94-4 lac for amplification of 315-bp product from DNA or cells of *Xac*.
5. Use 20 cycles at the same temperature profile mentioned above for the second round PCR.

6. Resolve the amplified products by electrophoresis through 2% NuSieve (FMC Bioproducts, USA) agarose (3:1) gels and staining with ethidium bromide.

Appendix 11: Detection of *Xylella fastidiosa* (*Xf*) in Xylem Exudates of Grapevine By IC-PCR Technique (Guo and Lu 2004)

Extraction of Xylem Exudates

1. Adopt the pressure chamber method as follows: collect 10–12 cm sections of infected and healthy shoots from vineyard between 09 and 10 AM; wrap them in moist paper towels and place them in plastic bags separately.
2. Extract the xylem fluids from the shoots or leaf petioles using a pressure chamber (PMS Instrument CO., USA) as per the manufacturer's instructions; plate 1 µl xylem exudates onto PD3 medium and count the number of colonies after 3 weeks.

Preparation of Xf Suspension by Immunocapture Procedure

1. Add 100 µl of PBS-BSA (PBS plus 0.2% BSA) to 5–10 µl of xylem exudates; add 100 µl diluted whole rabbit serum (1/1,000 in PBS-BSA) and incubate the mixture for 45 min.
2. Centrifuge at 12,000 rpm for 5 min; reject the supernatant; add 5 µl of sheep anti-rabbt antibody-coated beads (Dynal, Oslo, Norway) to the pellet; incubate at 4°C for 45 min and wash the beads four or five times using PBS-BSA.
3. Recover the beads by centrifugation at 12,000 rpm and resuspend in 10 µl of sterile water.
4. Alternatively, centrifuge the xylem exudates at 12,000 rpm for 5 min; wash the pellet four or five times as described above and resuspend in 10 µl of sterile distilled water.

PCR Amplification

1. Use two oligonucleotide primer pairs XF766f/XF686r and RST31/RST33.
2. Perform PCR amplification using a volume of 25 µl containing 1 × reaction buffer, 1.5 mM $MgCl_2$, 160 µM each dNTP, 2 µM of each primer, 1 U of *Taq* DNA polymerase (Promega, Madison, USA) and 2 µl of bacterial suspension as the DNA template.

3. Provide the following conditions using a DNA termocycler (MJ Research, USA): initial denaturing at 92°C for 5 min, followed by 35 PCR amplification cycles; denaturing at 92°C for 30 s, annealing at 50°C for 30 s, extension at 72°C for 1 min, final extension at 72°C for 8 min.
4. Resolve the PCR products on a 1.5% agarose gel; stain with ethidium bromide and photograph on a UV transilluminator (Polaroid Co., Cambridge, USA).

Appendix 12: Detection of *Xanthomonas axonopodis* pv. *citri* by Real Time PCR Assay (Mavrodieva et al. 2004)

Sample Preparation Using the IT 1-2-3 RAPID DNA Purification Kit

1. Collect the field samples from freshly infected lesions from leaves and fruits, in addition to uninfected samples; swip the leaves and fruits with moistened swabs provided with the kits; place the swabs inside tubes containing ceramic beads suspended in water provided with the kit and store the samples up to 1 week at 4°C prior to further processing.
2. Vortex the samples in 5 min in DNA extraction buffer provided in the kit; centrifuge and run the supernatant over a filtration spin column; wash the column once; elute the DNA in 100 μl of elution buffer and perform real-time PCR analysis.

Real-Time PCR Assay

1. Perform the assay on a mobile RAPID 7200 System (Idaho Technology, USA) using SYBR Green I (Molecular Probes, Eugene, UK) fluorescent dye detection system; run each sample in duplicates in glass capillary tubes (Roche Diaganostics, USA); include a negative (without target DNA) and positive (with target DNA) controls.
2. Add 2 μl of sample supernatant or purified DNA extracted from culture-grown cells or plant tissues to 18 μl of reaction mix containing 1 × PCR buffer with bovine serum albumin (BSA), 2mM $MgCl_2$, 0.5 μM each primer, 0.2 mM dNTPs, 2 μl of 1:3,000 SYBR Green I and 0.8 units of *Taq* polymerase (Roche Diganostics).
3. For hot-start PCR, incubate *Taq* polymerase with TaqStart Antibody (Clontech, CA) 1:28 (v/v) for 10 min at room temperature before adding to the reaction mixture.
4. Provide conditions as per manufacturer's recommendations.
5. Detect total SYBR Green fluorescence after each amplification cycle which provides a measure of PCR product formation; use a melting curve analysis as a second criterion to determine the presence or absence of target, as it helps

in detecting false positive PCR product formation; each PCR product shows a characteristic peak at its melting temperature (Tm) maxium.

Appendix 13: Detection of Phytoplasmal Pathogens by Histological Techniques (Galvis et al. 2007)

DAPI Fluorescence Test

1. Fix the petioles (5–8 mm long) healthy and diseased leaves in 2.5% glutaraldehyde dissolved in 0.1 M phosphate buffer, pH 7.0 and store at 4°C till required for staining.
2. Rinse the petiole tissues with water to remove the excess fixative; immerse in 4′6-diamidino-2 phenylindole [(DAPI), Sigma] at 0.4 g/ml; mount the tissues into carrots and prepare sections (15–20 μm thick).
3. Examine the sections under epifluorescence microscope (Zeiss Jenamed) with a set of one BP365/11 excitation filter and an LP395 suppressor filter.

Transmission Electron Microscopy (TEM)

1. Cut the petioles of healthy and diseased leaves into 1 × 2 mm pieces and prefix in 2.5% glutaraldehyde (0.1 M phosphate buffer, pH 7.2) for 24 h.
2. Prepare sections (500 nm) with a diamond knife using an MT 6000 Sorvall Ultramicrotome; post-fix and precontrast in acetate; dehydrate in an ethanol series of 25, 50, 75, 90 and 100% (20 min each) and continue with pure acetone (20 min, three times).
3. Embed in Spurr's resin (60°C for 8 h), after a 60-min infiltration with acetone-Spurr (1:1) to facilitate entry of resin into the tissues.
4. Mount the utrathin sections on copper grids (1 mesh) and observe on a JEOL 100 CX transimission electron microscope; note the presence of pleomorphic bodies in the phloem cells.

Appendix 14: Detection of Spiroplasma Infecting Carrot by Nested PCR Assay (Green et al. 1999; Lee et al. 2006)

Extraction of Pathogen DNA

1. Prepare plant tissues (leaf petioles and midribs, scion wood or root tissues) for total DNA extraction from samples of 0.5–0.6 g; place them in a sample bag

(Agadia) suitable for use with a Homex 5 homogenizer–extractor and homogenize with 5 ml of CTAB extraction buffer consisting of 100 mM Tris, pH 8.0, 1.4 M NaCl, 50 mM EDTA, pH 8.0, 2.5% (w/v), CTAB, 1% polyvinyl pyrrolidone (PVP-40) and 0.2% 2-mercaptoethanol (added just before homogenization).

2. Transfer 0.5 ml homogenate to a 1.5 ml microcentrifuge tube; mix with 22 µl RNase A (20 mg/ml) and incubate at 65°C with gentle shaking (800 rpm) using an Eppendorf Thermomixer for 25–35 min.
3. Add 162 µl AP2 buffer (Qiagen), mix by inversion; place on ice for 5 min; transfer the entire contents to QIA shredder (Qiagen) column sitting in a 2-ml collection tube and microcentrifuge at a maximum speed for 2 min in an Eppendorf mcirocentrifuge.
4. Transfer the column flow through (450 µl) without disturbing the cell debris pellet to a new 1.5 ml microcentrifuge tube and mix by inversion with 225 µl of AP 3 buffer (Qiagen) and 450 µl of 95% ethanol.
5. Transfer 650 µl of the mixture to a DNeasy (Qiagen) column; microcentrifuge at 8,000 × g for 1 min; discard the flowthrough and repeat the microcentrifugation with the remaining sample.
6. Transfer the column to a clean 2-ml collection tube; add 500 µl of AW (Qiagen) buffer to the column; wash throughly by microcentrifugation at 8000 × g for 1 min; discard the flowthrough and add additional 500 µl of AW buffer to the column.
7. Microcentrifuge at maximum speed for 2 min; transfer the column to a new 1.5 ml microcentrifuge tube; elute the DNA with 100 µl of pre-warmed (60°C) AE (Qiagen) buffer and microcentrifuge at 8,000 × g for 1 min.
8. Heat the DNA extracts at 95°C for 5 min; place them on ice for 5–10 min and store at −80°C until needed for PCR analysis.

Nested PCR and RFLP Analysis for Phytoplasmas

1. Perform the first amplification using primer pair P1/16S-Sr, followed by R16F2n/R162n in the second amplification and maintain a negative control devoid of DNA templates in the reaction mix in all PCR assays.
2. Use an automotive thermocycler for 38 cycles of amplification with AmpliTaq Gold polymerase (Applied Systems); carry out the reaction in 25 µl reaction mixtures containing 1 µl undiluted nucleic acid preparation (100–200 ng), 200 µM of each dNTP, and 0.4 µM of each primer.
3. Maintain the following conditions: denaturation at 94°C for 1 min (11 min for the first cycle to activate AmpliTaq Gold polymerase), annealing for 2 min at 55°C and primer extension for 3 min (7 min in the final cycle) at 72°C.
4. Use 1 µl of diluted (1:30) PCR product from the amplification as a template in the reaction mixture (50 µl) for nested PCR format.
5. Electrophorese the products (5 µl) on a 1% agarose gel; stain in ethidium bromide and visualize with UV-transilluminator.

6. Digest the PCR product (6 μl) individually with restriction enzymes *Alu*I, *Hha*I *Mse*I and *Sau*3AI (New England Biolabs) according to the manufacturer's instructions; separate the restriction products by electrophoresis through a 5% polyacrylamide gel for 1 h at 150v; stain with ethidium bromide and visualize with UV transilluminator.

Nested PCR and RFLP Analysis for Spiroplasmas

1. Use the primer pair ScR1F1/ScR16R1 in the first amplification followed by ScR16F1A/ScR16R2 in the second amplification.
2. Adopt PCR conditions as described for nested PCR format except the annealing temperature of 50°C.
3. Digest PCR products (1.5 kb) obtained from second amplification with *Mse*I, *Alu*I, *Hha*I and *Hae* III.
4. Use appropriate reference strain(s) for comparison with test samples.
5. Follow the rest of the steps as described above.

Appendix 15: Detection of Phytoplasmas by Oligonucleotide Microarray-Based Assay (Nicolaisen and Bertaccini 2007)

PCR Amplification

1. Perform all PCR amplification using *Taq* DNA polymerase (Promega) according to the manufacturer's recommendations, providing the following conditions: 94°C for 2.5 min followed by 30 cycles of 94°C for 15 s, 62°C for 30s and 72°C fir 15 min and finally 72°C for 10 min.

Labeling of Hybridization Probes

1. Label the PCR products (1 μl PCR) for hybridization using the BioPrime Array CGH Labeling Kit (Invitrogen) with Cy-3 dCTP (Amersham Biosciences) according to the manufacturer's instructions (volume reduced to one fourth of the recommended quantity to reduce the costs).
2. Purify labeling reactions using MinElute Reaction Clean up Kit (Qiagen) according to the manufacturer's recommendations.

Printing and Post-processing of Oligonucleotide Arrays

1. Prepare a 384-well plate with 10 μl of 20 μM capture oligonucleotide (CO) solutions in 30% DMSO, 1.5 × SSC and 0.005% SDS and spot the samples with

a QArray Minispotter (Genetix) on NextronE slides (Schott UK Ltd) using one aQu printing pin K2801 (Genetix) at 45% humidity.

2. Spot each oligonucleotide in quadruplicate in each of the 16 fields and post-process as per the manufacturer's recommendations.

Microarray Hybridization

1. Attach a FAST Frame (Schleicher & Schuell) array grid with 16 wells to printed slides to allow 16 individual hybridization reactions on each slide.
2. Heat the labeling reactions in 80 μl of 10% formamide, 5 × SSC, 0.1% SDS and 0.1 μg/μl sheared salmon sperm DNA to 95°C for 3 min before each hybridization mix is applied to individual wells; cover the slides and place them in a moist chamber with saturated NaCl solution and incubate at 48°C overnight.
3. After detaching the array grid, wash the slides in a 2 × SSC, 0.1% SDS at 42°C for 5 min, repeat washing twice in 1 × SSC, 0.1% SDS for 1 min at room temperature and dry the slides using compressed oil-free air.

Microarray Scanning

1. Scan the hybridized arrays using an ArrayWoRx® Biochip Reader (Applied Precision) as per the manufacturer's instructions with Cy3 settings (595 nm) and 1 s exposure; locate DNA spots on the array using ARRAYWoRx software and define the spots to be within a circle with a diameter of 200 μm.
2. Define a circumscribing square (300 μm) as a region of interest (ROI) and use it to determine local background of spots.
3. Process the results as mean pixel intensity within the defined spot area subtracted from local mean background as defined by the ROI.

Appendix 16: Detection of *Clavibacter michiganensis* subsp. *michiganensis* (*Cmm*) by Immunomagnetic Separation (IMS) – Plating Technique (de León et al. 2006)

Coating Magnetic Beads

1. Coat the immunomagnetic beads (IMBs) precoated with sheep anti-rabbit IgG (Dynabeads M-280, Dynal, Norway) with specific antisera for target pathogen (*Cmm*); wash the beads three times with phosphate buffer saline 0.1 M, pH 7.2

(PBS); collect the IMBs using magnetic particle concentrator (MPC-s, Dynal) and resuspend in PBS to have the desired concentration of IMBs.
2. Incubate the IMBs with anti-*Cmm* antiserum for 24 h at 4°C with gentle shaking as per the instruction of the manufacturer; wash three times with PBS containing 0.1% bovine serum albumin (PBS-BSA) and 0.05% Tween-20 to remove unattached antibodies and resuspend the beads in PBS-BSA to obtain a stock suspension (10^3 IMBs/ml) and store at 4°C.

Standardization of IMS Technique

1. Prepare required dilutions of the pathogen (*Cmm*)-specific antiserum; add aliquots of anti-*Cmm*-coated IMBs to 1 ml of *Cmm* suspension (5.7×10^3 cfu/ml) to obtain 10^5 IMBs/ml and incubate for 1 h at room temperature with gentle shaking.
2. Wash the IMBs three times with 1 ml of PBS-BSA containing 0.05% Tween X-20 for 10 min; resuspend in 1 ml of PBS and spread 0.1 ml of suspension on petriplates containing appropriate nutrient medium (YPGA); incubate for 5 days at 25°C and enumerate the pathogen (*Cmm*) colonies.

Isolation of Target Pathogen from Naturally Infected Seeds by IMS

1. Soak 2 g of seeds in 20 ml of PBS + 0.1% Tween X-20; gently shake in a rotary shaker for 15 min; incubate for 18 h at 4°C; remove the supernatant and keep it separately.
2. Deposit the seeds in an extraction bag provided with a synthetic intermediate mesh (Bioreba); crush the seeds with a pestle and resuspend in the same supernatant.
3. Inoculate the extract (large debris is retained in the extraction bag) with the pathogen to obtain different dilutions; plate 100 µl of seed extract on appropriate medium (YPGA) before and after IMS and examine the plates for the development of the target pathogen.
4. Perform the assays in quadruplicates.

Appendix 17: Detection of *Xanthomonas campestris* pv. *carotae* by PCR Assay (Meng et al. 2004)

Preparation of Seed-wash

1. Incubate 10 g seed samples (about 10,000 seeds) in 100 ml 0.85% NaCl at 4°C for 16–18 h in a 250 ml flask; add two drops of Tween 20; place the flasks on a

rotary shaker (250 rpm) for 2 min and filter through four layers of sterilized cheese cloth.

2. Centrifuge the filtrate at 1,000 × g for 10 min; resuspend the pellet in 10 ml 0.85% NaCl; add 0.5 or 1 ml of the suspension to a 1.5 ml Eppendorf tube and centrifuge at 10,000 × g for 10 min and use the pellet for DNA extraction.

DNA Extraction from Seed-Wash

1. Use the CTAB method for extraction by resuspending the pellet in 800 μl of 2% CTAB buffer and use 2 μl of this extract in 50 μl PCR.
2. Use the PCR primer pair 3SF/9BF designed on the sequences common to all strains of the pathogen species for amplification of ~350-bp target fragment.

Appendix 18: Detection of *Acidovorax avenae* subsp. *citrulli* by Immuno-capture PCR Assay (Xiao et al. 2007)

Selection of Antibodies and Primers

1. Select suitable polyclonal antibodies (Ab and Ab1) specific to target pathogen raised in New Zealand white rabbits.
2. Design specific primers (WFB1 and WFB2) based on the sequences of 16S rRNA gene from the standard strain of the pathogen.

Detection by Standard PCR Assay

1. Use aliquots of 25 μl PCR mixture containing 200 μmol/l of dNTP, 0.25 μmol/l of each primer (WFB1 and WFB2), 1 μmol/l of *Taq* DNA polymerase, 2.5 μl 10 × PCR buffer with $MgCl_2$ and 2–4 μl of bacterial cell lysates as template.
2. Perform amplification in a thermal cycler providing the following conditions: initial denaturation at 95°C for 5 min followed by 30 cylcles of denaturation at 90°C for 30 s, primer annealing at 65°C for 30 s and elongation at 72°C for 30 s and incubation of tubes at 72°C for 5 min.
3. Resolve the PCR amplicons by electrophoresis on 1% agarose gels.

Detection by Immuno-capture (IC)-PCR Assay

1. Dilute the antiserum specific to the target pathogen to 1:2,000 with double sterilized distilled water; transfer 200 μl aliquots into each 0.25 ml Eppendorf tube and incubate for 3 h at 57°C.

2. Rinse thrice with washing buffer containing 0.01 mol/l phosphate-buffered saline (PBS) plus 0.2% Tween 20 (PBST); add 200 µl of 0.5% dried skimmed milk powder and incubate for 30 min at 37°C.
3. Add 200 µl bacterial cell lysates after removing the milk powder; incubate overnight at 4°C and rinse thrice with PBST followed by rinsing once with sterile distilled water.
4. All reagents except the template and all steps as in standard PCR protocol (Section B above) are to be adopted.

Appendix 19: Detection of *Burkholderia glumae* in Rice Seeds by Real-time PCR Assay (Sayler et al. 2006)

Preparation of Seed-wash

1. Wash 2 g seed samples (approximately 20 seeds) in 2 ml distilled water containing 0.1% Tween-20; mix the aqueous suspensions on a rotary shaker at 150 rpm for 30 min and test 1 µl aliquots of wash solution in duplicates for each sample.

Real-time PCR Assay

1. Perform the assay using a Thermo-Fast 96-well reaction plate and Thermo-Fast Caps (AB gene Epsom, UK) in the Mx 3000 P real-time PCR machine (Stratagene Corporation, USA).
2. Transfer to each well a 25-µl reaction mixture containing 12.5 µl 2 × SYBR Green PCR Master Mix (Qiagen) and 12.5 µl primer (3 pmol each of forward and reverse primer and 1 µl of bacterial cells or seed wash/ plant extract).
3. Provide the following conditions for amplification: 95°C for 15 min; 40 cycles each of 95°C for 15 s, 60°C annealing for 30 s and 72°C for extension for 15 s, 95°C for 1 min and 55°C for 30 s.
4. Construct a standard curve to calculate the correlation between bacterial cell numbers and PCR Ct values by assaying a series of dilutions of bacterial culture; determine the bacterial cell number by turbidimetric assay using a spectrophotometer, followed by plate counting.

References

Agarwal PC, Mortensen CN, Mathur SB (1989) Seedborne diseases and seed health testing of rice. Phytopathological Papers No. 30, Danish Government Institute of Seed Pathol Devloping Countries, Copenhagen-CAB International Mycological Institute, UK.

Agrios GN (2005) Plant Pathology, fifth edition, Elsevier/Academic, Amsterdam, The Netherlands.

Ahrens U, Seemüller F (1992) Detection of DNA of plant pathogenic mycoplasma-like organisms by a polymerase chain reaction that amplifies a sequence of the 16S rRNA gene. Phytopathology 82: 828–832.

Aldaghi M, Massart S, Roussel S, Jijakli MH (2006) Establishment of a new method for rapid and precise estimation of apple proliferation phytoplasama concentration in periwinkle. Comm Appl Biol Sci Ghent Univ 71: 853–857.

Aldaghi M, Massart S, Roussel S, Jijakli MH (2007) Development of a new probe for specific and sensitive detection of '*Candidatus* Phytoplasma mali' in inoculated apple trees. Ann Appl Biol 151: 251–258.

Aldaghi M, Massart S, Dutrecq O, Bertaccini A, Jijakli MH (2009) A simple and rapid protocol of crude DNA extraction from apple trees for PCR and real-time PCR detection of '*Candidatus* Phytoplasma mali'. J Virol Meth 156: 96–101.

Aljanabi SM, Paramessure Y, Moutia Y, Saumtally A, Dookun A (2001) Further evidence of the association of a phytoplasma and a virus with yellow syndrome in sugarcane. Plant Pathol 50: 628–636.

Alvarez AM (2001) Differentiation of bacterial populations in seed extracts by flow cytometry. In: De Boer SH (ed), Plant Pathogenic Bacteria, Kluwer, Dordrecht, The Netherlands, pp. 393–396.

Alvarez AM (2004) Integrated approaches for detection of plant pathogenic bacteria and diagnosis of bacterial diseases. Annu Rev Phytopathol 42: 339–366.

Alvarez AM, Benedict AA, Mizumoto CY, Hunter JE, Gabriel DW (1994) Serological, pathological and genetic diversity among strains of *Xanthomonas campestris* infecting crucifers. Phytopathology 84: 1449–1457.

Alvarez AM, Schenck S, Benedict AA (1996) Differentiation of *Xanthomonas albilineans* strains with monoclonal antibody reaction patterns and DNA fingerprints. Plant Pathol 45: 358–366.

Angelini E, Bianchi GL, Filippin L, Morassutti C, Borgo M (2007) A new TaqMan method for the identification of phytoplasmas associated with grapevine yellows by real-time PCR assay. J Microbiol Meth 68: 613–622.

Anon (2005) EPPO standards PM7/44(1) Quarantine procedure for *Xanthomonas axonopodis* pv. *citri.* OEPP/EPPO Bull 35: 289–294.

Anovich D, Garnett HM (1989) The use of immunogold staining technique for detection of a bacterium associated with greening diseased citrus. Phytopathology 79: 382–384.

Arocha Y, Almeida R, Jones P (2008) Using a ribosomal probe for the detection of the phytoplasma associated with 'bunchy top symptom of papaya' (BTS) in plant and insect hosts. Rev Prot Veg 23: 16–20.

Audy P, Braat CE, Saindon G, Huang HC, Laroche A (1996) A rapid and sensitive PCR-based assay for concurrent detection of bacteria causing common and halo blights in bean seed. Phytopathology 86: 361–366.

Audy P, Laroche A, Saindon G, Huang HC, Gilbertson RC (1994) Detection of bean common blight bacteria *Xanthomonas campestris* pv. *phaseoli* and *X. campestris* pv. *phaseoli* var. *fuscans* using polymerase chain reaction. Phytopathology 84:1185–1192.

Avila FJ, Burton BD, Fletcher J, Sherwood JL, Pair SD, Melcher U (1998) Polymerase chain reaction of an agent associated with yellow vein disease of cucurbits. Phytopathology 88: 428–436.

Avramov Z, Gillet J, Langinova M (2008) First detection of stolbur phytoplasma in grapevine (*Vitis vinifera* cv. Merlot) affected with grapevine yellows in Bulgaria. J Phytopathol 156: 112–114.

Baer D, Mitzel E, Pasche J, Gudmestad NC (2001) PCR detection of *Clavibacter michinganensis* subsp. *sepedonicus*-infected tuber samples in a plate capture assay. Am J Potato Res 78: 269–277.

Baharuddin B, Rudolph K, Niepold F (1994) Production of monospecific antiserum against the blood disease bacterium affecting banana and plantain. Phytopathology 84: 570–575.

Ball S, Reeves J (1992) Application of rapid techniques to seed health testing – prospects and potential. In: Duncan JM, Torrance L (eds), Techniques for the Rapid Detection of Plant Pathogens, Blackwell Scientific Publications, Oxford, pp. 193–207.

Baric S, Dalla-Via J (2004) A new approach to apple proliferation detection: a highly sensitive real-time PCR assay. J Microbiol Meth 57: 135–145.

Baric S, Kerschabamer C, Via JD (2006) TaqMan real-time PCR versus four conventional PCR assays for detection of apple proliferation phytoplasma. Plant Mol Biol Rep 24: 169–184.

Benedict A, Alvarez AM, Berestecky J, Imanaka W, Mizumoto C, Pollard L, Mew T, Gonzalez C (1989) Pathovar-specific monoclonal antibodies for *Xanthomonas campestris* pv. *oryzae*. Phytopathology 79: 322–328.

Bentsink L, Leone GOM, van Beckhoven JRCM, van Schijndel MB, van Gemen B, van der Wolf JM (2002) Amplification of RNA by NASBA allows direct detection of viable cells of *Ralstonia solanacearum* in potato. J Appl Microbiol 93: 647–655.

Bereswill S, Bugert P, Volksch B, Ulrich M, Bender CL (1994) Identification and relatedness of coronate-producing *Pseudomonas syringae* pathovars by PCR analysis and sequence determination of amplification products. Appl Environ Microbiol 60: 2924–2930.

Bereswill S, Geider K (1996) Identification of *Erwinia amylovora* by PCR analyses. Acta Hortic 411: 57–62.

Berg M, Davies DL, Clark MF, Vetten HJ, Maier G, Marcone C, Seemüller E(1999) Isolation of the gene encoding one immunodominant membrane protein of the apple proliferation phytoplasma and expression and characterization of the gene product. Microbiology (Reading) 145: 1937–1943.

Berg T, Tesoriero L, Hailstones DL (2005) PCR-based detection of *Xanthomonas campestris* pathovars in Brassica seed. Plant Pathol 54: 416–426.

Bertaccini A, Davis RE, Lee I-M (1992) *In vitro* micropropagation for maintenance and mycoplasma-like organisms in infected plant tissues. Hortic Sci 27: 1041–1043.

Bertolini E, Olmos A, López MM, Cambra M (2003) Multiplex nested-RT-PCR in a single closed tube for sensitive and simultaneous detection of four RNA viruses and *Pseudomonas savastanoi* pv. *savastanoi* in olive trees. Phytopathology 93: 286–292.

Bertolini E, Penyalver R, García A, Olmos A, Quesada JM, Cambra M, López MM (2003) Highly sensitive detection of *Pseudomonas savastanoi* pv. *savastanoi* in asymptomatic olive plants by nested PCR in single closed tube. J Microbiol Meth 52: 261–266.

Bertolini E, Torres E, Olmos A, Martin MP, Bertaccini A, Cambra M (2007) Cooperational PCR coupled with dot blot hybridization for detection and 16Sr X grouping of phytoplasma. Plant Pathol 56: 677–682.

Bextine BR, Miller TA (2004) Comparison of whole-tissue and xylem fluid collection techniques to detect *Xylella fastidiosa* in grapevine and oleander. Plant Dis 88: 600–604.

Bhat AI, Madhubala R, Hareesh PS, Anandaraj M (2006) Detection and characterization of the phytoplasma associated with a phyllody disease of balck pepper (*Piper nigrum* L.) in India. Sci Hortic 107: 200–204.

Biosca EG, Gonzalez R, Lopez-Lopez MJ, Soria S, Montón C, Peréz-Laorga E, López MM (2003) Isolation and characterization of *Brennaria quercina*, causal agent of bark canker and drippy nut of *Quercus* spp. in Spain. Phytopathology 93: 485–492.

Boudon-Padieu E, Larrue J, Clair D, Hourdel J, Jeanneau A, Sforza R, Collin E (2004) Detection and prophylaxis of elm yellows phytoplasma in France. Invest Agrar Sist Recur For 13: 71–80.

Bouzar H, Jones JB, Minsavage GV, Stall RE, Scott JE (1994) Proteins unique to phenotypically distinct groups of *Xanthomonas campestris* pv. *vesicatoria* revealed by silver staining. Phytopathology 64: 39–44.

Bricker JS, Stutz JC (2004) Phytoplasmas associated with ash decline. J Arboricult 30: 193–199.

Brinkerhoff LA (1960) Variability for pathogenicity of *Xanthomonas malvacearum*. Emp Cotton Grow Rev 37: 235.

Bukhalid RA, Chung SY, Loria R (1998) *Nec1*, a gene conferring a necrogenic phenotype, is conserved in plant pathogenic *Streptomyces* spp. and linked to a transposase pseudogene. Mol Plant Microbe Interact 11: 960–967.

Bultreys A, Gheysen I (1999) Biological and molecular detection of toxic lipodepsipeptide-producing *Pseudomonas syringae* strains and PCR identification in plants. Appl Environ Microbiol 65: 1904–1909.

Burdman S, Kots N, Kritzman G, Kopelowitz J (2005) Molecular, physiological and host range characterization of *Acidovorax avenae* subsp. *citrulli* isolates from watermelon and melon in Israel. Plant Dis 89: 1339–1347.

Burokiene D. (2006) Early detection of *Clavibacter michiganensis* subsp. *michiganensis* in tomato seedlings. Agron Res 4: 151–154.

Buzkan N, Kocsis L, Walker MA (2005) Detection of *Xylella fastidiosa* from resistant and susceptible grapevine by tissue sectioning and membrane entrapment immunofluorescence. Microbiol Res 160; 225–231.

Buzombo P, Jaimes J, Lam V, Cantrell K, Harkness M, McCullough D, Morano L (2006) An American hybrid vineyard in the Texas Gulf Coast: analysis within a Pierce's disease hot zone. Am J Enol Vitic 57: 347–355.

Candlish AAG, Taylor JD, Cameron J (1988) Immunological methods as applied to bacterial pea blight. Proc Brighton Crop Prot Conf 2: 787–794.

Caruso P, Gorris MT, Cambra M, Palomo JL, Collar J, López MM (2002) Enrichment double-antibody enzyme-linked immunosorbent assay that uses a specific monoclonal antibody for sensitive detection of *Ralstonia solanacearum* in asympomatic potato tubers. Appl Environ Microbiol 68: 3634–3638.

Casati P, Quaglino F, Tedeschi R, Spiga FM, Alma A, Spadone P, Bianco PA (2010) Identification and molecular characterization of '*Candidatus* Phytoplasma mali' isolates in north-western Italy. J Phytopathol 158: 81–87.

Caudwell A, Meigno SR, Kuszala C, Schneider C, Larrue J, Fleury A, Boudon E (1982) Serological purification and visualization in the electron microscope of grapevine Flvescence dorée pathogen (MLO) in diseased plants and infectious vector extracts. Yale J Biol Mediterr 56: 936–937.

Cazelles D, Ruchta A, Schwarzel R (1995) Bacterial soft rot of potato tubers: progress in detection of latent infection of *Erwinia chrysanthemi* and forecasting field outbreaks. Rev Suisse Agric 27: 17–22.

Ceramic-Zagorac P, Hiruki C (1996) Comparative molecular studies on aster yellows phytoplasmas. Acta Hortic 377: 266–276.

Chen J, Groves R, Civerolo EL, Viveros M, Freeman M, Zheng Y (2005) Two *Xylella fastidiosa* genotypes associated with almond leaf scorch disease on the same location in California. Phytopathology 95: 708–714.

Chen J, Pu X, Deng D, Liu X, Li H, Civerolo E (2009) A phytoplasma related to '*Candidiatus* Phytoplasma asteris' detected in citrus showing huanglongbing (yellow shoot disease) symptoms in Guangdon, P.R. China. Phytopathology 99: 236–242.

Chen KH, Credi R, Loi N, Maixner M, Chen TA (1994) Identification and grouping of mycoplasma-like organisms associated with grapevine yellows and clover phyllody disease based on immunological and molecular analyses. Appl Environ Microbiol 60: 1905–1913.

Chitaara LG, Breeuwer P, Abee T, van den Bulk RW (2006) The use of fluorescent probes to assess viability of the plant pathogenic bacterium *Clavibacter michiganensis* subsp. *michiganensis* by flow cytometry. Fitopatol Bras 31: 349–356.

Chitaara LG, Langerak CJ, Begervoet JH, van den Bulk RW (2002) Detection of the plant pathogenic *Xanthomonas campestris* pv. *campestris* in seed extracts of *Brassica* sp. applying fluorescent antibodies and flow cytometry. Cytometry 47: 118–126.

Chitaranjan S, De Boer SH (1997) Detection of X*anthomonas campestris* pv. *pelargonii* in geranium and greenhouse nutrient solution by serological and PCR techniques. Eur J Plant Pathol 103: 553–563.

Christensen NM, Nicolaisen M, Hansen M, Schulz A (2004) Distribution of phytoplasmas in infected plants as revealed by real-time PCR and bioimaging. Mol Plant-Microbe Interact 17: 1175–1184.

Cimmerman A, Arnaud G, Foissac X (2006) Stolbur phytoplasma genome survey achieved using a suppression subtractive hybridization approach with high specificity. Appl Environ Microbiol 72: 3274–3283.

Clark MF, Morton A, Buss SL (1989) Preparation of mycoplasma immunogens from plants and a comparison of polyclonal and monoclonal antibodies made against primula yellows MLO associated with antigens. Ann Appl Biol 114: 111–124.

Colletta-Filho HD, Takita MA, de Souza AA, Neto JR, Destefano SAL, Hartung JS, Machado MA (2006) Primers based on the *rpf* gene region provide improved detection of *Xanthomonas axonopodis* pv. *citri* in naturally and artificially infected citrus plants. J Appl Microbiol 100: 279–285.

Coletta-Filho HD, Takita MA, Targon MLPN, Machado MA (2005) Analysis of 16S rDNA sequences from citrus huanglongbing bacteria reveal a different '*Ca.* Liberibacter strain associated with citrus disease in São Paulo. Plant Dis 88: 848–652.

Coletto-Filho HD, Targon MLPN, Takita MA, de Negri JD, Pompedu J Jr. do Amaral AM, Muller GW, Machado MA (2004) First report of the causal agent of huanglongbing (*Candiatus* Liberibacter asiaticus) in Brazil. Plant Dis 88: 1382.

Comstock JC, Irey MS, Lockhart BEL, Wang ZK (1998) Incidence of yellow leaf syndrome in CP cultivars based on polymerase chain reaction and serological reactions. SugarCane 4: 21–24.

Conci L, Meneguzzi N, Galdeano E, Torres L, Nome C, Nome S (2005) Detection and molecular characterization of an alfalfa phytoplasma in Argentina that represents a new group in the 16S rDNA ash yellows group (*Candidatus* Phytoplasma fraxini). Eur J Plant Pathol 113: 255–265.

Cordova J, Jones P, Harrison NA, Oropeza C (2003) In situ PCR detection of phytoplasma DNA in embryos from coconut palms with lethal yellowing disease. Appl Microbiol 4: 99–108.

Costa HS, Raetz E, Pinckard TR, Gispert C, Hernandez-Martinez R, Dumnyo CK, Cooksey DA (2004) Plant hosts of *Xylella fastidiosa* in and near southern California vineyards. Plant Dis 88: 1255–1261.

Cottyn B, Bautista AT, Nelson RJ, Leach LE, Swings J, Mew TW (1994) Polymerase chain reaction amplification of DNA from bacterial pathogens of rice using specific oligonucleotide primers. Int Rice Res Notes 19: 30–32.

Cousin MT, Roux J, Boudon-Padieu E, Berges R, Seemüller E, Hiruki C (1998) Use of heteroduplex mobility analysis (HMA) for differentiating phytoplasma isolates causing wiches' broom disease on *Populus nigra* cv. Italica and stolbur or big bud symptoms on tomato. J Phytopathol 146: 97–102.

Crosslin JM, Vandemark GJ, Munyaneza JE (2006) Development of a real-time, quantitative PCR for detection of the Columbian Basin potato purple top phytoplasma in plants and leafhoppers. Plant Dis 90: 663–667.

Cubero J, Graham JH (2002) Genetic relationship among worldwide strains of *Xanthomonas* causing canker in citrus species and design of new primers for their identification by PCR. Appl Environ Microbiol 68: 1257–1264.

Cubero J, Graham JH (2005) Quantitative real-time polymerase chain reaction for bacterial enumeration and allelic discrimination to differentiate *Xanthomonas* strains on citrus. Phytopathology 95: 1333–1340.

Cubero J, Martinez MC, Llop P, Lopez M (1999) A simple and efficient PCR method for the detection of *Agrobacterium tumefaciens*. J Appl Microbiol 86: 591–602.

Cubero J, van der Wolf J, van Beckhoven J, López MM (2002) An intenal control for the diagnosis of crown gall by PCR. J Microbiol Meth 51: 387–392.

Dai Q, He FT, Lue PY (1997) Elimination of phytoplasma by stem culture from mulberry plants (*Morus alba*) with dwarf disease. Plant Pathol 46: 56–61.

Dale JL (1988) Rapid compression technique for detecting mycoplasma-like organisms in leaf midrib sieve tubes by fluorescence microscopy. Phytopathology 78: 118–120.

Da Silva ACR, Ferro JA, Reinach FC, Farah CS, Furlan LR, Quaggio RB, Monteiro-Vitorella CB, Sluys MAV (2002) Comparisons of genomes of two *Xanthomonas* pathogens with different host specificities. Nature 417: 459–463.

Davies DL (1998) The occurrence of two distinct phytoplasma isolates associated with *Rubus* species in the UK. Acta Hortic 471: 63–65.

Davies RE, Dally EL, Gundersen DE, Lee I-M, Habili N (1997) '*Candidatus* Phytoplasma australiense' a new phytoplasma taxon associated with Australian grapevine yellows. Int J Syst Bacteriol 47: 262–269.

Davis RE, Flectcher J (1983) *Spiroplasma citri* in Maryland: isolation from field-grown plants of horseradish with brittle root symptoms. Plant Dis 67: 900–973.

Davis RE, Lee I-M (1982) Pathogenicity of spiroplasma, mycoplasma-like organisms and vascular-limited fastidious walled bacteria. In: Mount GS, Lacy GH (eds), Phytopathogenic Prokaryotes, Vol 1, Academic, New York, pp. 492–513.

Davis RE, Lee I-M, Douglas SM, Dally EL (1990) Molecular cloning and detection of chromosomal and extrachromosomal DNA of the mycoplasma-like organisms associated with little leaf disease in periwinkle (*Catharanthus roseus*). Phytopathology 80: 789–793.

Davis RE, Lee I-M, Douglas SM, Dally EL, Dewitt N (1988) Cloned nucleic acid hybridization probes in detection and classification of mycoplasma-like organisms. Acta Hortic 234: 115–122.

Davis RE, Sinclair WA (1998) Phytoplasma identity and disease etiology. Phytopathology 88: 1372–1376.

Davis RE, Sinclair WA, Lee I-M, Dally E (1992) Cloned DNA probes specific for detection of a mycoplasma-like organism associated with ash yellows. Mol Plant-Microbe Interact 5: 163–169.

Davis RE, Dally EL, Jomantiene R, Zhao Y, Roe B, Lin SP, Shao J (2005) Cryptic plasmid pSKU146 from wall-less plant pathogen *Spiroplasma kunkelii* encodes an adhension and components of a type IV translocation-related conjugation system. Plasmid 53:179–191

De Boer SH, Cupples DA, Gitaitis RD (1996) Detecting latent bacterial infections. Adv Bot Res 23: 27–57.

De Boer SH, Hall JW (2000) Proficiency testing in a laboratory accreditation program for bacterial ring rot pathogen of potato. Plant Dis 84: 649–653.

De Boer SH, McNaughton ME (1987) Monoclonal antibodies to the lipopolysaccharides of *Erwinia carotovora* subsp. *atroseptica* serogroup I. Phytopathology 77: 828–832.

De Boer SH, Stead DE, Alivizatos AS, Janse JD, van Vaerenbergh J, De Haan TL, Mawhinney J (1994) Evaluation of serological tests for detection of *Clavibacter michiganensis* subsp. *sepedonicus* in composite stem and tuber samples. Plant Dis 74: 1431–1434.

De Boer SH, Ward LJ (1995) PCR detection of *Erwinia carotovora* subsp. *atroseptica* associated with potato tissue. Phytopathology 85: 854–858.

Deepa James, Girija D, Mathew SK, Nazeem PA, Babu TD, Varma AS (2003) Detection of *Ralstonia solanacearum* race 3 causing bacterial wilt of solanaceous vegetables in Kerala using random amplified polymorphic DNA (RAPD) analysis. J Trop Agric 41: 33–37.

de León L, Rodriguez A, Lopez MM, Siverio F (2007) Evaluation of the efficacy of immunomagnetic separation for the detection of *Clavibacter michiganensis* subsp. *michiganensis* in tomato seeds. J Appl Microbiol 104: 776–786.

de León L, Siverio F, Rodriguez A (2006) Detection of *Clavibacter michiganensis* subsp. *michiganensis* in tomato seeds using immunomaganetic separation. J Microbiol Meth 67: 141–149.

De Parasis J, Roth DA (1990) Nucleic acid probes for identification of genus specific 16S rRNA sequences. Phytopathology 80: 618.

Derrick KS, Brlansky RH (1976) Assay for viruses and mycoplasma using serologically specific electron microscopy. Phytopathology 66: 815–820.

Dianese JC, Schaad NW (1982) Isolation and characterization of inner and outer membranes of *Xanthomonas campestris* pv. *campestris*. Phytopathology 72: 1284–1289.

Doi Y, Ternaka M, Yora K, Asuyama H (1967) Mycoplasma and PLT group-like microorganisms infected with mulberry dwarf, potato witches' broom, aster yellows and paulownia witches' broom. Ann Phytopathol Soc Jpn 33: 259–266.

Dong LC, Sun CW, Thies KL, Luther DS, Graves CH Jr (1992) Use of polymerase chain reaction to detect pathogenic strains of *Agrobacterium*. Phytopathology 82: 434–439.

Dreo T, Gruden K, Manceau C, Janse JD, Ravnikar M (2007) Development of a real-time PCR based method for detection of *Xylophilus ampelinus*. Plant Pathol 56: 9–16.

Dreier J, Bermpohl A, Eichenlaub R (1995) Southern hybridization and PCR for specific detection of phytopathogenic *Clavibacter michiganensis* subsp. *michiganensis*. Phytopathology 85: 462–468.

Duduk B, Bertaccini A (2006) Corn with symptoms of reddening: new host of stolbur phytoplasma. Plant Dis 90: 1313–1319.

Duduk B, Bott S, Ivanovic M, Krstic B, Dukic N, Bertaccini A (2004) Identification of phytoplasmas associated with grapevine yellows in Serbia. J Phytopathol 152: 575–579.

Duduk B, Calari A, Paltrinieri S, Duduk N, Bertaccini A (2009) Multigenic analysis for differentiation of aster yellows phytoplasamas infecting carrots in Serbia. Ann Appl Biol 154: 219–229.

Duduk B, Perić P, Mařuć D, Drobnjaković T, Picuau L, Alma A, Bertaccini A (2008) Phytoplasma in carrots: disease and potential vectors in Serbia. Bull Insectol 61: 327–331.

Dye DW (1962) The inadequacy of the usual determinative tests for the identification of *Xanthomonas* spp. New Zealand J Sci 5: 393–416.

Ebisugi H, Ooishi S, Goda T, Kubo H, Sakiyama K (1988) Effectiveness of membrane filtration for phage technique for detection of *Xanthomonas campestris pv. citri*. Res Bull Plant Protect Serv 34: 113–115.

Edwin BT, Mohankumar C (2007) Kerala wilt disease phytoplasma: phylogenetic analysis and identification of vector *Proutista moesta*. Physiol Mol Plant Pathol 71: 41–47.

Ed Stover, Riaz S, Walker MA (2008) PCR screening for *Xylella fastidiosa* in grape genebank accessions collected in the southeastern United States. Am J Enol Vitic 59: 437–439.

Elibox W, Umaharan P (2007) A green fluorescent protein-based screening method for identification of resistance in anthurium to systemic infection by *Xanthomonas axonopodis* pv. *diffenbachiae*. Plant Pathol 56: 819–827.

Elphinstone JG, Stead DE, Caffer D, Janse JD, López MM, Mazzucchi U, Müller P, Persson P, Rauscher E, Schiessendoppler E, Sousa Santos M, Stefani E, van Vaerenbergh J (2000) Standardization of methods for detection of *Ralstonia solanacearum* in potato tuber extract. Bull OEPP 30: 391–395.

Errampalli D, Fletcher J (1993) Production of monospecific polyclonal antibodies against aster yellows mycoplasma-like organisms-associated antigen. Phytopathology 83: 1279–1282.

Fang CT, Allen ON, Riker AJ, Dickson JG (1950) The pathogenic physiological and serological reactions of the form species *Xanthomonas translucens*. Phytopathology 40: 44–64.

Fegan M, Croft BJ, Teakle DS, Hayward AC, Smith GR (1998) Sensitive and specific detection of *Clavibacter xyli* subsp. *xyli* causal agent of ratoon stunting disease of sugarcane with a polymerase chain reaction based assay. Plant Pathol 47: 495–504.

Fessehaie A, De Boer SH, Lévesque CA (2003) An oligonucleotide array for the identification and differentiation of bacterial pathogenic on potato. Phytopathology 93: 262–269.

Fessehaie A, Wydra K, Rudolph K (1999) Development of a new semiselective medium for isolating *Xanthomonas campestris* pv. *manihotis* from plant material and soil. Phytopathology 89: 591–597.

Fogliano V, Gallo M, Vinale F, Ritieni A, Randazzo G, Greco ML, Lops R, Graniti A (1999) Immunological detection of syringopeptins produced by *Pseudomonas syringae* pv. *lachrymans*. Physiol Mol Plant Pathol 55: 255–261.

Firrao G, Gobbi E, Locel R (1994) Rapid diagnosis of apple proliferation mycoplasma-like organism using a polymerase chain reaction procedure. Plant Pathol 43: 669–674.

Fraaije BA, Birnbaum Y, Franken AAJM, van den Bulk RW (1996) The development of a conductimetric assay for the automated detection of metabolically active soft rot *Erwinia* spp. in potato tuber peel extracts. J Appl Bacteriol 81: 375–382.

Francis M, Lin H, Rosa JC, Doddapaneni H, Civerolo EL (2006) Genome-based PCR primers for specific and sensitive detection and quantification of *Xylella fastidiosa*. Eur J Plant Pathol 115: 203–213.

Franken AAJM (1992) Application of polyclonal and monoclonal antibodies for the detection of *Xanthomonas campestris* pv. *campestris* in crucifer seeds using immunofluorescence microscopy. Neth J Plant Pathol 98: 95–106.

Franken AAJM, Zilverentant JF, Boonekamp PM, Schots A (1992) Specificity of polyclonal and monoclonal antibodies for the identification of *Xanthomonas campestris* pv. *campestris*. Neth J Plant Pathol 98: 81–94.

Freigoun SO, El Faki HI, Gelie B, Schirmer M, Lemattre M (1994) Phage sensitivity in relation to pathogenicity and virulence of the cotton bacterial blight pathogen of Sudan. Plant Pathol 43: 493–497.

Frommel MI, Pazos G (1994) Detection of *Xanthomonas campestris* pv. *undulosa* infested wheat seeds by combined liquid medium enrichment and ELISA. Plant Pathol 43: 589–596.

Frosini A, Casati P, Bianco PA, Bordoni R, Consolandi C, Catiglioni B, Mezzelani A, Rizzi E, Battaglia C, Belli G, Bernardi R, DeBellis G (2002) Ligase detection reaction and universal array as a tool to detect grapevine-infecting phytoplasmas. Minerva Biotechnol 14: 265–267.

Fukui H, Alvarez AM, Fukui R (1998) Differential susceptibility of anthurium cultivars to bacterial blight in foliar and systemic infection phases. Plant Dis 82: 800–806.

Gaitaitis R, Sumner D, Gay D, Smittle D, McDonald G, Maw B, Johnson WC III, Tolner B, Hung Y (1997) Bacterial streak and bulb rot of onion. I. A diaganostic medium for semiselective isolation and enumeration of *Pseudomonas viridiflava*. Plant Dis 87: 897–900.

Galetto L, Bosco D, Marzachi C (2005) Universal and group-specific real-time PCR diagnosis of flavescence doree (16Sr-V), bois noir (16Sr-Xii) and apple proliferation (16Sr-XII) phytoplasmas from field-collectded plant hosts and insect vectors. Ann Appl Biol 147: 191–201.

Gallo M, Fogliano V, Ritieni A, Peluso L, Greco ML, Lops R, Graniti A (2000) Immunoassessment of *Pseudomonas syringae* lipdepsipeptides (syringomycins and syringopeptins). Phytopathol Mediterr 39: 410–416.

Galvis CA, Leguizamon JE, Gaitan AL, Mejia JF, Alvarez E, Arroyave J (2007) Detection and identification of a group 16SrIII-related phytoplasma associated with coffee crispiness disease in Colombia. Plant Dis 91: 248–252.

Gao R, Zhang G-M, Lan Y-F, Zhu T-S, Yu X-Q, Zhu X-P, Li X-D (2008) Molecular characterization of phytoplasma associated with rose witches' broom in China. J Phytopathol 156: 93–98.

Gardan L, Moreno B, Guinebretiere JP (1984) The use of a micromethod of auxanogram, and data analysis for identification of *Pseudomonas*. Proceedings of the 2nd Working Group, Athens Greece, pp. 37–39.

Gibb KS, Padovan AC, Mogen BD (1995) Studies on sweet potato little leaf phytoplasma detected in sweet potato and other species growing in Northern Australia. Phytopathology 85: 169–174.

Gibb KS, Persley DM, Schneider B, Thomas JE (1996) Phytoplasmas associated with papaya diseases in Australia. Plant Dis 80: 174–178.

Gibson UEM, Heid CA, Williams PM (1996) A novel method for real-time quantitative RT-PCR. Genome Res 6: 995–1001.

Gilbertson RL, Maxwell DP, Hagedorn DJ, Leong SA (1989) Development and application of a plasmid DNA probe for detection of bacterial causing common bacterial blight of bean. Phytopathology 79: 418.

Gillings MR, Fahy PC, Broadbent P, Barnes D (1995) Rapid identification of a second outbreak of Asiatic citrus canker in the Northern Territory using polymerase chain reaction and genomic fingerprinting. Aust Plant Pathol 24: 104–111.

Giunchedi L, Pollini CP, Biondi S, Babini AR (1993) PCR detection of MLOs in quick decline affected trees in Italy. Ann Appl Biol 124: 399–403.

Gnanamanickam SS, Shigaki T, Medalla ES, Mew T, Alvarez AM (1994) Problems in detection of *Xanthomonas oryzae* pv. *oryzae* in rice seeds and potential for improvement using monoclonal antibodies. Plant Dis 78: 173–178.

Golmohammadi M, Cubero J, Peñyaler J, Quesada JM, López MM, Llop P (2007) Diagnosis of *Xanthomonas axonopodis* pv. *citri*, causal agent of citrus canker in commercial fruits by isolation and PCR-based methods. J Appl Microbiol 103: 2309.

Goodwin PH, Nassuth A (1993) Detection and characterization of plant pathogens. In: Glick BR, Thomson JC (eds), Methods in Plant Molecular Biology and Biotechnology, CRC Press Inc., Boca Raton, FL, pp. 303–319.

Goodwin PH, Xue BG, Kuske CR, Sears MK (1994) Amplification of plasmid DNA to detect plant pathogenic mycoplasma-like organisms. Ann Appl Biol 124: 27–36.

Gouda KA, Baranwal VK, Ahlawat YS (2006) Simplified DNA extraction and improved PCR-based detection of greening bacterium in citrus. J Plant Biochem Biotechnol 15: 117–121.

Green MJ, Thompson DA, MacKenzie DJ (1999) Easy and efficient DNA extraction from woody plants for the detection of phytoplasma by polymerase chain reaction. Plant Dis 83: 482–485.

Griep RA, van Twisk C, van Beckhoven JRCM, van der Wolf J, Schots A (1998) Development of specific recombinant monoclonal antibodies against the lipopolysaccharide of *Ralstonia solanacearum* race 3. Phytopathology 88: 795–803.

Griffiths HM, Sinclair WA, Davis RE, Lee I-M, Dally EL, Guo YK, Chen TA, Hibben CR (1994) Characterization of mycoplasma-like organisms from *Fraxinus*, *Syringa* and associated plants from geographically diverse sites. Phytopathology 84: 110–126.

Grisham MP, Pan YB, Richard EP Jr (2007) Early detection of *Leifsonia xyli* subsp. *xyli* in sugarcane leaves by real-time polymerase chain reaction. Plant Dis 91: 430–434.

Groves PL, Chen J, Civerolo EL, Freeman MW, Viveros MA (2005) Spatial analysis of almond leaf scorch disease in the San Joaquim Valley of California: factors affecting pathogen distribution and spread. Plant Dis 89: 581–589.

Gudmestad NC, Mallik I, Pasche JS, Anderson NR, Kinzer K (2009) A real-time PCR assay for the detection of *Clavibacter michiganensis* subsp. *sepedonicus* based on cellulase A gene sequences. Plant Dis 93: 649–659.

Guo X, Lu J (2004) Use of a pressure chamber to isolate and detect *Xylella fastidiosa* in xylem exudates of grapevines. Am J Enol Vitic 52: 202–204.

GuoYH, Cheng Z-M, Walla JA (2003) Rapid PCR-based detection of phytoplasmas from infected plants. Hort Sci 38: 1134–1136.

Guo YH, Cheng Z-M, Walla JA, Zhang Z (1998) Diagnosis of X-disease phytoplasma in stone fruits by a monoclonal antibody developed directly from a woody plant. J Eniviron Hortic 16: 33–57.

Ha Y, Fessehaie A, Ling KS, Wechter WP, Keinath AP, Walcott RR (2009) Simultaneous detection of *Acidovorax avenae* subsp. *citrulli* and *Didymella bryoniae* in cucurbit seedlots using magnetic capture hybridization and real-time polymerase chain reaction. Phytopathology 99: 666–678.

Haas JH, Moore LW, Ream W, Manulis S (1995) PCR primers for detection of phytopathogenic *Agrobacterium* strains. Appl Environ Microbiol 61: 2879–2884.

Hadas R, Kritzman GK, Kleitman F, Gefen T, Manulis S (2005) Comparison of extraction procedures and determination of the detection threshold for *Clavibacter michinganensis* ssp. *michiganensis* in tomato seeds. Plant Pathol 54:643–649

Hale CN, Taylor RK (1999) Effect of cool storage on survival of *Erwinia amylovora* in apple calyxes. Acta Hortic 489: 139–143.

Han S (2005) Specific primer for detection of jujube witches' broom phytoplasm group (16SrV) in Korea. Plant Pathol J 21: 55–58.

Han S-S, Hiurki C, Kim S-M (1998) Rapid analysis of genetic relationship of phytoplasma isolates by a DNA heteroduplex mobility assay. Korean J Plant Pathol 14: 382–385.

Hara H, Hoga K, Tanaka H (1995) An improved selective medium for quantitative isolation of *Pseudomonas solanacearum.* Ann Phytopathol Soc Jpn 61: 255.

Harris-Baldwin A, Gudmestad NC (1996) Identification of phytopathogenic coryneform bacteria using the Biolog automated microbial identification system. Plant Dis 80: 874–878.

Harrison NA, Bourne CM, Cox RL, Tsai JH, Richardson PA (1992) DNA probes for detection of mycoplasma-like organisms associated with lethal yellowing disease of plants in Florida. Phytopathology 82: 216–224.

Hartung F, Werner R, Mühlbach HP, Büttner C (1998) Highly specific PCR diagnosis to determine *Pseudomonas solanacearum* strains of different geographic origins. Theor Appl Genet 96: 797–782.

Hartung JS, Civerolo EL (1989) Restriction fragment length polymorphisms distinguish *Xanthomonas campestris* strains isolated from Florida citrus nurseries from *X. campestris* pv. *citri.* Phytopathology 79: 793–799.

Hartung JS, Pruvost OP, Villeniot I, Alvarez A (1996) Rapid and sensitive colorimetric detection of *Xanthomonas axonopodis* pv. *citri* by immunocapture and nested polymerase chain reaction assay. Phytopathology 86: 95–101.

Henrich M, Botti S, Caprara L, Arthofer W, Strommer S, Hanzer V, Katinger H, Bertaccini A, Laimer M, Machado DC (2001) Improved detection methods for fruit tree phytoplasmas. Plant Mol Biol Rep 19: 169–179.

Helias V, Roux AC, le Bertheau Y, Andrivon D, Gauthier JP, Jouan B (1998) Characterization of *Erwinia carotovora* subspecies and detection of *Erwinia carotovora* subsp. *atroseptica* in potato plants, soil or water extracts with PCR-based methods. Eur J Plant Pathol 104: 685–699.

Henriquez NP, Kenyone L, Quiroz L (1999) Corn stunt complex mollicutes in Belize. Plant Dis 83: 77.

Hernandez-Martinez R, Costa HS, Dumenyo CK, Cooksey DA (2006) Differentiation of strains of *Xylella fastidiosa* infecting grape, almonds and oleander using a multiprimer PCR assay. Plant Dis 90: 1382–1388.

Hiruki C, da Rocha A (1986) Histochemical diagnosis of mycoplasma infections in *Catharanthus roseus* by means of a fluorescent DNA binding agent 4,6-diamidino-2-phenylindole-dihydrochloride. Can J Plant Pathol 8: 185–188.

Hobbs HA, Reddy DVR, Reddy AS (1989) Detection of a mycoplasma-like organism in peanut plants with witches' broom using indirect enzyme-linked immunosorbent assay (ELISA). Plant Pathol 36: 164–167.

Hodgetts J, Ball T, Boonham N, Mumford R, Dickinson M (2007) Use of terminal restriction fragment length polymorphism (T-RFLP) for identification of phytoplasmas in plants. Plant Pathol 56: 357–365.

Hodgetts J, Boonham N, Mumford R, Harrison N, Dickinson M (2008) Phytoplasma phylogenetics based on anlysis of *secA* and 23S rRNS gene sequences for improved resolution of candidate species of '*Candidatus* Phytoplasma'. Int J Syst Evol Microbiol 58: 1826–1837.

Hooker ME, Lee RF, Civerolo EL, Wang SY (1993) Reliability of gentisic acid, a fluorescent marker for diagnosis of citrus greening disease. Plant Dis 77: 174–180.

Horita M, Yano K, Tsuchiya K (2004) PCR-based specific detection of *Ralstonia solanacearum* race 4 strains. J Gen Plant Pathol 70: 278–283.

Hren M, Boben J, Rotter A, Kralj P, Gruden K, Ravnikar M (2007) Real-time PCR detection systems for Flavescence dorée and Bois noir phytoplasms in grapevine: comparison with conventional PCR detection and application in diagnostics. Plant Pathol 56: 785–796.

Hseu SH, Lin CY (2000) Serological detection of *Xanthomonas axonopodis* pv. *diffenbachiae* in Taiwan. Plant Prot Bull (Taipei) 42: 97–106.

Hu X, Lai F-M, Reddy ANS, Ishimaru CA (1995) Quantitative detection of *Clavibacter michiganensis* subsp. *sepedonicus* by competitive polymerase chain reaction. Phytopathology 85: 1468–1473.

Huang BC, Zhu H, Hu GG, Li QX, Gao JL, Goto M (1993) Serological studies of *Xanthomonas campestris* pv. *oryzae* with polyclonal and monoclonal antibodies. Ann Phytopathol Soc Jpn 59: 123–127.

Huang Q, Bentz J, Sherald JL (2006) Fast, easy and efficient DNA extraction and one-step polymerase chain reaction for the detection of *Xylella fastidiosa* in potential vectors. J Plant Pathol 88: 77–81.

Hung TH, Wu ML, Su HJ (1999) Detection of fastidious bacteria causing citrus greening disease by non-radioactive DNA probes. Ann Phytopathol Soc Jpn 65: 140–146.

Hyman LJ, Birch PRJ, Dellagi A, Avrova AO, Toth IK (2000) A competitive PCR-based method for detection and quantification of *Erwinia carotovora* subsp. *atroseptica* on potato tubers. Lett Appl Microbiol 30: 330–335.

Ishiie T, Doi Y, Yora K, Asuyama H (1967) Suppressive effects of an antibiotics of tetracycline group on symptom development of mulberry dwarf disease. Ann Phytopathol Soc Jpn 33: 267–275.

Ito S, Ushijima Y, Fujii T, Tanaka M, Kameya-Iwaki S, Yoshiwara S, Kishi F (1998) Detection of viable cells of *Ralstonia solanacearum* in soil using a semiselective medium and PCR technique. J Phytopathol 146: 379–384.

Jagoueix S, Bové JM, Garnier M (1994) The phloem-limited bacterium of greening disease of citrus is a member of the alpha subdivision of the *Proteobacteria*. Int J Syst Bacteriol 44: 379–386.

Jagoueix S, Bové JM, Garnier M (1996) PCR detection of the two '*Candidatus* Liberibacter' species associated with greening disease of citrus. Mol Cell Probes 10: 43–50.

Jarausch W, Lansac M, Saillard C, Broquaire JM, Dosba F (1998) PCR assay for specific detection of European stone fruit yellows phytoplasmas and its use for epidemiology. Eur J Plant Pathol 104: 17–27.

Ji P, Allen C, Sanchez-Perez A, Yao J, Elphinstone JG, Jones JB, Momol MT (2007) New diversity of *Ralstonia solanacearum* strains associated with vegetable and ornamental crops in Florida. Plant Dis 91: 195–203.

Jin K-S, Kang I-B, Ko K-I, Lee E-S, Heo J-Y, Kang Y-K, Kim B-K (2001) Detection of *Xanthomonas axonopodis* pv. *citri* on citrus fruits using enzyme-linked immunosorbent assay. Plant Pathol J 17: 62–66.

Jones JB, van Vuurde JWL (1996) Immunomagnetic isolation of *Xanthomonas campestris* pv. *pelargonii.* J Appl Bacteriol 81: 78–82.

Jones R, Barnes L, Gonzalez C, Leach J, Alvarez A, Benedict A (1989) Identification of low virulence strains of *Xanthomonas campestris* pv. *oryzae* from rice in the United States. Phytopathology 79: 984–990.

Jones WT, Harvey D, Zhao YF, Mitchell RE, Bender CL, Reynolds PHS (2001) Monoclonal antibody based immunoassays for the phytotoxin coronatine. Food Agric Immunol 13: 19–32.

Joppa B, Li S, Cole S, Gallagher S (1992) Pulsed field gel electrophoresis for separation of large DNA. Probe 2(3): Fall, nal.usda.gov/pgdic/Probe/v2n3/puls/html

Joshi M, Rong S, Moll S, Kers J, Franco C, Loria R (2007) *Streptomyces turgidiscabies* secretes a novel protein Nec1 which facilitates infection. Mol Plant-Microbe Interact 20: 599–608.

Kado CI (1971) Methods in Plant Bacterioloty-Laboratory Manual. Department of Plant Pathology, University of California, Davis

Kamińska M, Śliwa H (2005) Detection of '*Candidatus* Phytoplasma asteris' in ashleaf maple trees with shoot prolifieration and decline. J Fruit Ornament Plant Res 13: 123–134.

Kanamori H, Sugimoto H, Ochiar K, Kaku H, Tsuyumu S (1999) Isolation of *hrp* cluster from *Xanthomonas campestris* pv.*citri* and its application for RFLP analyses of xanthomonads. Ann Phytopathol Soc Jpn 65: 110–115.

Kawabe K, Truc NTN, Lan BTN, Hong LTT, Onuki M (2006) Quantification of DNA of citrus huanglongbing pathogen in diseased leaves using competitive PCR. J Gen Plant Pathol 72: 355–359.

Kazmi MR, Farooq SI, Fateh FS, Hameed S, Fahmeed F, Ashraf M, Ahmed I (2007) Identification of phytoplasma from mango trees showing sudden death phenomenon through molecular techniques. Pak J Bot 39: 2153–2157.

Kelman A (1954) The relationship of pathogenicity of *Pseudomonas solancacearum* to colony appearance in a tetrazolium medium. Phytopathology 44: 693–895.

Khadhair A-H, Kawchuk LM, Taillon RC, Botar G (1998) Detection and molecular characterization of an aster yellows phytoplasma in parsley. Can J Plant Pathol 20: 55–61.

Khan AA, Furuya N, Matsumoto M, Matsuyama N (1999) Trial for rapid identification of pathogens from blasted pear blossoms and rotted radish leaves by direct TLC and whole cellular fatty acid analysis. J Fac Agric Kyushu Univ 43: 327–335.

Khan AJ, Bonti S, Al-Subhi AM, Gundersen-Rindal DE, Bertaccini AF (2002) Molecular identification of a new phytoplasma associated with alfalfa witches' broom in Oman. Phytopathology 92: 1038–1047.

Khan JA, Srivastava P, Singh SK (2009) Sensitive detection of a phytoplasma associated with little leaf symptoms in *Withania somnifera*. Eur J Plant Pathol 115: 401–408.

Khoodoo MHR, Sahin F, Jaufeerally-Fakim Y (2005) Sensitive detection of *Xanthomonas axonopodis* pv. *diffenbachiae* on *Anthurium andreanum* by immunocapture PCR (IC-PCR) using primers designed from sequence characterized amplified regions (SCAR) of the blight pathogen. Eur J Plant Pathol 112: 379–390.

Kievits T, van Gemen B, van Strijp D, Schukkink R, Dircks M, Adriaanse H, Malek L, Sooknanan R, Lens P (1991) NASBA™ isothermal enzymatic *in vitro* nucleic acid amplification optimized for the diagnosis of HIV-1 infection. J Virol Meth 35: 273–286.

Kirkpatrick BC, Stenger DC, Morris TJ, Purcell H (1987) Cloning and detection of DNA from a nonculturable plant pathogenic mycoplasma-like organism. Science 238: 197–200.

Klinger JM, Stowe RP, Obenhuber DC, Groves TO, Mishra SK, Pierson DL (1992) Evaluation of the Biolog automated microbial identification system. Appl Environ Microbiol 58: 2089–2092.

Ko HC, Lin CP (1994) Development and application of cloned DNA probes for mycoplasma-like organism associated with sweet potato witches' broom. Phytopathology 84: 468–473.

Koike ST, Barak JD, Handerson DM, Gilbertson RL (1999) Bacterial blight of leek: a new disease in California caused by *Pseudomonas syringae*. Plant Dis 83: 165–170.

Kokošková B, Mraz I, Hýblová J (2007) Comparison of specificity and sensitivity of immunochemical and molecular techniques for reliable detection of *Erwinia amylovora*. Folia Microbiol 52: 175–182.

Kokošková B, Panková I, Krejzar V (2000) Characteristics of polyclonal antisera for detection and determination of *Clavibacter michiganensis* subsp. *insidiosus*. Plant Prot Sci 36: 46–52.

Krell RK, Boyd EA, Nay JE, Park Y-L, Perring TM (2007) Mechanical and insect transmission of *Xylella fastidiosa* to *Vitis vinifera*. Am J Enol Vitic 58: 211–216.

Krell RK, Perring TM, Hashim-Buckey JM, Pinckard TR (2008) Susceptibility of *Vitis vinifera* L. cv. Redglobe and Thompson seedless to Pierce's disease. Am J Enol Vitic 59: 61–66.

Krieg NR, Holt JG (eds) (1984) Bergy's Manual of Systematic Bacteriology, Vol I, Williams & Wilkins, Baltimore, MD.

Kuflu KM, Cuppels DA (1997) Development of a diagnostic cDNA probe for Xanthomonads causing bacterial leaf spot of peppers and tomatoes. Appl Environ Microbiol 63: 4462–4470.

Kunkel LO (1926) Studies on aster yellows. Am J Bot 23: 646–705.

Kuo TT, Huang TC, Wu RY, Yang CM (1967) Characterization of three bacteriophages of *Xanthomonas oryzae* (Uyeda et Ishiyama)Dowson. Bot Bull Acad Sinica 8: 246–254.

Kutin RK, Alvarez A, Jenkins DM (2009) Detection of *Ralstonia solanacearum* in natural substrates using phage amplification integrated with real-time PCR assay. J Microbiol Meth 76: 241–246.

Lamka GL, Hill JH, McGee DC, Brawn EJ (1991) Development of an immunosorbent assay for seedborne *Erwinia stewartii* in corn seeds. Phytopathology 81: 839–846.

Lang JM, Gent DH, Schwartz HF (2007) Management of Xanthomonas leaf blight of onion with bacteriophages and plant activator. Plant Dis 91: 871–878.

Leach JE, White FE (1991) Molecular probes for disease diagnosis and monitoring. In: Khush GS, Toenniessen GH (eds), Rice Biotechnology, CAB International, UK; International Rice Research Institute, Philippines, pp. 281–307.

Lee I-M, Bottner KD, Munyaneza JE, Davis RE, Crosslin JM, du Toit LJ, Crossby T (2006) Carrot purple leaf: a new spiroplasmal disease associated with carrots in Washington State. Plant Dis 90: 989–993.

Lee I-M, Davis RE (1983) Phloem-limited prokaryotes in sieve elements isolated by enzyme treatment of diseased plant tissues. Phytopathology 73: 1540–1543.

Lee I.M, Davis RE (1988) Detection and investigation of genetic relatedness among ater yellows and other mycoplasma-like organisms by using cloned DNA and RNA probes. Mol Plant-Microbe Interact 1: 303–310.

Lee I-M, Gunderson-Rindal DE, Bertaccini A (1998a) Phytoplasma: Ecology and genomic diversity. Phytopathology 88: 1359–1366.

Lee I-M, Gunderson-Rindal DE, Davis RE, Bartoszyk IM (1998b) Revised classification scheme of phytoplasmas based on RFLP analyses of 16S rRNA and ribosomal protein gene sequences. Int J Syst Bacteriol 48: 1153–1169.

Lee I-M, Hammond RW, Davis RE, Gunderson DE (1993) Universal amplification and analysis of pathogen 16S rDNA for classification and identification of mycoplasma-like organisms. Phytopathology 83: 834–842.

Lee I-M, Lukaesko LA, Maroon CJM (2001) Comparison of dig-labeled PCR, nested PCR and ELISA for the detection of *Clavibacter michiganensis* subsp. *sepedonicus* in field-grown potatoes. Plant Dis 85: 261–266

Lee I-M, Zhao Y, Davis RE, Wei W, Martini M (2007) Prospects of DNA-based systems for differentiation and classification of phytoplasmas. Bull Insectol 60: 239–244.

Lee ME, Grau CR, Lukaesko LA, Lee I-M (2002) Identification of aster yellows phytoplasma in Wisconsin based on RFLP analysis of PCR-amplified products. Can J Plant Pathol 24: 125–130.

Lehtonen MJ, Rantala H, Kreuze JF, Bang H, Kuisma L, Koski P, Virtanen E, Vihlman K, Valkonen JPT (2004) Occurrence and survival of potato scab pathogens (*Streptomyces* species) on tuber lesions: quick diagnosis based on PCR-based assay. Plant Pathol 53: 280–287.

Leite RP Jr, Minsavage GV, Bonas U, Stall RF (1994) Detection and identification of phytopathogenic *Xanthomonas* strains by amplifications of DNA sequences related to the *hrp* genes of *Xanthomonas campestris* pv. *vesicatoria*. Appl Environ Microbiol 60: 1068–1077.

Leu HH, Leu LS, Lin CP (1998) Development and application of monoclonal antibodies against *Xylella fastidiosa* the causal bacterium of pear leaf scorch. J Phytopathol 146: 31–37.

Lherminier J, Prensier G, Boudon-Padieu E, Caudwell A (1990) Immunolabeling of grapevine Flavescence dorée MLO in salivary glands of *Euscelidius variegatus*: a light and electron microscopy study. J Histochem Cytochem 38: 79–85.

Li W, Harung JS, Levy L (2007) Evaluation of DNA amplification methods for improved detection of '*Candiatus* Liberibacter species' associated with citrus huanglongbing. Plant Dis 91: 51–58.

Li W, Levy L, Hartung JS (2009) Quantitative distribution of '*Candiatus* Liberibacter asiaticus' in citrus plants with citrus huanglongbing. Phytopathology 99: 139–144.

Li WB, Pria WD Jr, Laeava PM, Qin X, Hartung JS (2003) Presence of *Xylella fastidiosa* in sweet orange fruit and seeds and its transmission to seedlings. Phytopathology 93: 953–958.

Li X, De Boer SH (1995) Selection of polymerase chain reaction primers from an RNA intergenic spacer region for specific detection of *Clavibacter michiganensis* subsp. *sepedonicus*. Phytopathology 85: 837–842.

Li X, De Boer SH, Ward LJ (1997) Improved microscopic identification of *Clavibacter michiganensis* subsp. *sepedonicus* by combining *in situ* hybridization with immunofluoresence. Lett Appl Microbiol 24: 431–434.

Li X, Wong WC, Hayward AC (1993) Production and use of monoclonal antibodies to *Pseudomonas andropogonis*. J Phytopathol 138: 21–30.

Liefting LW, Sutherland PW, Ward LI, Paice KL, Weir BS, Clover GRG (2009) A new '*Candidatus* Liberibacter' species associated with diseases of solanaceous crops. Plant Dis 93: 208–214.

Lima JEO de, Miranda VS, Hartung JS, Brlansky RH, Coutinho A, Roberto SR, Carlos EF (1998) Coffee leaf scorch bacterium: axenic culture, pathogenicity and comparison with *Xylella fastidiosa* in citrus. Plant Dis 82: 94–97.

Lin CP, Chen TA (1985) Production of monoclonal antibodies against *Spiroplasma citri*. Phytopathology 75: 845–851.

Lin CP, Chen TA (1986) Comparison of monoclonal antibodies and polyclonal antibodies in detection of aster yellows mycoplasma-like organism. Phytopathology 76: 45–50.

Lin CP, Shen WC, Ko HC, Chang FL, Yu YL, Chen MF, Wu FY (1993) Application of monoclonal antibodies and cloned DNA probes for the detection and differentiation of phytopathogenic mycoplasma-like organisms. Plant Pathol Bull 2: 161–168.

Llop Bonaterra A, Peñalver I, Lopez MM (2000) Development of a highly sensitive nested-PCR procedure using a single closed tube for detection in asymptomatic plant material. Appl Environ Microbiol 66: 2071–2078.

Loi N, Ermacora P, Carraro L, Osler R, Chen TA (2002) Production of monoclonal antibodies against apple proliferation phytoplasma and their use in serological detection. Eur J Plant Pathol 108: 81– 86.

Loi N, Ermacora P, Chen TA, Carraro L, Osler R (1998) Monoclonal antibodies for the detection of tagetes witches' broom agent. J Plant Pathol 80: 171–174.

Lorenz K-H, Schneider B, Ahrens U, Seemüller E (1995) Detection of the apple proliferation and pear decline phytoplasmas by PCR amplification of ribosomal and nonribosomal DNA. Phytopathology 85: 771–776.

Louws FJ, Beu J, Medina-Mora CM, Smart CD, Opgenorth D, Ishimaru CA, Hausbeck MK, de Bruijn FJ, Fulbright DW (1998) rep-PCR-mediated genomic fingerprinting: a rapid and effective method to identify *Clavibacter michiganensis*. Phytopathology 88: 862–868.

Louws FJ, Fulbright DW, Stephens CT, de Bruijn FJ (1994) Specific genomic fingerprints of phytopathogenic *Xanthomonas* and *Pseudomonas* cultivars and strains generated with repetitive sequences and PCR. Appl Environ Microbiol 60: 2286–2295.

Luo LX, Walters C, Bolkan H, Liu XL, Li JQ (2008) Quantification of viable cells of *Clavibacter michiganensis* subsp. *michiganensis* using a DNA binding dye and a real-time PCR assay. Plant Pathol 57: 332–337.

Lydon J, Patterson CD (2001) Detection of tabtoxin-producing strains of *Pseudomonas syringae* by PCR. Lett Appl Microbiol 32: 166–170.

Lyons NF, Cruz L, Santos MS (2001) Rapid field detection of *Ralstonia solanacearum* in infected tomato and potato plants using *Staphylococcus aureus* slide agglutination test. Bull OEPP 31: 91–93.

Lyons NF, Taylor JD (1990) Serological detection and identification of bacteria from plants by the conjugated *Staphylococcus aureus* slide agglutination test. Plant Pathol 39: 584–590.

Machado MA (2009) *Xylella fastidiosa* and the citrus variegated chlorosis (CVC), a new destructive citrus disease in Brazil. www.lbi.ic.unicampa.br/xylella _cvc-strain/home mmachado.html

Maes M, Garbera P, Crepel C (1996) Identification and sensitive endophytic detection of the fire blight pathogen *Erwinia amylovora* with 23S ribosomal DNA sequences and polymerase chain reaction. Plant Pathol 45: 1139–1149.

Mahmoudi E, Soleimani MJ, Taghavi M (2007) Detection of bacterial soft rot of crown imperial caused by *Pectobacterium carotovorum* subsp. *carotovorum*, using specific primers. Phytopathol Mediterr 46: 168–176.

Malandrin L, Huard A, Samson R (1996) Discriminant envelope profiles between *Pseudomonas syringae* pv. *pisi* and *P. syringae* pv. *syringae*. FEMS Microbiol Lett 41: 11–17.

Malandrin L, Samson R (1998) Isozyme analysis for the identification of *Pseudomonas syringae* pv. *strains*. J Appl Microbiol 84: 895–902.

Malandrin L, Samson R (1999) Serological and molecular size characterization of flagellins of *Pseudomonas syringae* pathovars and related bacteria. Syst Appl Microbiol 22: 534–545.

Malisano G, Firrao G, Locci R (1996) 16S rDNA oligonucleotide probes for the differential diagnosis of plum leptonecrosis and apple proliferation phytoplasmas. Bull OEPP 26: 421–428.

Manulis S, Kogan N, Valinsky L, Dror O, Kleitman F (1998) Detection of *Erwinia herbicola* pv. *gysophilae* in gysophila plants by PCR. Eur J Plant Pathol 104: 85–91.

Marcone C, Hergenhahn F, Ragozzino A, Seemüller R (1999) Dodder transmission of pear decline, European stone fruit yellows, rubus stunt, *Pieris echioides* yellows and cotton phyllody phtoplasmas to periwinkle. J. Phytopathol 147: 187–192.

Marcone C, Seemüller E (2001) A chromosome map of the European fruit yellows phytoplasma. Microbiology 174: 1213–1221.

Maringoni AC, Camara RC, Souza VL (2006) Semi-selective medium for *Curtobacterium flaccumfaciens* pv. *flaccumfaciens* isolation from bean seeds. Seed Sci Technol 34: 117–124.

Marzarei M, Hajimoral MR, Kerr A (1922) Specificity of polyclonal antibodies to different antigenic preparations of Pseudomonas syringae pv. pisi strain UQM551 and Pseudomonas syringae pv. syringae strain L. Plant Pathol 41:437–443

Matsuyama N (1995) Trials for rapid identification of phytopathogenic bacteria by HPLC and direct colony TLC. J Fac Agric Kyushu Univ 40: 87–91.

Matsuyama N, Daikohara M, Yoshimura K, Manabe K, Khan MAA, de Melo MS, Negishi H, Suyama K, Tsuchiya K, Furuya N (2009) Presumptive differentiation of phytopathogenic and non-pathogenic bacteria by improved rapid-extraction TLC method. J Fac Agric Kyushu Univ 54: 1–11.

Matsuyama N, Furuya N (1993) Application of the direct colony TLC for identification of phytopathogenic bacteria (II) chromatographic profile of *Erwinia* and *Pseudomonas* spp. J Fac Agric Kyushu Univ 38: 89–95.

Matsuyama N, Mian IH, Akanda AM, Furuya N (1993) On a rapid identification of phytopathogenic bacteria by direct colony thin-layer chromatography. Proc Assoc Plant Prot Kyushu 39: 60–63.

Matsuyama N, Ueda Y, Iiyama K, Furuya N, Ura H, Ashraf Khan A, Matusmoto M (1998) Rapid extraction – HPLC as a tool for presumptive identification of *Burkholderia gladioli*, *B. glumae* and *B. Plantarii*, causal agents of various rice diseases. J Fac Agri Kyushu Univ 42: 265–272.

Mavrodieva V, Levy L, Gabriel DW (2004) Improved sampling methods for real-time polymerase chain reaction diagnosis of citrus canker from field samples. Phytopathology 94: 61–68.

Mazarei A, Kerr A (1990) Distinguishing pathovars of *Pseudomonas syringae* pv. *syringae* on peas: nutritional pathogenicity and serological tests. Plant Pathol 39: 278–285.

Mc Garvey A, Denny TP, Schell MA (1999) Spatial-temporal and quantitative analysis of growth and EPS production by *Ralstonia solanacearum* in resistant and susceptible tomato cultivars. Phytopathology 89: 1233–1239.

McIntosh AH, Skowronski BS, Maramorosch K (1974) Rapid identification of *Spiroplasma citri* and its relation to other yellows agent. Phytopathol Z 80: 150–156.

McManus PS, Jones AC (1996) Detection of *Erwinia amylovora* by nested PCR and PCR-dot blot and reverse blot hybridization. Acta Hortic 411: 87–90.

Mello APOA, Bedendo IP, Camargo LEA (2006) Source heterogeneity in the 16S rDNA of tomato big bud phytoplasma belonging to group 16SrIII. J Phytopathol 154: 245–249.

Meng XQ, Umesh KC, Davis RM, Gilbertson R (2004) Development of PCR-based assays for detecting *Xanthomonas campestris* pv. *carotae*. Plant Dis 88: 1226–1234.

Merighi M, Sandrini A, Landini S, Ghini S, Girotti S, Malaguti S, Bazzi (2000) Chemiluminescent and colorimetric detection of *Erwinia amylovora* by immunoenzymatic determination of PCR amplicons from plasmid pEA29. Plant Dis 84: 49–54.

Mguni CM, Mortensen CN, Keswani CL, Hockenhull J (1999) Detection of the black rot pathogen *Xanthomonas campestris* pv. *campestris* and other xanthomonads in Zimbabwean and imported Brassica seed. Seed Sci Technol 27: 447–454.

Mills D, Russell BW (2003) Parameters for specific detection of *Clavibacter michiganensis* subsp. *sepedonicus* in potato stems and tubers by multiplexed PCR-ELISA. Am J Potato Res 80: 223–234.

Mills D, Russell BW, Hanus JW (1997) Specific detection of *Clavibacter michiganensis* subsp. *sepedonicus* by amplification of three unique DNA sequences isolated by subtraction hybridization. Phytopathology 87: 853–861.

Milne RG (1992) Immuno-electron microscopy of plant viruses and phytoplasmas. Adv Dis Vector Res 9: 283–312.

Milne RG, Ramasso E, Lenzi R, Masenga V, Sarindu N, Clark MF (1995) Pre- and post-embedding immunogold labeling and electron microscopy in plant host tissues of three antigenically unrelated MLOs: primula yellows, tomato big bud and bermudagrass white leaf. Eur J Plan Pathol 101: 57–67.

Mitchell RE, Young H (1978) Identification of chlorosis inducing toxin of *Pseudomonas glycinea* as coronatine. Phytochemistry 17: 20–28.

Mitrev S, Kovačevic B (2006) Characterization of *Xanthomonas axonopodis* pv. *vesicatoria* isolated from peppers in Macedonia. J Plant Pathol 88: 321–324.

Miyoshi T, Sawada H, Tachibana Y, Matsuda I (1998) Detection of *Xanthomonas campestris* pv. *citri* by PCR using primers from the spacer region between the 16S and 23S rRNA genes. Ann Phytopathol Soc Jpn 64: 249–254.

Moline HE (1985) Differentiation of postharvest soft rotting bacteria with two-dimensional polyacrylamide gel electrophoresis. Phytopathology 75: 549–553.

Moretti C, Silvestri FM, Rossini E, Natalini G, Buonaurio R (2007) A protocol for rapid identification of *Brennaria nigrifluens* among bacteria isolated from bark cankers in Persian walnut plants. J Plant Pathol 89: 211–218.

Mosqueda-Cano G, Herrera-Estrella L (1997) A simple and efficient PCR method for the specific detection of *Pseudomonas syringae* pv. *phaseolicola* in the bean seeds. World J Microbiol Biotechnol 13: 463–467.

MrázL, Pánková I, Petrzik K, Šíp M (1999) *Erwinia* and *Pseudomonas* bacteria can be reliably screened by an improved serological agglutination test. J Phytopathol 147: 429–431.

Musetti R, Favalli MA (2004) Microscopy techniques applied to the study of phytoplasma diseases: traditional and innovative methods. www.formatex.org/microscopy2/papers/72-80.pdf

Myrie W, Harrison N, Dollet M, Been B (2007) Molecular detection and characterization of phytoplasmas associated with coconut palms in Jamaica. Bull Insectol 60: 159–160.

Myung I-S, Hyn J-W, Cho W-D (2006) Detection of *Xanthomonas axonopodis* pv. *citri* on Satsuma Mandrin orange fruits using phage technique in Korea. Plant Pathol J 22: 314–317.

Myung I-S, Nam K-W, Kwon H-M (2001) Dispersal of citrus bacterial canker caused by *Xanthomonas axonopodis* pv. *citri* in nursery plots of Unshiu orange. Plant Pathol J 19: 205–209.

Najar A, Bouachem S, Danet JL, Saillard C, Garnier M, Bové JM (1998) Presence of *Spiroplasma citri*, the pathogen responsible for citrus stubborn disease and its vector leafhopper *Circulifer haemotoceps* in Tunisia: contamination of both *C. haemotoceps* and *C. opacipennis*. Fruits (Paris) 53: 391–396.

Nakamura H, Yoshikawa N, Takahashi T, Sahashi N, Kubono T, Shoji T (1996) Evaluation of primer pairs for the reliable diagnosis of paulownia witches' broom disease using a polymerase chain reaction. Plant Dis 80: 302–305.

Nakashima K, Hayashi T (1995) Extrachromosomal DNAs of rice yellow dwarf and sugarcane white leaf phytoplasmas. Ann Phytopathol Soc Jpn 61: 451–462.

Nakashima K, Kato S, Iwanami S, Murata N (1991) Cloning and detection of chromosomal and extrachromosomal DNA from mycoplasma-like organisms that cause yellow dwarf disease of rice. Appl Environ Microbiol 57: 3570–3575.

Narayanasamy P (2001) Plant Pathogen Detection and Disease Diagnosis, Second Edition, Marcel Dekker, New York.

Narayanasamy P (2008) Molecular Biology in Plant Pathogenesis and Disease Management, Volumes 1, 2, and 3, Springer, Heidelberg, Germany.

Naumann K, Karl H, Zielke R, Schmidt A, Griesbach E (1988) Comparative analysis of bean seeds on the incidence of causal agent of halo blight *Pseudomonas syringae* pv. *phaseolicola* by different identification methods. Zentr Mikrobiol 143: 487–498.

Navrátil M, Válova P, Fialová R, Priblova J, Špak J, Kubelkova D, Karešová R (2001) Occurrence of phytoplasmas in red and white currant in the Czech Republic. Acta Hortic 656: 119–124.

Nicolaisen M, Bertaccini A (2007) An oligonucleotide microarray-based assay for identification of phytoplasma 16S ribosomal groups. Plant Pathol 56: 332–336.

Niepold F (1999) A simple and fast extraction procedure to obtain amplifiable DNA from *Ralstonia solanacearum* and *Clavibacter michiganensis* subsp. *michiganensis*-inoculated potato tubers to conduct a polymerase chain reaction (PCR). J Phytopathol 147: 249–256.

Nipah JO, Jones, P, Dickinson MJ (2007) Detection of lethal yellowing phytoplasma in embroys from coconut palms infected with Cape St. Paul wilt disease in Ghana. Plant Pathol 56:777–784

Nishiyama K, Sakai R, Ezuka A, Ichihara A, Shiraishi K, Sakamura S (1977) Phytotoxic effect of coronatine in haloblight lesions of Italian ryegrass. Ann Phytopathol Soc Jpn 42: 219.

Norman DJ, Yuen JMF (1998) A distinct pathotype of *Ralstonia* (*Pseudomonas*) *solanacearum* race 1, biovar 1 entering Florida in pothos (*Epiremnum aureum*) cuttings. Can J Plant Pathol 20: 171–175.

Oh CS, Heu SG, Choi YC (1999) Sensitive and pathovar-specific detection of *Xanthomonas campestris* pv. *glycines* by DNA hybridization and analysis. Plant Pathol J 15:58–61

Okuda M, Matsumoto M, Tanaka Y, Subandiyah S, Iwananu T (2005) Characterization of the *tuft*B-secE-*nus*G-*rpl*KAJL-*rpo* gene cluster of the citrus greening organism and detection by loop-mediated isothermal amplification. Plant Dis 89: 705–711.

Olsen M, Bigelow D, Pryor B, Ramussen S (2002) Detection of *Xanthomonas translucens* on barley seed. In Ottmann M (ed), Forage and Grain, A College of Agriculture and Life Sciences Report Series P-132. Publ. No. AZ1301. University of Arizona, College of Agriculture and Life Sciences, Cooperative Extension, Tucson, AZ, pp. 1–3.

Opgenorth DC, Smart CD, Louws PJ, Bruijn FJ, de Kirkpatrick BC (1996) Identification of *Xanthomonas fragariae* field isolates by rep-PCR genomic fingerprinting. Plant Dis 80: 868–873.

Ou SH (1985) Rice Diseases, Second Edition, CMI, Surrey, UK.

Ovod VV, Knirel YA, Samson R, Krohn KJ (1999) Immunochemical characterization and taxonomic evaluation of the O-polysaccharides of the lipopolysaccharides of *Pseudomonas syringae* serogroup 01 strain. J Bacteriol 181: 6937–6947.

Ozakman M, Schaad NW (2004) A real-time BIO-PCR assay for the detection of *Ralstonia solanacearum* race 3, biovar 2, in asymptomatic potato tubers. Can J Plant Pathol 26: 232–239.

Ozdemir Z (2005) Development of a multiplex PCR assay for concurrent detection of *Clavibacter michiganensis* subsp. *michiganensis* and *Xanthomonas axonopodis* pv. *vesicatoria*. Plant Pathol J 4: 133–137.

Paltrinieri S, Bertaccini A (2007) Detection of phytoplasmas in plants grown from different batches of seed potatoes. Bull Insectol 60: 379–380.

Pan YN, Grisham MP, Burner DM, Damann KE Jr, Wei Q (1998) A polymerase chain reaction protocol for the detection of *Clavibacter xyli* subsp. *xyli*, the causal bacterium of sugarcane ratoon stunting disease. Plant Dis 82: 285–290.

Paret ML, de Silva AS, Alvarez AM (2009) Bioindicators for *Ralstonia solanacearum* race 4: plants in the Zingiberaceae and Costaceae families. Aust Plant Pathol 38: 6–12.

Park DS, Hyun JW, Park YJ, Kim JS, Kang HW, Hahn JH, Go SJ (2006) Sensitive and specific detection of *Xanthomonas axonopodis* pv. *citri* by PCR using pathovar specific primers based on *hrp*^{W} gene sequences. Microbiol Res 161: 145–149.

Park DS, Shim JK, Kim JS, Lim CK, Shreshtha R, Hahn JH, Kim HG (2009) Sensitive and specific detection of *Xanthomonass campestris* pv. *vesicatoria* by PCR using pathovar-specific primers based on *rhs* family gene sequences. Microbiol Res 164: 36–42.

Park YJ, Lee BM, Ho-Hahn J, Lee GB, Park DS (2004) Sensitive and specific detection of *Xanthomonas campestris* pv. *campestris* by PCR using species-specific primers based on *hrpF* gene sequences. Microbiol Res 159: 419–423.

Pasichnyk LA (2000) Antigenic properties of bacteria of pathovars of *Pseudomonas syringae* affecting cereals. Mikrobiol Z 62: 18–22.

Pastrik KH, Elphinstone JG, Pukall R (2002) Sequence analysis and detection of *Ralstonia solanacearum* by multiplex PCR amplification of 16S–23S ribosomal intergenic spacer region with internal positive control. Eur J Plant Pathol 108: 831–842.

Pataky JK, Block CC, Michener PM, Shepherd LM, McGee DC, White DG (2004) Ability of an ELISA-based seed health test to detect *Erwinia stewartii* in maize seed treated with fungicides and insecticides. Plant Dis 88: 633–640.

Patel MK, Dhande GM, Kulkarni YS (1951) Studies on some species of *Xanthomonas*. Ind Phytopathol 4: 123–140.

Peacock SJ, de Silva GD, Justice A, Cowland CE, Moore CE, Winerals CG, Day NP (2002) Comparison of multilocus sequence typing and pulsed-field gel electrophoresis as tools for typing *Staphylococcus aureus* isolates in a microepidemiological setting. J Clin Microbiol 40: 3764–3770.

Peñyalver R, García A, Ferrer A, Bertolini E, López MM (2000) Detection of *Pseudomonas savastanoi* pv. *savatanaoi* in olive plants by enrichment and PCR. Appl Environ Microbiol 66: 2673–2677.

Permar TA, Gottwald TR (1989) Specific recognition of a *Xanthomonas campestris* Florida citrus nursery strain by a monoclonal antibody probe in a microinfiltration enzyme immunoassay. Phytopathology 79: 780–783.

Pilkington LJ, Gibb KS, Gurr GM, Fletcher MJ, Nikandrow A, Elliott E, van de Ven R, Read DMY (2003) Detection and identification of a phytoplasma from lucerne with Australian lucerne yellows disease. Plant Pathol 52: 754–762.

Pollini CP, Giunchedi L, Bissani R (1997) Immunoenzymatic detection of PCR products for the identification of phytoplasmas in plants. J Phyotpathol 145: 371–374.

Pooler MR, Myung IS, Bentz J, Sherald J, Hartung JS (1997) Detection of *Xylella fastidiosa* in potential insect vectors by immunogenetic separation and nested polymerase chain reaction. Lett Appl Microbiol 25: 123–126.

Pradel KS, Ullrich CI, Santa Cruz S, Opar Ka KJ (1999) Symplastic continuity in *Agrobacterium tumefaciens*-induced tumors. J Exp Bot 50: 183–192.

Přibylova J, Petrzik K, Špak J (2009) The first detection of '*Candidatus* Phytoplasma trifolii' in *Rhododendron hybridum*. Eur J Plant Pathol DOI 10.1007/s.10658-008-9391-1

Priou S, Gutarra L, Aley P (1999) Highly sensitive detection of *Ralstonia solanacearum* in latently infected potato tubers by post-enrichment enzyme-linked immunosorbent assay on nitrocellulose membrane. Bull OEPP 29: 117–125.

Prossen D, Hatziloukas E, Panopoulos NJ, Schaad NW (1991) Direct detection of the halo blight pathogen *Pseudomonas syringae* pv. *phaseolicola* in bean seeds by DNA amplification. Phytopathology 81: 1159.

Puopolo G, Raio A, Zoina A (2007) Early detection of *Agrobacterium tumefaciens* in symptomless artificially inoculated chrysanthemum and peach plants using PCR. J Plant Pathol 89: 185–190.

Purcell AH, Saunders SR, Hendson M, Grebus ME, Henry MJ (1999) Causal role of *Xylella fastidiosa* in oleander leaf scorch disease. Phytopathology 89: 53–58.

Qu X, Wanner LA, Christ BJ (2008) Using the TxtAB operon to quantify pathogenic *Streptomyces* in potato tubers and soil. Phytopathology 98: 405–412.

Rabenstein F, Nachtigall M, Boll E (1999) Production and characterization of monoclonal antibodies against *Xanthomonas campestris* pv. *campestris*. Beit Zucht Bunde Zucht Kultur 5: 4–6.

Rademaker JLW, Janse D (1994) Detection and identification of *Clavibacter michiganensis* subsp. *sepedonicus* and *C. michiganens* subsp. *michiganensis* by nonradioactive hybridization, polymerase chain reaction and restriction enzyme analysis. Can J Microbiol 40: 1007–1018.

Rademaker JLW, Norman RL, Forster RL, Louws FJ, Schultz MH, de Bruijn FJ (2006) Classification and identification of *Xanthomonas translucens* isolates including those pathogenic to ornamental asparagus. Phytopathology 96: 876–884.

Rajeshwari N, Shylaja MD, Krishnappa M, Shetty HS, Mortensen CN, Mathur SB (1998) Development of ELISA for the detection of *Ralstonia solanacearum* in tomato: its application in seed health testing. World J Microbiol Biotechnol 14: 697–704.

Rico A, López R, Asensio C, Aizpún T, Asensio-S-Manzanera MC, Murillo J (2003) Nontoxigenic strains of *Pseudomonas syringae* pv. *phaseolicola* are a main cause of halo blight of beans in Spain and escape current detection methods. Phytopathology 93: 1553–1559.

Robéne-Soustrade I, Laurent P, Gagnevin L, Jouen E, Pruvost O (2006) Specific detection of *Xanthomonas axonopodis* pv. *diffenbachiae* in anthurium (*Anthurium andreanum*) tissues by nested PCR. Appl Environ Microbiol 72: 1072–1078.

Rodrigues JLM, Silva-Stencio ME, de Souza AN, Lopes JRS, Tsai SM (2006) *In situ* probing of *Xylella fastidiosa* in honeydew of a xylem sap-feeding insect using 16S rRNA-targeted fluorescent oligonucleotides. Environ Microbiol 8: 747–754.

Rodrigues JLM, Silva-Stencio ME, Gomes JE, Lopes JRS, Tsai sM (2003) Detection and diversity assessment of *Xylella fastidiosa* in field-collected plant and insect samples by using 16S rRNA and *gyrB* sequences. Appl Environ Microbiol 69: 4249–4255.

Romeiro RS, Silva HSA, Beriam LOS, Rodrigues Neto J, Carvalho MG de (1999) A bioassay for the detection of *Agrobacterium tumefaciens* in soil and plant material. Summ Phytopathol 25: 359–362.

Rott P, Davis MJ, Baudin P (1994) Serological variability in *Xanthomonas albilineans*, causal agent of leaf scald disease of sugarcane. Plant Pathol 43: 344–359.

Sachadyn P, Kur J (1997) A new PCR system for *Agrobacterium tumefaciens* detection based on amplification of T-DNA fragment. Acta Microbiol Polo 46: 145–156.

Saeed EM, Roux J, Cousin MT (1993) Studies on polyclonal antibodies for the detection of MLOs associated with faba bean (*Vicia faba* L.) using different ELISA methods and dot blot. J Phytopathol 137: 33–43.

Sagaram US, DeAngelis KM, Trivedi P, Andersen GL, Lu S-N, Wang N (2009) Bacterial diversity analysis of huanglongbing pathogen-infected citrus, using PhyloChip arrays and 16S rRNA gene clone library sequencing. Appl Environ Microbiol 75: 1566–1574.

Samuitiene M, Navalinskiene M (2006) Molecular detection and characterization of phytoplasma infecting *Celosia argentea* L. plants in Lithuania. Agron Res 4: 345–348.

Sayler RJ, Cartwright RD, Yang Y (2006) Genetic characterization and real-time PCR detection of *Burkholderia glumae*, a newly emerging bacterial pathogen of rice in the United States. Plant Dis 90: 603–610.

Scally M, Schuenzel EL, Stouthamer RA, Nunney L (2005) Multilocus sequence type system for the plant pathogen *Xylella fastidiosa* and relative contributions of recombination and point mutation of clonal diversity. Appl Environ Microbiol 71: 8491–8499.

Schaad NW, Azad H, Peet RC, Panopoulos NJ (1989) Identification of *Pseudomonas syringae* pv. *phaseolicola* by a DNA hybridization probe. Phytopathology 79: 903.

Schaad NW, Berthier-Schaad Y, Knorr D (2007) A high throughput membrane BIO-PCR technique for ultra-sensitive detection of *Pseudomonas syringae* pv. *phaseolicola*. Plant Pathol 56: 1–8.

Schaad NW, Berthier-Schaad Y, Sechler A, Knorr D (1999) Detection of *Clavibacter michiganensis* subsp. *sepedonicus* in potato tubers by BIO-PCR and an automated real-time fluorescence detection system. Plant Dis 83: 1095–1100.

Scaad NW, Cheong SS, Tamaki E, Hatziloukas E, Panopoulos NJ (1995a) A combined biological and enzymatic amplification (BIO-PCR) technique to detect *Pseudomonas syringae* pv. *phaseolicola* in bean seed extracts. Phytopathology 95: 243–248.

Schaad NW, Cheong SS, Tamaki S, Hatziloukas E, Panopoulos NJ (1995b) A combined biological and enzymatic amplification (BIO-PCR) technique to detect Pseudomonas syringae pv. phaseolicola from wheat seeds. Phytopathology 85: 260–263.

Schaad NW, Forster RL (1985) A semi-selective agar medium for isolating *Xanthomonas campestris* pv. *translucens* from wheat seeds. Phytopathology 75: 260–263.

Schaad NW, Jones JB, Chun W (2002a) Laboratory Guide for Identification of Plant Pathogenic Bacteria. American Phytopathol Society, St. Paul, MN.

Schaad NW, Opgennorth D, Gaush P (2002b) Real-time polymerase chain reaction for one-hour on site diagnosis of Pierce's disease of grape in early season asymptomatic vines. Phytopathology 92: 721–728.

Schaad NW, Takatsu A, Dianese JC (1978) Serological identification of strains of *Pseudomonas solanacearum* in Brazil. Proceedings of the 4th International Conference on Plant Pathogenic Bacteria, Angers, France, 1, pp. 295–300.

Schmidt T (1994) Fingerprinting bacterial genomes using ribosomal RNA genes and operons. Meth Mol Cell Biol 5: 3–12.

Schneider B, Ahrens U, Kirkpatrick BC, Seemüller E (1993) Classifiction of plant pathogenic mycoplasma-like organisms using restriction-site analysis of PCR-amplified 16S rDNA. J Gen Microbiol 139: 519–527.

Schneider B, Seemüller E (1994) Studeis on taxonomic relationships of mycoplasma-like organisms by Southern blot analysis. J Phytopathol 141: 173–185.

Schneider B, Torres E, Martin MP, Schroder M, Behnke HD, Seemüller E (2005) '*Candiatus* Phytoplasma pini', a novel taxon from *Pinus silvestris* and *Pinus halepensis*. Int J Syst Evol Microbiol 55: 303–307.

Schneider BJ, Zhao JL, Orser CS (1993) Detection of *Clavaibacter michiganensis* subsp. *sepedonicus* by DNA amplification. FEMS Microbiol Lett 109: 207–212.

Scholberg P, O'Gorman D, Bedford K, Levesque CA (2005) Development of DNA macroarray for detection and monitoring of economically important apple diseases. Phytopathology 89: 1143–1150.

Schubert TS, Rivui SA, Sun XA, Gottwald TR, Graham JH, Dixon WN (2001) Meeting the challenge of eradicating citrus canker in Florida again. Plant Dis 85: 340–356.

Schumann GL (1991) Plant Diseases: Their Biology and Social Impact. American Phytopathological Society, St. Paul, MN.

Schwartz DC, Cantor CR (1984) Separation of yeast chromosome-sized DNA by pulsed field gradient gel elelctrophoresis. Cell 37: 67–75.

Schwarz RE (1968) Indexing of greening and excortis through fluorescent marker substance. Proc 4th Conf Citrus Virol Univ Florida, pp. 118–123.

Seal SE, Taghavi M, Fegan N, Hayward AC, Fegan M (1999) Determination of *Ralstonia* (*Pseudomonas*) *solanacearum* rDNA subgroups by PCR tests. Plant Pathol 48: 115–120.

Sechler A, Schuenzel EL, Cooke S, Donnua S, Thaveechai N, Postnikova E, Stone AL, Schneider WL, Damsteegt D, Schaad NW (2009) Cultivation of '*Candiatus* Liberibacter asiaticus', '*Ca.* L. africanus' and '*Ca.* L. americanus' associated with Huanglongbing. Phytopathology 99: 480–486.

Secor GA, Rivera VV, Abad JA, Lee I-M, Clover GRG, Liefting LW, Li X, De Boer SH (2009) Association of '*Candidatus* Liberibacter solanacearum' with Zebra Chip disease of potato established by graft and psyllid transmission, electron microscopy and PCR. Plant Dis 93: 574–583.

Seddas A, Meignoz R, Daire X, Boudon-Padieu E (1996) Generation and characterization of monoclonal antibodies to flavescence dorée phytoplasma: serological relationships and differences in electroblot immunoassay profiles of flavescence dorée and elm yellows phytoplasmas. Eur J Plant Pathol 102: 757–764.

Seo ST, Furuya N, Lim CK, Takanami Y, Tsuchiya K (2003) Phenotypic and genetic characterization of *Erwinia carotovora* from mulberry (*Morus* spp.) Plant Pathol 52: 140–146.

Seo ST, Tsuchiya K, Murakami R (2004) Characterization and differentiation of *Erwinia carotovora* strains from mulberry trees. J Gen Plant Pathol 70: 120–123.

Shan ZH, Liao BS, Tan YJ, Li D, Lei Y, Shen MZ (1997) ELISA technique used to detect latent infection of groundnut by bacterial wilt (*Pseudomonas solanacearum*). Oilcrops China 19: 45–47.

Shao J, Jomantiene R, Dally EL, Zhao Y, Lee I-M, Nuss DL, Davis RE (2006) Phylogeny and characterization of phytoplasmal NusA and use of the *nusA* gene in detection of group 16SrII strains. J Plant Pathol 88: 193–201.

Shigaki R, Nelson SC, Alvarez AM (2000) Symptomless spread of blight-inducing strains of *Xanthomonas campestris* pv. *campestris* on cabbage seedlings in misted seed beds. Eur J Plant Pathol 106: 339–346.

Siddique ABM, Guthrie JN, Walsh KB, White DT, Scott PT (1998) Histopathology and within-plant distribution of the phytoplasma associated with Australian papaya dieback. Plant Dis 82: 1112–1120.

Sinclair WA, Griffiths HM, Davis RE, Lee I-M (1992) Detection of ash yellows mycoplasma-like organisms in different tree organs and in chemically preserved specimens by cDNA probe vs DAPI. Plant Dis 76: 154–158.

Sinclair WA, Iuli RT, Dyer AT, Larsen AD (1989) Sampling and histological procedures for diagnosis of ash yellows. Plant Dis 73: 432–435.

Singh U, Trevors CM, De Boer SH, Janse JD (2000) Fimbrial-specific monoclonal antibody-based ELISA for European potato strains of *Erwinia chrysanthemi* and comparison to PCR. Plant Dis 84: 443–448.

Sinha RC (1988) Serological detection of mycoplasma-like organisms from plants affected with yellows disease. In: Hiruki C (ed), Tree Mycoplasmas and Mycoplasma Diseases, University of Alberta Press, Edmonton, AB, pp. 143–156.

Sinha RC, Benhamou N (1983) Detection of mycoplasma-like organism antigens from aster yellows-diseased plants by two serological procedures. Phytopathology 73: 1199–1202.

Slack SK, Drennan, JL, Westra AAG, Gudmestad NC, Oleson AE (1996) Comparison of PCR, ELISA and DNA hybridization for the detection of Clavibacter michiganensis susbsp. sepedonicus in field-grown potatoes. Plant Dis 80:519–524

Sledz W, Jafra S, Waleron M, Malinowska A, Lojkowska E (1999) Diagnostics of pectinolytic *Erwinia* isolated from infected potato plants and tubers by PCR. Petria 9: 73–76.

Smith DS, De Boer SH (2009) Implementation of an artificial reaction control in a TaqMan method PCR detection of *Ralstonia solanacearum* race 3 biovar 2. Eur J Plant Pathol 124: 405–412.

Smith DS, De Boer SH, Gourley J (2008) An internal reaction control for routine detection of *Clavibacter michiganensis* subsp. *sepedonicus* using real-time TaqMan PCR-based assay. Plant Dis 92: 684–693.

Solomon JJ, Govindan Kutty MP (1991) Etiology-mycoplasma-like organisms. In: Nair MK, Nambiar KKN, Koshy PK, Jayasankar NP (eds), Coconut Root (wilt) Disease, CPCRI, Kasaragod, India, pp. 35–40.

Špak J, Přibylová J, Kubelková D, Špaková V (2004) The presence of phytoplasma in black currant infected with the black currant reversion disease. J Phytopathol 152: 600–605.

Srinivasan N (1982) Simple diagnostic technique for plant diseases of mycoplasma etiology. Curr Sci 51: 883–885.

Srinivasulu B, Narayanasamy P (1995) Serological detection of phyllody disease in sesamum and leafhopper *Orosius albicinctus*. J Mycol Plant Pathol 25: 154–157.

Stöger A, Schaffer J, Ruppitsch W (2006) A rapid and sensitive method for direct detection of *Erwinia amylovora* in symptomatic and asymptomatic plant tissues by polymerase chain reaction. J Phytopathol 154: 469–473.

Streten C, Conde B, Herrington M, Moulden J, Gibb K (2005) *Candidatus* Phytoplasma australiense is associated with pumpkin yellow leaf curl disease in Queensland, Western Australia and the Northern Territory. Aust Plant Pathol 34: 103–105.

Sun X, Stall RE, Jones JB, Cubero J, Gottwald TR, Graham JH, Dixon WN, Schubert TS, Chaloux PH, Stromberg VK, Lacy GH, Sutton BD (2004) Detection and characterization of a new strain of citrus canker bacteria from Key/Mexican lime and Alemow in South Florida. Plant Dis 88: 1179–1188.

Takahashi Y, Omura T, Hibino H, Sato M (1996) Detection and identification of *Pseudomonas syringae* pv. *atropurpurea* by PCR amplification of specific fragments from an indigenous plasmid. Plant Dis 80: 783–788.

Tambong JT, Mwange KN, Bergeron M, Ding T, Mandy F, Reid LM, Zhu X (2008) Rapid detection and identification of the bacterium *Pantoea stewartii* in maize by TaqMan® real-time PCR targeting the *cpsD* gene. J Appl Microbiol 104: 1525–1537.

Tatineni S, Sagaram US, Gowda S, Robertson CJ, Dawson WO, Iwanami T, Wang N (2008) *In planta* distribution of '*Candidatus* Liberibacter asiaticus' as revealed by polymerase chain reaction (PCR) and real-time PCR. Phytopathology 98: 592–599.

Taylor RK, Guilford PJ, Clark RG, Hale CN, Forster RLS (2001) Detection of *Erwinia amylovora* in plant material using novel polymerase chain reaction (PCR) primers. NZ J Crop Horti Sci 29: 35–43.

Taylor RK, Tyson JL, Fullerton RA, Hale CN (2002) Molecular detection of exotic phytopathogenic bacteria: a case study involving canker-like symptoms on citrus. NZ Plant Prot 55: 53–57.

Teixeira DC, Saillard C, Eveilard S, Danet JL, Ayers AJ, Bové JM (2005) '*Candidatus* Liberibacter americanus' sp. nov., associated with citrus huanglongbing (greening disease) in Sao Paulo, Brazil. Int Syst Bacteriol 55: 18857–18662.

Thaveechai N, Schaad NW (1986) Immunochemical characterization of a subspecies-specific antigenic determinant of a membrane protein complex of *Xanthomonas campestris* pv. *campestris*. Phytopathology 76: 148–153.

Thomas S, Balasundaram M (2001) Purification of sandal spike phytoplasma for the production of polyclonal antibody. Curr Sci 80: 1489–1494.

Torres E, Bertolini E, Cambra M, Monton C, Martin MP (2005) Real-time PCR for simultaneous and quantitative detection of quarantine phytoplasmas from apple proliferation (16SrX) group. Mol Cell Probes 19: 334–340.

Toth IK, Hyman LJ, Wood JR (1999) A one-step PCR-based method for the detection of economically important soft rot *Erwinia* species in micropropagated potato plants. J Appl Microbiol 87: 158–166.

Toussaint V, Morris CE, Carisse O (2001) A new semi-selective medium for *Xanthomonas campestris* pv. *vitians*, the causal agent of bacterial spot of lettuce. Plant Dis 85: 131–136.

Tran-Nguyen LTT, Gibb KS (2007) Optimizing phytoplasma DNA purification for genome analysis. J Biomol Tech 18: 104–112.

Trindade LC, Marques E, Lopes DB, Ferreira MASV (2007) Development of a molecular method for detection and identification of *Xanthomonas campestris* pv. *viticola.* Summa Phytopathol 33: 16–23.

Triplett LR, Zhao Y, Sundin GW (2006) Genetic differences between blight-causing *Erwinia* species with differing host specificities identified by suppression subtractive hybridization. Appl Environ Microbiol 72: 7359–7364.

Trujillo GE, Saettler AW (1979) A combined semi-selective medium and serology test for the detection of *Xanthomonas* blight bacteria in bean seed. J Seed Sci Technol 4: 35–41.

Tsuchiya K, d'Ursel CCM (2004) Development of a sensitive ELISA using three monoclonal antibodies against lipopolysaccharides to detect *Xanthomonas campestris* pv. *vesicatoria*, the causal agent of bacterial spot of tomato and pepper. J Gen Plant Pathol 70: 21–26.

Tsuchiya K, d'Uresel CCM, Nozu Y (2004) Production and preliminary characterization of monoclonal antibodies raised against *Xanthomonas campestris* pv. *mangiferaeindicae.* J Gen Plant Pathol 70: 27–33.

Tsuchiya K, Yano K, Horita M, Morita Y, Kawada Y, d'Ursel CM (2005) Occurrence and epidemic adaptation of new strains of *Ralstonia solanacearum* associated with Zingiberaceae plant under agro-ecosystem in Japan. In: Allen C, Prior Ph, Hayward AC (eds), Bacterial Wilt Disease and the *Ralstonia solanacearum* species complex, American Phytopathological Society, St. Paul, MN, pp. 463–469.

Tsushima S, Narimatsu C, Mizuno A, Kimura R (1994) Cloned DNA probes for detection of *Pseudomonas glumae* causing bacterial grain rot of rice. Ann Phytopathol Soc Jpn 60: 576–584.

Tymon AM, Jones P, Harrison NA (1997) Detection and differentiation of African coconut phytoplasma: RFLP analysis of PCR-amplified 16S rDNA and DNA hybridization. Ann Appl Biol 131: 91–102.

Umesha S, Richardson PA, Kong P, Hong CX (2008) A novel indicator plant to test the hypersensitivity of phytopathogenic bacteria. J Microbiol Meth 72: 95–97.

Ura H, Matsumoto M, Iiyama K, Furuya N, Matsuyama N (1998) PCR-RFLP analysis for distinction of *Burkholderia* species, causal agents of various rice diseases. Proc Assoc Plant Prot Kyushu 44: 5–8.

Urabana Vičienė L, Jomantienė R, Valiūnas D, Davis RE (2004) Molecular detection of phytoplasmas in oats, barley and *Triticosecale* and their classification based on 16S rRNA gene polymorphisms. Zemes Ukio Mokslai 4: 15–19.

Urabana Vičienė D, Valiūnas D, Jomatienė R (2006) Detection of aster yellows group (subgroup 16SrI-B) phytoplasma in oats based on nested PCR and RFLP in Lithuania. Agron Res 4: 417–420.

Urasaki N, Kawano S, Mukai H, Uemori T, Takeda O, Sano T (2008) Rapid and sensitive detection of '*Candiatus* Liberibacter asiaticus' by cycleave isothermal and chimeric primer-initiated amplification of nucleic acids. J Gen Plant Pathol 74: 151–155.

Valiūnas D, Samuitiene M, Navalinskiene, Davis RE (2008) Identification of viral and phytoplasmal agents causing diseases in *Gaillardia* Foug plants in Lithuania. Agron Res 6: 109–118.

Van der Vliet GME, Schepers P, Schukkink RAF, van Gemen B, Klaster PR (1994) Assessment of mycobacterial viability by RNA amplification. Antimicrob Agents Chemotherap 38: 1959–1965.

Van der Wolf JM, Hyman LJ, Jones DAC, Grevesse C, Beckhoven JRCM, van Vuurde JWL, Pérombelon MCM (1996) Immunomagnetic separation of *Erwinia carotovora* subsp. *atroseptica* from potato peel extracts to improve detection sensitivity on a crystal violet pectate medium or by PCR. J Appl Bacteriol 80: 487–495.

Van der Wolf JM, Schoen CD (2004) Bacterial Pathogens: Detection and Identification Methods, Marcel Dekker, New York.

Van der Wolf JM, Van Beckhoven JRCM, De Haan EG, Van den Bovenkamp GW, Leone GOM (2004) Specific detection of *Ralstonia solanacearum* 16S rRNA sequences by AmpliDet RNA. Eur J Plant Pathol 110: 25–33.

Van Vuurde JWL, Van den Bovenkamp GW (1981) Routine methods for the detection of halo-blight (*Pseudomonas phaseolicola*) in beans. Report on the 17th International Workshop on Seed Pathology, International Seed Testing Association, Zurich, Switzerland, p. 22.

Veena MS, van Vuurde JWL (2002) Indirect immunofluorescence colony staining method for detecting bacterial pathogens of tomato. J Microbiol Meth 49: 11–17.

Veera C, Milne RG (1994) Immunosorbent electron microscopy and gold label antibody decoration of MLOs from crude preparations of infected plants and insects. Plant Pathol 43: 190–199.

Verdier V, Mosquera G, Assigbetse K (1998) Detection of the cassava bacterial blight pathogen *Xanthomonas axonopodis* pv. *manihotis* by polymerase chain reaction. Plant Dis 82: 79–83.

Verniere C, Devaux M, Pruvost O, Couteau A, Luisetti J (1991) Studies on the biochemical and physiological variations among strains of *Xanthomonas campestris* pv. *citri*, the causal agent of citrus bacterial canker disease. Fruits 46: 162–170.

Verniere C, Pruvost O, Civerolo EL, Gambin O, Jacquemoud-Collet JP, Luisetti J (1993) Evaluation of the Biolog substrate utilization system to identify and assess metabolic variation among strains of *Xanthomonas campestris* pv. *citri*. Appl Environ Microbiol 59: 243–249.

Verreault H, Lafond M, Asselin A, Banville G, Bellemare G (1998) Characterization of two DNA clones specific for identification of *Corynebacterium sepedonicum*. Can J Microbiol 34: 993.

Vicente JG, Everett B, Roberts SJ (2006) Identification of isolates that cause a leaf spot disease of Brassicas as *Xanthomonas campestris* pv. *raphani* and pathogenic and genetic comparison with related pathovars. Phytopathology 96: 735–745.

Viswanathan R (1997a) Detection of ratoon stunting disease (RSD) bacterium by ELISA. Madras Agric J 84: 374–377.

Viswanathan R (1997b) Detection of phytoplasma associated with grassy shoot disease by ELISA techniques. Z Pflanzenkrank 104: 9–16.

Viswanathan R (2000) Detetcion of phytoplasmas causing grassy shoot disease in sugarcane by immunofluorescence technique. Ind Phytopathol 53: 475–477.

Wakimoto S (1957) A simple method for comparison of bacterial populations in a large number of samples by phage technique. Ann Phytopathol Soc Jpn 22: 159–163.

Wakimoto S (1960) Classification of strains of *Xanthomonas oryzae* on the basis of their susceptibility against bacteriophages. Ann Phytopathol Soc Jpn 25: 193–198.

Walcott RR, Gitaitis RD (2000) Detection of *Acidovorax avenae* subsp. *citrulli* in watermelon seed using immunomagnetic separation and the polymerase chain reaction. Plant Dis 84: 470–474.

Walcott RR, Gitaitis RD, Castro AC (2003) Role of blossom in watermelon seed infection by *Acidovorax avenae* subsp. *citrulli*. Phytopathology 93: 528–534.

Walcott RR, Gitaitis RD, Castro AC, Sanders FH Jr, Diaz-Perez JC (2002) Natural infestation of onion seed by *Pantoea anantis* causal agent of center rot. Plant Dis 86: 106–111.

Walcott RR, Langston DB Jr, Sanders FH, Gitaitgis RD (2000) Investigating intraspecific variation of *Acidovorax avenae* subsp. *citrulli* using DNA fingerprinting and whole cell fatty acid analysis. Phytopathology 90: 191–196.

Wang K, Hiruki C (2001) Molecular characterization and classification of phytoplasmas associated with canola yellows and a new phytoplasma strain associated with dandelions. Plant Dis 85: 76–79.

Wang K, Hiruki C (2005) Distinctions between phytoplasmas at the subgroup level detected by heteroduplex mobility assay. Plant Pathol 54: 625–633.

Wang ZK, Luo HH, Yi SZ (1997) Application of DIA (dot immunobinding assay) for rapid detection of *Xanthomonas axonopodis* pv. *citri*. J Southwest Agric Univ 19: 529–532.

Ward LJ, De Boer SH (1984) Specific detection of *Erwinia carotovora* subsp. *atroseptica* with digoxigenin-labeled DNA probe. Phytopathology 74: 180–186.

Ward LJ, De Boer SH (1990) A DNA probe specific for serologically diverse strains of *Erwinia carotovora*. Phytopathology 80: 665.

Warokka JS, Jones P, Dickinson MJ (2006) Detection of phytoplasmas associated with Kalimantan wilt disease of coconut by the polymerase chain reaction. J Liitri 12: 154–160.

Webb DR, Bonfiglioli RG, Carraro L, Osler R, Symons RH (1999) Oligonucleotides as hybridization probes to localize phytoplasmas in host plants and insect vectors. Phytopathology 89: 894–901.

Wei W, Davis RE, Lee I-M, Zhao Y (2007) Computer simulated RFLP analysis of 16S rRNA genes: identification of ten new phytoplasma groups. Int J Syst Evol Microbiol 57: 1855–1867.

Wei W, Opgenorth DC, Davis RE, Chang CJ, Summers CG, Zhao Y (2006) Characterization of a novel adhesion-like gene and design of a real-time PCR for rapid sensitive and specific detection of *Spiroplasma kunkelii*. Plant Dis 90: 1233–1238.

Weller SA, Elphinstone JG, Smith NC, Boonham N, Stead DE (2000) Detection of *Ralstonia solanacearum* strains with a quantitative, multiplex, real-time, fluorogenic PCR (TaqMan) assay. Appl Environ Microbiol 66: 2853–2858.

Wu WC, Lee ST, Kuo HP, Wang LY (1993) Use of phase for identifying the citrus canker bacterium *Xanthomonas campestris* pv. *citri* in Taiwan. Plant Pathol 42: 389–395.

Wullings BA, Beuningen AR, van Janse JD, Akkermans ADL (1998) Detection of *Ralstonia solanacearum* which causes brown rot of potato by fluorescent *in situ* hybridization with 23S rRNA-targeted probes. Appl Environ Microbiol 64: 4546–4554.

Xiao W, Le Z, Fu-Shou X, Li-Han Z, Guan-Lin X (2007) Immunocapture PCR method for detecting *Acidovorax avenae* subsp. *citrulli* from watermelon. Chin J Agric Biotechnol 4: 173–183.

Xie GL, Mew TW (1998) A leaf inoculation method for the detection of *Xanthomonas oryzae* pv. *oryzicola*. Plant Dis 82: 1007–1011.

Yakrus M, Schaad NW (1979) Serological relationships among strains of *Erwinia chrysanthemi*. Phytopathology 69: 517–522.

Yamada T, Kawasaki T, Nagata S, Fujiwara A, Usami S, Fujie M (2007) New bacteriophages that infect the phytopathogen *Ralstonia solanacearum*. Microbiology 153: 2630–2639.

Yashitola J, Krishanaveni D, Reddy APK, Sonti RV (1997) Genetic diversity within populations of *Xanthomonas oryzae* pv. *oryzae* in India. Phytopathology 87: 760–765.

Young JM, Takikawa Y, Gardan L, Stead DE (1992) Changing concepts in the taxonomy of plant pathogenic bacteria. Annu Rev Phytopathol 30: 67–105.

Yu J, Wayadande AC, Fletcher J (2000) *Spiroplasma citri* surface protein p89 implicated in adhesion to cells of vector *Circulifer tenellus*. Phytopathology 90: 716–722.

Zaccardelli M, Spasiano A, Bazzi C, Merighi M (2005) Identification and *in planta* detection of *Pseudomonas syringae* pv. *tomato* using PCR amplification of *hrpZpst*. Eur J Plant Pathol 111: 85–90.

Zhang Y, Merighi M, Bazzi C, Geider K (1998) Genomic analysis by pulsed field gel electrophoresis of *Erwinia amylovora* strains from the Mediterranean region including Italy. J Plant Pathol 80: 225–232.

Zhao W-J, Chen H-Y, Zhu S-F, Xia M-X, Tan T-W (2007) One-step detection of *Clavibacter michiganensis* subsp. *michiganensis* in symptomless tomato seeds using a TaqMan probe. J Plant Pathol 89: 349–351.

Zhao Y, Damicone JP, Bender CL (2002) Detection, survival and sources of inoculum for bacterial diseases of leafy crucifers in Oklahoma. Plant Dis 86: 883–888.

Zhu H, Gao JL, Li QX, Hu GX (1988) Detection of *Xanthomonas oryzae* in rice by monoclonal antibodies. J Jiangsu Agric Coll 9: 41–43.

Zreik L, Carle P, Bové JM, Garnier M (1995) Characterization of the mycoplasma-like organism associated with witches' broom disease of lime and proposition of a Candidatus taxon for the organism, '*Candidatus* Phytoplasma aurantifolia'. Int J Syst Bacteriol 45: 449–453.

Chapter 3
Detection of Bacterial and Phytoplasmal Pathogens in the Environment

Abstract Plant pathogenic bacteria, except a few, can exist outside plant hosts for short or long periods in soil, water or air and infect the healthy susceptible plants, when they are available. Various methods of detection of bacterial pathogens in the environmental samples have been followed. Biological methods depending on their ability to develop in cell-free media and pathogenicity are the basic and essential methods, although they may be time-consuming and labor intensive. Immunological assays and nucleic acid-based techniques provide more sensitive and specific detection and quantification of pathogen population in the environment especially in the soil samples. These methods are useful in establishing the identity of bacterial pathogens conclusively, when many other saprophytic bacteria are present. Diagnostic methods are very useful in assessing the host range of bacterial pathogens and this information will be required for epidemiological studies.

Phytoplasmas, the cell wall-less bacteria-like organisms require living cells for their existence. They can infect other plant species and natural insect vectors. Both may be able to serve as sources of infection of the primary crop hosts. Various diagnostic techniques especially nucleic acid-based procedures have been applied for the detection, quantification and differentiation as well as for the classification of phytoplasmas in the plant hosts and also in the vector insects. Concerted attempts of researchers have recently succeeded in culturing one of the phytoplasmas in cell-free media, opening up the possibility of rapid advancement in understanding the biology of these pathogens and consequent development of effective management methods for the phytoplasmal diseases.

Bacterial plant pathogens can exist, outside plant hosts, in the soil, water and air. On the other hand, phytoplasmal pathogens are obligate pathogens and hence, they are incapable of existing outside the host. Both bacterial and phytoplasmal pathogens exhibit biological relationships with various insect species that carry them even to long distances breaking the geographical barriers and making the efforts of various countries to contain the spread of these pathogens ineffective. However, consistent attempts have been made to detect these pathogens in different habitats to develop suitable disease management systems.

P. Narayanasamy, *Microbial Plant Pathogens-Detection and Disease Diagnosis: Bacterial and Phytoplasmal Pathogens, Vol. 2*, DOI 10.1007/978-90-481-9769-9_3,

3.1 Detection of Bacterial Pathogens in the Environment

3.1.1 Detection of Bacterial Pathogens in Soil

Bacterial plant pathogens may reach the soil, when infected aerial plant parts left in the soil decompose. In addition, the bacteria present on the aerial plant parts in a standing crop may be washed down by rain water and may move to other fields along with running surface water. They may remain as saprophytes living on dead organic matter available in the soil till they find suitable host plant species. Three different kinds of methods depending on the biological, immunological and genomic characteristics of bacterial pathogens have been taken into consideration for the detection and diagnosis of the diseases caused by them.

3.1.1.1 Biological Methods

Bacterial pathogens present in the soil may be detected by using highly sensitive indicator plants that develop characteristic symptoms of infection rapidly. In the case of *Ralstonia solanacearum* (*Rs*), causing wilt and brown rot disease of potato, sensitive plants such as potato have been used as indicator plants to detect the presence of *Rs* in the soil samples (Graham and Lloyd 1978). A tomato bioassay system was applied for the detection of *Rs*. This bioassay consistently detected populations of *Rs* at or greater than 7.5×10^5 cfu/g of soil (Pradhanang et al. 2000). *Agrobacterium tumefaciens* (*At*) causing tumors in large number of plant species could be detected by a simple, rapid and reliable bioassay method. Detached leaves of the highly susceptible *Kalanchoe tubiflora* were used as baits for trapping *At* present in the soil. This bacterial pathogen alone was retrieved from the soil in spite of the presence of a large and heterogeneous microflora in the soil. *At* could be isolated from the leaf baits and brought into pure culture (Romeiro et al. 1999). Relatively large volumes of soil, greenhouse space and long time will be required for performing the bioassays, although they may be able to give an estimate of viable pathogen propagules that can cause disease.

Quantitative assessments can be made, by adopting dilution plating on a semi-selective medium that differentially favors the development of the target pathogen. However, the high populations of non-target bacteria present in the soil easily tend to overgrow the target bacterial species. Detection limit of 10^4 cfu/g of soil may be obtained (Graham and Lloyd 1979). Use of semi-selective medium like SMSA may improve the limit of detection of *R. solanacearum* by dilution plating procedure (Elphinstone 1996). In a later investigation, it was shown that a modified SMSA medium was able to detect *R. solanacearum* (*Rs*) at concentration of less than 10 viable cfu /ml in inoculated soil suspension (or $<10^2$ cfu/g of soil), suggesting that the sensitivity of the pathogen detection in soil could be improved by using this modified SMSA medium. However, a number of saprophytic soilborne bacteria were observed to develop on SMSA medium, which could adversely affect detection under natural conditions due to competition or antagonism. Soilborne *Rs* populations could be recovered on SMSA agar up to 18 months after harvest of an infected potato crop, indicating that residual population can remain viable over this period. Under natural conditions the lowest detectable populations of *Rs* on SMSA was found to be 5×10^2 cells/g of soil (Pradhanang et al. 2000).

Bacteriophages cause lysis of cells of susceptible bacterial species. The specificity of phages to certain bacterial species has been exploited for detection and quantification of certain phytopathogenic bacteria. *Ralstonia solanacearum* causing bacterial wilt diseases, can survive in the soil for several years during the absence of plant hosts. Four bateriophages øRSL, øRSA, øRDM and øRSS capable of infecting *R. solanacearum* were isolated. These phages had relatively wide host ranges and induced larger clear plaques (clear zone formed in bacterial cultures due to lysis of bacterial cells). The phage øRSA1 could infect all 15 strains of *R. solanacearum* of different races or biovars tested. The results indicated that the phages may be useful not only for the detection of *R. solanacearum*, but also for its control as a disease management measure (Yamada et al. 2007).

3.1.1.2 Immunoassays

Detection of soilborne bacterial plant pathogens by employing immunoassays has been shown to be more sensitive and rapid compared to biological methods.

Immunofluorescence Tests

Immunofluorescence colony-staining (IFC) test which combines bacterial colony growth and immunoassay, allows sensitive and quantitative detection of several phytopathogenic bacteria in complex backgrounds. Both monoclonal and polyclonal antibodies (MAbs and PAbs) generated against *R. solanacearum* (*Rs*) exhibited cross-reaction with taxonomically related bacteria such as the causative agent of banana blood disease and *R. syzygii*, but also with more distantly related soil saprophytes (Griep et al. 1998). Hence, the results of immunoassay have to be interpreted carefully, especially quarantine pathogens like *Rs* have to be detected in the soil, where populations of *Rs* may be low and high numbers of nontarget soil bacterial may be present, some of which might cross-react with antibodies giving false positive results.

A specific and sensitive quantitative technique for detection of *R. solanacearum* (*Rs*) biovar 2 (race 3) was developed, based on IFC followed by confirmation of the fluorescent colonies by PCR-amplification or dilution plating on a semi-selective SMSA medium. Addition of sucrose and antibiotics cycloheximide and crystal violet to the non-selective tryticase soy broth agar resulted in increased colony size and staining intensity of *Rs* in IFC test. The IFC test was able to detect consistently ca. 100 cfu/g of soil, a detection level similar, but less laborious, whereas the bioassay using tomato plants had a detection limit of 10^4–10^5 cfu/g of soil (van der Wolf et al. 2000).

Enzyme-Linked Immunosorbent Assay

Ralstonia solanacearum (*Rs*) can survive in the soil for a long time, in addition to its ability to be transmitted through infected potato tubers in which it causes the brown rot disease. The population of *Rs* in the field soil has to be determined to reduce incidence of bacterial wilt and brown rot diseases in potatoes. An indirect ELISA format was found to be sensitive in detecting as few as 10^4 cfu of *Rs*/g of soil, when the bacterial suspension was incubated in a modified semi-selective medium prior to applying the ELISA test (Pradhanang et al. 2000). A double antibody sandwich (DAS)-ELISA test was applied after selective enrichment for the detection of *R. solanacearum* (*Rs*) in the soil. *Rs* was successfully detected by post-enrichment DAS-ELISA at low population levels in inoculated soil. The DAS-ELISA showed a satisfactory level of sensitivity and there was no cross-reaction following enrichment of soil extracts inoculated with soil saprophytes. Similar absence of cross-reaction was also demonstrated in samples from naturally infested field soil. This study revealed the reliability and sensitivity of post-enrichment DAS-ELISA format suitable for disease management and epidemiological investigations on bacterial wilt disease (Priou et al. 1999). International Potato Center (CIP) developed a new semi-selective broth containing potato tuber infusion which differentially favored the development of *R. solanacearum* (*Rs*). Potato isolates of *Rs* (273) belonging to five different biovars (BV) originating from 33 countries were incubated in this new semi-selective broth prior to the detection by DAS-ELISA procedure. By using specific antibodies, the presence of *Rs* in samples with low pathogen population could be detected, after incubation of soil suspensions for 48 h at 30°C in the broth. The detection thresholds for BV1 and BV2A were 20 and 200 cfu/g of inoculated soil respectively (Priou et al. 2006).

3.1.1.3 Nucleic Acid-Based Techniques

Polymerase Chain Reaction-Based Assays

Polymerase chain reaction (PCR)-based assays may be employed either alone or in combination with other techniques which may precede or succeed PCR amplification. PCR assays based on amplification of species-specific genes such as 16S rRNA, 16S-23SS ITS regions and the nec1 gene governing virulence of *Streptomyces* spp. causing potato scab disease have been employed for the detection of bacterial pathogens. The DNA recovered from the soils usually contains potential inhibitors of PCR, such as humic acids rendering the measured values insufficiently reliable. Many procedures have been developed to remove/inactivate the PCR inhibitors from soil samples (Kuske et al. 1998; Miller 2001). However, removal of PCR inhibitors from soil samples completely is yet to be achieved and still remains problematic.

Ralstonia solanacearum (*Rs*) is soilborne and well distributed in tropical and subtropical countries. A DNA fragment (0.7 kb) amplified by the randomly amplified polymorphic DNA (RAPD) procedure from the total DNA extract of *Ralstonia solan-*

acearum (*Rs*) was cloned and evaluated as a specific DNA probe. This 0.7-kb DNA fragment hybridized to a 2.7-kb fragment in the *Eco*RI-digested total DNA of *Rs* and this fragment was sequenced. Specific oligonucleotide primers were designed based on the sequences of the 2.7-kb fragment for PCR amplification. The primers amplified the expected product from all strains of *Rs*, but not from any other plant pathogenic strains tested. The detection limit of the PCR assay was about 20 bacterial cells. This PCR format was successfully employed to detect *Rs* in the soil samples, using the specific primers developed in this investigation (Yung and Chichung 2000).

Although *R. solanacearum* (*Rs*) could be detected by immunofluorescence colony (IFC) staining method, it was considered necessary to confirm the results by PCR assay, especially when the *Rs* population was low in the samples. PCR amplification based on primer pair D2/B, allowed specific and rapid confirmation of *Rs* biovar 2 taken from IFC-positive colonies. Primers D2/B reacted with all *Rs* division 2 strains. The success rates of rapid verification of IFC-positive results by PCR assay were 86% and 96% with spiked and naturally contaminated soil samples respectively (van der Wolf et al. 2000). A nested PCR assay was successfully developed for the detection of *Rs* in soil, since the conventional PCR format was unable to detect *Rs* in any of the inoculated soil suspensions. However, after adopting overnight enrichment, positive results were obtained on broth that had been inoculated with soil suspension that contained at least 10^6 cfu/g of soil. The primer pair OLI-1/OLI-2 was employed in the first round amplification followed by the amplification of the PCR product by primer pair JE2/Y2 in the nested PCR format. Detection limit of the assay was 7.5×10 cfu/ml of soil suspension, when the suspension was incubated in the SMSA broth for 60 h. A minimum population of 10^6 cells/ml of soil suspension was required for positive detection by nested PCR without enrichment in SMSA broth. As few as 5×10^2 cfu *Rs*/g of soil were recovered from naturally infested soil by nested PCR assay, whereas the indirect ELISA procedure required a concentration of 10^6 cfu/g of soil (Pradhanang et al. 2000).

Two primer sets were designed based on the sequence of polymorphic bands that were derived from repetitive sequences based PCR (rep-PCR) fingerprinting. These primer pairs specifically detected *R. solanacearum* (*Rs*) race 4 strains that infect ginger, mioga and curcuma plants. One primer set AKIF-AKIR, amplified a single band of 165-bp from genomic DNA of isolates of *Rs* infecting mioga and curcuma, whereas another primer set 21F-21R amplified a 125-bp band from ginger isolates of *Rs*. The rep-PCR assay along with standard PCR was applied for the detection of *Rs* in soil samples. The detection limits of rep-PCR formats in pure culture and soil artificially inoculated with *Rs* were 2×10^2 cells/ml of cell suspension and 3×10^7 cfu/g of soil respectively (Horita et al. 2004).

A nested PCR assay was employed for the detection of *R. solanacearum* (*Rs*) in the soil. The primer pair OLI-1 and OLI-2 amplified a 410-bp from the *Rs* DNA in the first round and the primer pair Y2/JE2 produced a 220-bp amplicon in the nested PCR format. These primer pairs provided a very specific tool for the detection and differentiation of *Rs* strains in soil and water samples. This investigation emphasized the need for obtaining suitable DNA from the target pathogen for efficient detection of the pathogen concerned present in soil samples (Khakvar et al. 2008).

The effectiveness of BIO-PCR and direct plating method was evaluated for the detection of *R. solanacearum* race 1 strains in field soils of Hsinchu and Tainan counties, Republic of China. A total of 320 soil samples collected from eight fields were tested. The positive detection frequencies by growth on MSM-1 selective medium and BIO-PCR methods were 39.7% and 66.6% respectively with a 3-day enrichment period in MSM-1 broth followed by PCR detection using the primer pair AU 759/760. The BIO-PCR protocol could detect as few as 1.9 cfu/ml from pure suspensions or 17 cfu/g of infested soil from Asian Vegetable Research and Development Center (AVRDC), Taiwan (Fig. 3.1). No clear relationship could be seen between the positive detection and bacterial population present in the field. The spatial distribution of the pathogen was not even regardless of the presence or absence of the disease and the differences in the agroecosystems where the sampled fields were located (Table 3.1). The degree of unevenness was more striking, when tomato was absent from the field (Lin et al. 2009) [Appendix 1].

The members of the genus *Burkholderia* constitute an important group among soil microbial community. However, no direct method to detect and assess the diversity of this genus was available. Hence, a DNA-based PCR denaturing gradient gel electrophoresis (DGGE) method was developed for the detection and analysis of *Burkholderia* diversity in soil samples. Primers specific for the genus *Burkholderia* were developed based on the sequences of 16S rRNA gene.

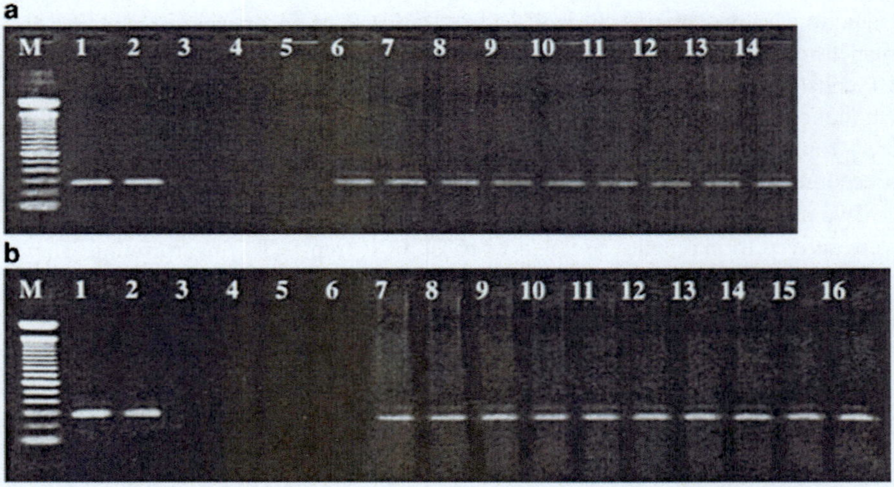

Fig. 3.1 Detection of *Ralstonia solanacearum* (*Rs*) by PCR in infested soil suspensions incubated in MSM-1 broth for 3 days. Note the presence of a 282-bp PCR product amplified using the primer pair AU759/AU760 on ethidium bromide-stained agarose gels **a** and **b**. M: 100 bp DNA ladder; Lane 1 and 2: positive control (pure Rs DNA from suspension of 10^8 cfu/ml); Lanes 3 and 4: negative control (water). Gel **a**: initial log population of *Rs* at 0.27 (lanes 5 and 6); 1.27 (lanes 7 and 8); 2.27 (lanes 9 and 10); 3.27 (lanes 11 and 12); 4.27 (lanes 13 and 14). Gel **b**: initial log population of *Rs* at 0.23 (lanes 5 and 6); 1.23 (lanes 7 and 8); 2.23 (lanes 9 and 10); 3.23 (lanes 11 and 12); 5.23 (lanes 15 and 16) (Courtesy of Lin et al. 2009 and with kind permission of Springer Science and Business Media, Heidelberg, Germany)

Table 3.1 Detection of *Ralstonia solanacearum* (*Rs*) in field soils collected from tomato production fields using MSM-1 medium and BIO-PCR method (Lin et al. 2009)

Field no.	Pathogen population[a]	MSM-1 plating[b]	BIO-PCR assay[b]
1	0.8–5.1	13/20	20/20
2	0.9–4.3	15/20	19/20
3	0.8–3.7	13/20	17/20
4	0.8–4.3	17/20	16/20
5	1.0–5.1	17/20	18/20
6	1.9–5.8	17/20	17/20
7	ND	0/20	8/20
8	0.8–5.0	13/20	18/20
9	ND	0/20	11/20
10	0.8–5.2	2/20	12/20
11	0.8–4.6	4/20	11/20
12	2.0–4.5	4/20	10/20
13	ND	0/20	4/20
14	0.8–2.0	2/20	12/20
15	1.4–1.5	7/20	12/20
16	1.1–2.1	3/20	8/20

[a]Range of means of four duplicate determinations of *Rs* population using selective MSM-1 medium

[b] Number of positive samples/total samples tested

ND Not determined

The primer system exhibited the required levels of specificity and sensitivity for the majority of established species of *Burkholderia.* DGGE analyses of the PCR products obtained showed that there were sufficient differences in the migration behavior to distinguish the majority of the 14 *Burkholderia* species tested. Sequence analysis of amplicons generated with soil DNA exclusively revealed sequences affiliated with sequences of *Burkholderia* species, demonstrating that the PCR-DGGE method is suitable for detecting and studying the diversity of *Burkholderia* spp. in natural settings (Salles et al. 2002).

Burkholderia (*Pseudomonas*) *cepacia* was reported to cause the onion slipper skin disease and later *B. cepacia* has emerged as a human pathogen, causing numerous outbreaks especially among patients suffering from cystic fibrosis (CF) disease (Holmes et al. 1978). The species composition of *Burkholderia cepacia* complex population naturally occurring in the maize rhizosphere was examined by using both culture-dependent and culture-independent methods. DNA extracted from root slurry was investigated by applying nested PCR to amplify *recA* genes with species-specific *B. cepacia* complex primers and to obtain a library of PCR-amplified *recA* genes. The culture-dependent method allowed assessment of greater diversity of *B. cepacia* complex population naturally occurring in the rhizosphere of maize plants under field conditions. But culture-independent methods detected additional species like *B. vietnamiensis.* Nonculture-based methods provide more reliable picture of the diversity of the *B. cepacia* complex population naturally occurring on the maize root system than culture-based methods (Pirone et al. 2005).

A novel real-time PCR technique using fluorescent quenching-based probes/primers was developed. This method requires only a single dye that interacts with a nuclease leading to the decrease in fluorescence intensity. Hence, they are cost-effective compared with other real-time PCR methods such as TaqMan PCR which needs two dyes (reporter and quencher dyes) for intramolecular fluorescence resonance energy transfer (FRET) (Crockett and Wittwer 2001). For the reliable, rapid and precise detection of *Streptomyces* spp. causing potato scab disease and residing in soil, a new competitive quenching probe PCR (QCQP-PCR) was developed. The virulence gene of pathogenic *Streptomyces* spp. *nec1* was selected as the target for QCQP-PCR assay. A specific primer set to amplify *nec1* gene and a fluorescently labeled probe that hybridized specifically with *nec1* amplicon were designed. An internal standard DNA (ISDNA) that was identical to the *nec1* amplicon, but has a 4-base mismatch in the probe-hybridizing region was included. In addition, a fluorescently labeled probe IS which specifically hybridized with ISDNA at the mutaginized region was also synthesized by PCR amplification. The target *nec1* gene was co-amplified with the known copy number of ISDNA by PCR using the same primer set in the presence of the specific probes. The PCR products were monitored in real-time by measuring the fluorescence intensity (quenching) of each probe. The initial amount of the *nec1* gene was quantified based on the ratio of the PCR products of the same PCR cycle. QCQP-PCR format could be used to accurately quantify *nec1* gene, even in the presence of PCR inhibitors in soil samples tested. The lower limit of quantification was 20 copies per tube which corresponded to 1,500 copies per g of dry soil. The results of quantification of pathogen population could be obtained within 5 h, indicating the rapidity and sensitivity of the protocol developed in this investigation that may be useful for monitoring pathogenic *Streptomyces* species in soil (Manome et al. 2008).

The phytotoxin thaxtomin produced by *Streptomyces* spp. has been shown to be a pathogenicity determinant. Genes encoding thaxtomin synthetase (*txtAB*) are located in a pathogencity island characteristic of genetically diverse plant pathogenic *Streptomyces* spp. The SYBR Green quantitative real-time PCR format using primers designed to anneal to the *txtAB* operon of *Streptomyces* was developed to quantify the pathogenic bacterial populations present in the soil samples. The assay specific for pathogenic *Streptomyces* strains had a detection limit of 10 fg of the target DNA or one genome equivalent. A standard curve was generated to quantify pathogenic *Streptomyces* in soil samples. Based on the standard curve, the numbers of pathogenic *Streptomyces* CFUs were extrapolated to range from 10^3 to 10^6/g of soil from potato fields where the disease incidence was recorded earlier. This real-time PCR protocol was found to be rapid, accurate and cost-effective to obtain quantitative estimates of the pathogenic *Streptomyces* strains present the soils (Qu et al. 2008).

3.1.2 Detection of Bacterial Pathogens in Water

Bacterial pathogens in or on the plant organs may be washed off by rain water or by water from sprinkler irrigation system and reach the soil. They may be carried to other parts of the same or different fields by rain water or irrigation water. The pathogen movement within or outside the field is largely aided by water. In a few cases river water may be involved in the dissemination of bacterial pathogens. Bacterial pathogens like *Erwinia* spp. infecting potato tubers may be carried in water from washing plants (Wicks et al. 2007). Presence of bacterial pathogens may be detected by employing different procedures.

3.1.2.1 Biological Methods

Phytopathogenic bacteria present in water may be isolated by dilution plating procedure using suitable media that will differentially favor the development of the target bacterial species. Many factors such as the population of the bacterial species, presence of antagonists and saprophytes may affect the isolation of bacterial pathogens concerned. Detection of bacterial pathogens in water has been accomplished by employing bacteriophages that are pathogenic to the bacterial species specifically. *Xanthomonas campestris* pv. *oryzae* (*Xco*) was detected in irrigation water by using the bacteriophages OP1 (Wakkimoto 1957) Xf (Kuo et al. 1967) that exhibited specific pathogenicity on Xco.

The presence of *Ralstonia solanacearum* (*Rs*), causative agent of bacterial wilt disease affecting several crops like tomato and potato has been detected in waterways. *R. solanacearum* biovar 2 has been shown to have the ability to survive in waterways at low temperatures in the UK and the Netherlands (Elphinstone et al. 1998; Janse and Schans 1998) and later in several European countries. Survival of *Rs* cells in water for variable periods depended on the temperature and inoculum density. The pathogen was recovered at low densities (10–80 cfu/ml) by direct plating on modified SMSA agar from water samples at 14°C or higher. Waterways can be major dissemination routes of *Rs* which is able to survive for long periods in sterilized water, but its survival in natural water in the presence of other microorganisms was not clearly understood. The fate of a Spanish strain of *Rs* inoculated in water microcosms from Spanish river containing different micobiota fractions at 24°C and 14°C was studied by plating the samples in YPGA and SMSA media. The presence of lytic phages and protozoa had adverse effects on the populations of *Rs*. In water microcosm, the temperature of 14°C was more favorable for the survival of *Rs* than at 24°C, since biotic interactions were slower at lower temperatures (Álvarez et al. 2007). Surveys for three seasons of irrigation, drainage and artesian well water were undertaken throughout the major potato-growing areas of Egypt to assess the population levels of *R. solanacearum* bv. 2 race 3 (phylotype II sequevar) causative agent of potato brown rot disease. The pathogen was limited to the canals of traditional potato-growing areas in the Nile Delta region. The pathogen populations in the canals of Delta (~100–200 cfu/l) were generally variable throughout the year with presence limited to potato cultivation in the immediate vicinity. In vitro experiments showed that the pathogen survival in the canal water depended on temperature and microbial activity. The results suggested the potential for long-distance spread of the pathogen in Egyptian surface waters from sources of contamination (Tomlinson et al. 2009).

The presence of soft rot pathogen *Erwinia* spp. infecting potato tubers was detected in wash water used for washing tubers at various sites in the washing plants and from ponded recycled water. The water samples, after serial dilutions, were plated on crystal violet pectate medium kept in petri dishes (Hyman et al. 2001). The pathogen population and soft rot disease incidence and severity were determined. Concentrations of *Erwinia* spp. around 10^4 cfu/ml and occasionally 10^6 cfu/ml were present in wash water at various sites. The most severe amount of rotting developed when tubers were immersed in water containing *Erwinia* spp. at concentration of 10^4 cfu/ml or greater. The results suggested that potato wash water should be replaced frequently with clean water to reduce the bacterial population and consequently to reduce the disease incidence (Wicks et al. 2007).

3.1.2.2 Immunoassays

The enzyme-linked immunosorbent assay (ELISA) has been shown to be efficient in detecting bacterial pathogens in water samples. *Ralstonia solanacearum* (*Rs*) was recovered from river waters by an enrichment step. The modified Wilbrink broth (MWB) was more effective than the modified SMSA broth for enrichment of *Rs* which was detected by indirect ELISA test. Enrichment in MWB at 29°C and 35°C allowed recovery of *Rs* cells that were unculturable on solid media. When this technique was applied for the detection of *Rs* during the cold months, best detection results were obtained after enrichment at 35°C with MWB. The enrichment protocol was combined with indirect DAS-ELISA and validated by cooperative (Co)-PCR assay to detect both naturally and artificially starved and cold-stressed cells of *Rs* in water which remained infective after treatments (Caruso et al. 2005).

3.1.2.3 Nucleic Acid-Based Methods

Potato and tomato crops may be infected by soilborne bacterial pathogens such as *Ralstonia solanacerum* (*Rs*) through the use of contaminated surface water for irrigation. Concentrations of *Rs* in surface water were found to vary between 10^3 and 10^6 cfu/l. Rapid and reliable detection of *Rs* in water was considered to be an effective approach to avoid the transfer of the pathogen to potato plants via irrigation (Elphinstone 1996; Wennecker et al. 1999). AmpliDet RNA is a sensitive procedure based on nucleic acid sequence-based amplification (NASBA) of RNA sequences and homogeneous real-time detection of NASBA amplicons with a molecular beacon. This procedure performed in sealed tubes, reduces the risks for carry over contamination that is usually associated with conventional PCR assay. The RNA was extracted from 200 ml of contaminated water samples, after removing coarse particles by filtration, followed by centrifugation to concentrate the bacteria by about 200 times. AmpliDet RNA protocol could detect *Rs* at a concentration of 10 cfu/ml in surface water and 1 cfu/ml in demineralized water (Table 3.2). The ability to detect and quantify the viable cells of *Rs* is the distinct advantage of AmpliDet RNA technique over other real-time

Table 3.2 Detection of *Ralstonia solanacearum* in contaminated and demineralized water by AmpliDet RNA technique (van der Wolf et al. 2004)

	Contaminated water		Demineralized water	
Cells per reaction	Average Rn value	S.D.	Average Rn value	S.D.
10,000[a]	*8.88*	2.57	*13.91*	3.13
1,000	*5.61*	1.70	*9.74*	4.72
100	2.52	1.72	*5.96*	4.04
10	1.54	0.76	1.96	0.63
0	1.60	0.85	0.94	0.62

Italic figures represent average values significantly positive ($p \geq 0.005$)
[a]Bacterial cells from 200 ml concentrated into 1 ml by centrifugation

Table 3.3 Detection of *Ralstonia solanacearum* in water samples from tomato production fields (Lin et al. 2009)

Technique/field no.	Entrance point	Standing water	Exit point
Plating on MSM-1 Medium			
1	0/5[a]	0/5[a]	1/5[a]
2	0/5	3/5	2/5
3	0/5	NT	2/5
4	0/5	NT	NT
5	2/5	4/5	4/5
6	0/5	NT	0/5
7	0/5	NT	0/5
8	0/5	NT	NT
BIO-PCR			
1	0/5	0/5	1/5
2	0/5	3/5	3/5
3	0/5	NT	2/5
4	0/5	NT	NT
5	1/5	2/5	3/5
6	4/5	NT	2/5
7	2/5	NT	4/5
8	0/5	NT	NT

[a] Number of positive detection/total number tested
NT Not tested

PCR methods (van der Wolf et al. 2004). *R. solanacearum* was detected by employing dilution plating on SMS-1 selective medium and BIO-PCR assay in irrigation water, standing water in the field planted with tomato and drainage water at the exit point of three tomato fields. Presence of *Rs* was monitored in water samples collected from the field (WL1) over several seasons. The frequency of positive detection by BIO-PCR was the highest in water samples collected from drainage water followed by standing water and the lowest was in irrigation water. The number of positive detection among 85 water samples was 18 (21.2%) and 27 (31.8%) by plating on MSM-1 medium and BIO-PCR procedures respectively (Table 3.3) (Lin et al. 2009).

3.1.3 Detection of Bacterial Pathogens in Air

Bacterial pathogens do not produce any kind of spores similar to that of fungal spores that can be dispersed by wind. But the combined action of wind and rain could be important for dispersal of the bacterial cells. Dynamics of dispersal of *Xanthomonas axonopodis* pv. *citri* (*Xac*) have been studied in some detail. When *Xac* lesions are wetted, slimy mass of bacterial cells begins to exude within a few minutes. This inoculum can be dispersed in gentle rain, rain with wind or rain storms, tropical storms and hurricanes. These weather factors become progressively more effective at dispersing the inoculum over long distances (Gottwald and Irey 2007). Dynamics of dispersal of *Xac* were assessed in simulated wind-driven rain splash. The wind-driven splash events were simulated using electric blowers to generate turbulent wind (15–20 m/s) and sprayer nozzles to produce water droplets entrained in the wind flow. The splash was blown at an inoculum source of canker-infected trees one meter downward. The splash downwind of the source of the infected trees was collected by vertical panel samplers and funnel samplers (Fig. 3.2). The number of bacterial cells collected declined with increase in time and distance from the source of infected citrus plants. Citrus canker was readily dispersed in wind-driven rain in large quantities immediately, after the stimulus occurred upon which wind-driven splash could disperse inoculum over a long period and over a substantial distance (Bock et al. 2005).

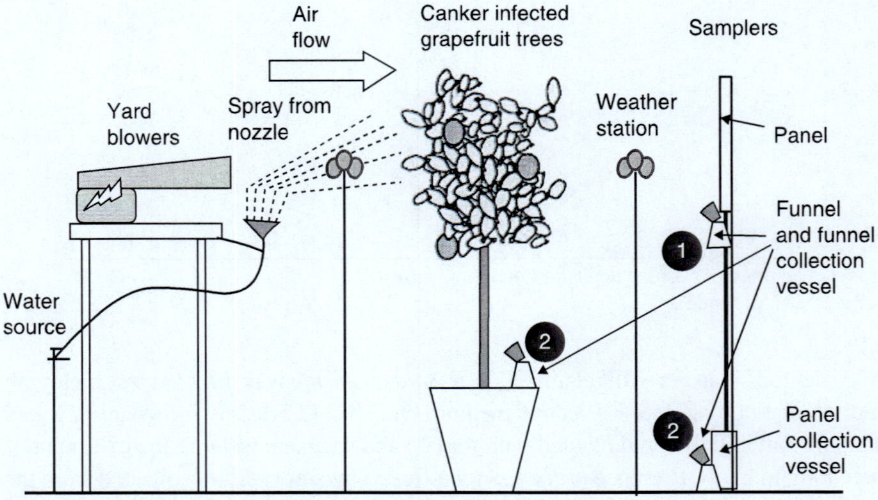

Fig. 3.2 Assessment of wind/rain-dispersed population of *Xanthomonas axonopodis* pv. *citri* (*Xac*) under simulated conditions using electric blowers and sprayer nozzles. Funnels are placed at position 1 for 52-h duration-of-dispersal experiments (1 m from inoculum source) and at position 2 for the 4-h duration-of-dispersal and distance-of-dispersal experiments (10, 4 and 6 m from inoculum source) (Courtesy of Bock et al. 2005; The American Phytopathological Soceity, MN, USA)

3.1.4 Detection of Bacterial Pathogens in Alternative Host Plants

The alternative host plant species of bacterial pathogens play a vital role for their survival in the absence of primary host plant species. They form a link between two crop seasons for the carryover of the bacterial pathogens. The range of plant species infected by bacterial pathogens may vary depending on their pathogenic potential. Techniques similar to those applied for the detection in their primary host plant species may be adopted for the detection of these pathogens in the alternative host plant species also. Depending on the level of sensitivity and specificity required and time available for obtaining results, conventional or modern molecular techniques may be chosen. It is well known that the infected plants, either primary or alternative hosts help build up of the inoculum. Hence, removal and proper disposal of infected plants and debris is important to avoid the chances of newly planted crops getting infected in the early growth stages. In addition, certain weed plant species like *Solanum dulcamera* growing along water courses, may serve as sources of inoculum for *Ralstonia solanacearum.* These plants allow the build up of pathogen population in the roots and the pathogen cells may be leached out from infected roots into the river water (Álvarez et al. 2007).

Potato haulms, infected tubers and wilted and volunteer plants have to be rogued to reduce the sources of inoculum and to avoid contamination of neighboring healthy potato tubers by *Ralstonia solanacearum* causing brown rot disease. The presence of *R. solanacearum* in asymptomatic tomato and pepper plants collected from fields with a history of tomato bacterial wilt disease was detected by immunocapture (IC)-PCR assay. This investigation indicated the possible role of latent infection of crop plants and weed plants in the survival of *R. solanacearum* between copping cycles (Dittapongpitch and Surat 2003).

R. solanacearum (*Rs*) is known to infect a wide range of plant species which can serve as reservoir of inoculum for the crop plant species. The alternative host plants, when remain symptomless, are more dangerous sources of infection. Asymptomatic plants of 13 weed species from tomato production fields in Taiwan were examined by employing BIO-PCR assay. A high positive detection of *Rs* in the root samples as well as in the rhizosphere soil samples was accomplished using BIO-PCR assay, compared to plating in MSM-1 selective medium. *R. solanacearum* was detected by BIO-PCR assay from roots of seven weed species compared to six species by MSM-1 plating method. Roots of *Ludwigia prostrata* (Onagraceae) harbored maximum population of *R. solanacearum* (Table 3.4) (Lin et al. 2009).

Xylella fastidiosa (*Xf*) is responsible for several diseases of crop plants including almond leaf scorch, phony peach, citrus variegated chlorosis and grapevine Pierce's disease. In order to assess the role of additional host plants of *Xylella fastidiosa* (*Xf*), in addition to grapevine, 96 native and ornamental plants from within or around the vineyards were examined for the presence of *Xf* first by ELISA test and confirmed by the PCR assay, using *Xylella*-specific primers RST31 and RST33. The plant species that harbored *Xf* included Yaupon holly (*Ilex vomitoria*), Mustang grape (*Vitis mustangensis*), American sycamore (*Platanus occidentalis*), southern dewberry

Table 3.4 Detection of *Ralstonia solanacearum* in the roots of weed species by BIO-PCR assay and plating on MSM-1 medium (Lin et al. 2009)

Weed species	MSM-1	BIO-PCR
Ageratum conyzoides	1/5[a]	1/5[a]
A. houstonianum	1/11	4/11
Eleucine indica	3/9	4/9
Ludwigia prostrata	4/7	6/7
Polygonum lapathifolium	2/6	2/6
Solanum nigrum	0/6	2/6

[a] Number of positive samples/total number tested

(*Rubus trivialis*), frogfruit (*Lippia nodiflora*) and Japaneses honeysuckle (*Lonicera japonica*). The first four plant species constituted a high percentage of total plant biomass around the vineyards under investigation and they could be the potential reservoirs of inoculum for infection of grapevines (Buzombo et al. 2006).

In another investigation native or ornamental plant species present around south Texas were tested by employing antibody- and nucleic acid-based techniques to identify the potential reservoirs of *X. fastidiosa* (*Xf*). For all plant species testing positive for *Xf* by ELISA, DNA from stem and petiole samples were extracted and the presence of *Xf* was tested by applying standard PCR, and quantitative real-time PCR (QRT-PCR) procedures. ELISA test detected *Xf* in 19 species belonging to 12 plant families. Of the 19 plant species tested ELISA-positive, 11 species were also positive for standard PCR and both QRT-PCR formats. All three wild vine species *Ampelopsis arborea*, *Vitis mustangensis* and *V. rotundifola* and two samples of *Tatibida columnifera* and two samples of *Magnolia grandiflora* tested positive for two or more diagnostic tests. This investigation revealed the existence of several additional host plant species that can serve as sources of inoculum of *X. fastidiosa* to perpetuate the pathogen in the perennial crops like grapevine (McGaha et al. 2007).

3.1.5 Detection of Bacterial Pathogens in Vector Insects

Dissemination through insects is important for bacterial pathogen to spread, but most difficult route that can be contained. Movement of bacterial pathogens through natural vector insects may occur across different countries and continents, though separated by natural barriers, making the quarantine and other restrictive measures ineffective. Several approaches have been made to detect the presence of bacterial pathogens in the insect tissues and to assess population of insects that can be potential vectors.

3.1.5.1 Biological Methods

The insect species suspected to carry bacterial pathogen(s) may be given access to infected plants for a definite period of time or insect species collected from the fields where there is high disease pressure may be tested on healthy plants for their ability to transmit the pathogen under investigation. These insects either in groups or individually may be allowed to feed on the highly susceptible host plant species which should be kept under observation in insect-proof glasshouse. Development of symptoms of the disease in question, will indicate the ability of the insect species to transmit the pathogen(s). This method requires a long time and large glasshouse space, in addition to being labor intensive, cumbersome, yielding sometimes inconclusive results.

3.1.5.2 Immunoassays

Immunoassays may be employed for the detection of bacterial pathogens in the insect vectors and estimation of their seasonal dynamics. *Pantoea stewartii* subsp. *stewartii*, causative agent of Stewart's wilt disease in corn, was successfully detected in the vector, corn beetles *Chaetocnema pulicaria* by employing ELISA test (Khan et al. 1997). The proportion of the beetles *Acalyamma vittata* harboring *Erwinia tracheiphila*, causing bacterial wilt of cucurbits was estimated. Further, the possibility of *A. vittata* serving as the primary overwintering reservoir of *E. tracheiphila* was indicated by this study. The bacterial pathogen might infect the cucurbits when the overwintered beetles feed on the healthy plants in the next season (Fleischer et al. 1999). A high percentage of individual beetles (60–70%), after feeding on infected plants for 24 h, tested positive for *E. tracheiphila* antigen by double antibody sandwich (DAS)-ELISA procedure. The beetles remained inoculative for about 30 days after acquiring the pathogen from infected source plants. (García-Salazar et al. 2000).

Pierce's disease (PD) has been found to be a chronic problem in many viticultural areas in California. *Xylella fastidiosa*, the causative agent, is spread by glassy winged sharpshooter (GWSS) aggravating this problem. The extracts of GWSS were tested by ELISA procedure. The total number of *Homalodisca coagulata* captured in vineyards increased in the month of June significantly. With the increase in the number of *H. coagulata* captured, there was a concomitant increase in the ELISA-positive insects also, some of them with readings threefold greater than the positive threshold level. Another smaller sharpshooter species *Oncometopia orbona* was present fairly in large numbers in June and several of these insects were also ELISA-positive. This investigation indicated the need for monitoring the vector population of sharpshooters capable of spreading the bacterial pathogens to uninfected plants in the same field, as well as to the crops that are far away from the vineyard concerned (Buzombo et al. 2006).

3.1.5.3 Nucleic Acid-Based Techniques

Nucleic Acid Hybridization Techniques

Fluorescence in situ hybridization (FISH) technique combines both specificity, adjustable to different phylogenetic levels with appropriate nucleotide probes and detection and enumeration without prior isolation and cultivation of the target bacterial pathogen. For the detection of *Xylella fastidiosa* that threatens grapevine and citrus-based industries, two fluorescent oligonucleotide probes complimentary to different regions of the 16S rRNA gene of the pathogen were developed. The probes S-S-X-fas-0067-a-A-18 and S-S-X-fas-1439-a-A-18 were shown to be specific by using them in FISH protocol for hybridization with *Xylella* isolates, 15 closely related bacteria and three plant endophytes. These probes were employed to detect and quantify *X. fastidiosa* cells in honey dew ($2.2 \pm 0.2 \times 10^4$ cells/ml) collected from the sharpshooter vector *Bucephalgonia xanthophis* during the acquisition access period on infected plant (Fig. 3.3). As the pathogenic bacterial cells may be detected also in nonvector insects, it is necessary to prove the ability of insects to transmit the pathogen to healthy plants (Rodrigues et al. 2006).

Polymerase Chain Reaction

Use of polymerase chain reaction (PCR) assay for the detection of *Xylella fastidiosa* (*Xf*), a xylem-limited bacteria (XLB), in insect vector was difficult because of lack of quick and easy DNA extraction protocols and the presence of PCR inhibitors in the insect tissues. Citrus variegated chlorosis (CVC), caused by *Xylella fastidiosa* is transmitted by several insect species. The immunomagnetic separation (IMS) procedure in combination with nested PCR format was employed for the detection of *Xf* in the leafhopper species *Graphocephala coccinea*, *G. versuta*, *Erythroneura* spp. and *Typhlocybia* spp. These insect species were identified as the vectors of bacterial leaf scorch disease of American elms (Pooler et al. 1997). The IMS-PCR

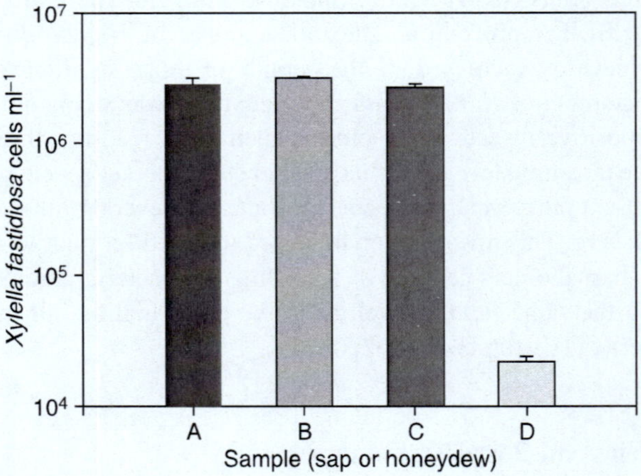

Fig. 3.3 Detection and quantification of *Xylella fastidiosa* (*Xf*) cells in citrus from three different orchards and honey dew of insect vectors by FISH technique employing species-specific probes.A: Gavião Peixoto; B: Neves Paulista; C: Paraíso and D: Bucephalogonia xanthophis. Standard error is indicated by error bar (Courtesy of Rodrigues et al. 2006; Society for Applied Microbiology/Blackwell Publishing Ltd, Oxford, UK)

technique requires the use of *Xf*-specific antibody, specific IgG-coated magnetic beads and a magnetic capture stand which increases the cost of tests. In addition, this protocol takes more than 1 day to yield the results. *Bucephalogonia xanthopis, Dilobopterus costalimai, Parathona gratiosa and Acerogonia citrina* were tested after feeding for 2 days on citrus infected with *Xf.* A multiplex PCR employing primers based on the sequences of 16S rRNA and *gyrB* genes was applied for the detection of *Xf* in field-collected vector insects. The assay yielded positive results only after a second round of PCR amplification. No PCR amplification occurred when DNA extracted from healthy laboratory-reared insects fed on noninfected plants was used as a template (Rodrgues et al. 2003).

Xylella fastidiosa (*Xf*) causes serious diseases in several crops including grapevine and citrus. *X. fastidiosa* has been detected in 39 species of sharpshooters collected from different plant species (Redak et al. 2004). However, the inoculativity of the insect species in which *Xf* was detected was not proved to determine their role in disease spread. In a later study, primers designed based on the sequences of 16s rRNA gene were employed for the detection of *Xf* in the honeydew of the sharpshooter insect *Bucephalogonia xanthophis* that had been fed on diseased citrus plants. Positive results were obtained only after two rounds of PCR amplification. An expected product of 1,348-bp was amplified. Partial sequencing of the PCR products resulted in 99% identity to the *Xf* 16S rRNA gene, confirming the presence of the pathogen in insect honeydew. No PCR amplification occurred in the honeydew samples from insects fed on the healthy laboratory-grown plants (Rodrigues et al. 2006). *X. fastidiosa* has been detected in 39 species of sharpshooters collected from different plant species (Redak et al. 2004).

A rapid, technically easy and convenient method was developed for efficient detection of *X. fastidiosa* (*Xf*) in potential insect vector species. This procedure combines the commercially available DNeasy tissue kit for the extraction of DNA from insect tissues with a one-step PCR protocol using *Xf*-specific 16S set C primers. Organic solvents, precipitation with ethanol, *Xf*-specific antibody and nested PCR are not required. Further, no additional purification or enrichment steps are needed and the results can be obtained in less than a day. Among the five species of leafhoppers species, *Graphocephala versuta* was selected, because of its abundance and positive detection of *Xf* in this insect. The nested PCR using primer set 272 and one-step PCR using 16S set C primers detected *Xf* respectively in 15 and 14 out of 30 *G. versuta* from which DNA was extracted by DNeasy tissue kit, but not by FastDNA kit. The presence of *Xf* was also detected in six of 13 of *Entilia concisa* (treehopper) and two of five of *Graphocephala coccinea*, but not other leafhoppers. DNA sequencing of selected PCR products cloned from nested PCR and one-step PCR assays confirmed that they were most similar to a partial 16S DNA or a 472-bp sequence of *X. fastidiosa* in GenBank. The PCR amplicons of 472 and 620-bp were detected by nested PCR and one-step PCR assays respectively. The one-step PCR could be employed for the detection of *Xf* in infected oleander plants also (Huang et al. 2006).

In another study, the glassy-winged sharpshooter (GWSS) *Homalodisca coagulata* was found to be a vector capable of transmitting *X. fastidiosa* to citrus plants which develop symptoms of citrus variegated chlorosis (CVC) disease. By employing real-time PCR SYBR Green I, *Xf* was detected in *H. coagulata*. The transmissibility of CVC strain of *Xf* by GWSS was tested by PCR assay and membrane entrapment immunofluorescence (MEIF) test after inoculating Madam Vinous sweet orange plants. Of the 16 inoculated plants, the presence of *Xf* was detected by PCR and MEIF analysis. The cibaria of 27 sharpshooters that had fed on infected plants revealed the presence of bacterial populations abundantly, when examined under scanning electron microscope (Damsteegt et al. 2006).

Detection of *X. fastidiosa* in vector insects was found to be difficult because of low concentration of *Xf* and the presence of PCR inhibitors in the insect tissues. A dual purpose standard PCR and quantitative PCR (TaqMan™) system was developed for generic detection of *Xf* strains. Primers HL5 and HL6 designed to amplify a unique region common to the sequenced genomes of four *Xylella* strains, amplified a 221-bp fragment from strains associated with Pierce's disease of grapes, almond leaf scorch and oleander leaf scorch diseases and from DNA from an *Xf* strain associated with citrus variegated chlorosis disease. The DNA extracted from transmission assayed- and field-collected sharpshooters (GWSS) was analyzed by real-time and standard PCR assays. The amount of bacteria per sample was estimated to be between 10 and 100 cells per whole insect, based on the Ct values. This bacterial population level was below the limit of detection by ELISA test. No amplification occurred from the DNA of endosymbiotic bacteria isolated from GWSS, indicating the specificity of the real-time quantitative PCR format in detecting *X. fastidiosa* (Francis et al. 2006).

Of the mechanisms of transmission of *X. fastidiosa* viz., naturally occurring graft transmission, mechanical transmission by pruning shears, transmission through the nymphs and adults of *Homalodisca liturata* (smoke tree sharpshooter) and adults of *Diceroprocta apache* (apache cicada), transmission by the insects is widespread and frequent. The insects were given access to infected plants to acquire the pathogen and then to healthy plants to transmit it. The presence of *Xf* in insects and plants was tested by PCR and ELISA tests respectively [Appendix 2]. In the test of nymphal transmission, three of 24 smoke tree sharpshooter nymphs transmitted *Xf* and one of the nymphs transmitted *Xf* to two different plants. Eight of nine nymphs were *Xf*-positive by PCR assay. None of the nymphs transmitted the pathogen after molting. Thirteen of the adult *H. liturata* insects tested PCR-positive for *Xf* and they were able to transmit *Xf* to test plants which tested positive for ELISA test. The presence of the pathogen could not be detected by PCR assay in these insects, possibly because the pathogen concentration required for positive transmission by the vector might be less than the threshold of detection by PCR assay. This investigation appears to be the first record of the transmission of *Xf* by nymphs of *H. liturata* and adults of *D. apache* (Krell et al. 2007).

Citrus huanglongbing (HLB or citrus greening) disease caused by '*Candiatus* Liberibacter' spp. has been steadily spreading to new areas in Texas and Mexico states, posing serious threat to citrus production. The pathogens could not be detected consistently in the infected plants consistently, presumably because of low concentration and irregular distribution in the plants. The bacterial species are spread by the psyllid vector *Diaphorina citri*. TaqMan-based real-time quantitative PCR assay was developed for the detection of '*Ca. L. asiaticus*' in *D. citri*. Psyllid adults and nymphs (over 1,200 samples) collected from various locations in Florida and HLB-symptomatic and asymptomatic trees at different times of the year were tested by the TaqMan QPCR procedure to monitor the incidence and spread of HLB disease. The results revealed the presence of HLB agent in the vectors which could inoculate healthy citrus plants on which symptoms might appear after several years. The spread of '*Ca. L. asiaticus*' in an area might be determined by employing efficient techniques. The role of discount garden centers and retail nurseries might play a significant role in the widespread distribution of psyllids and plants carrying HLB pathogens in Florida, as suggested by the results of this investigation (Manjunath et al. 2008).

3.2 Detection of Phytoplasmal Pathogens in the Environment

Phytoplasmal pathogens have not yet been successfully cultured on cell-free media with the techniques currently available. Hence, their existence outside living cells appears to be impossible and the chances of their presence in the soil, water or air are remote. The phytoplasmas may infect several plant species that may serve as sources of inoculum. They are spread from infected plants to healthy plants by the agency of insect vectors. However, the phytoplasmas may spread through infected propagative plant materials that are transported by growers from one location to another. Dissemination of phytoplasmas through planting materials has been discussed earlier.

3.2.1 Detection of Phytoplasmal Pathogens in Alternative Host Plants

3.2.1.1 Nucleic Acid-Based Techniques

Detection and identification of phytoplasma may be achieved by transmitting the phytoplasma under investigation to a convenient experimental plant species that reacts with the characteristic symptoms. The Italian clover phyllody phytoplasma (ICPh) was transmitted using dodder (*Cuscuta campestris*) from its original host *Chrysanthemum leucanthimum* to periwinkle (*Catharanthus roseus*) and then from periwinkle to *Chrysanthemum carinatum*. The ICPh was detected in *C. carinatum* by employing 20- to 24-mer oligonucleotides originally designed as PCR primers, as hybridization probes for non-radioactive detection of the phytoplasma. The Cy5-labeled oligonucleotide probes that hybridized to phytoplasmas present in plant tissues were visualized by confocal microscopy. By incorporating a tyramide signal-amplification procedure into the biotin or digoxigenin system, signal enhancement was observed in plant tissues. Presence of ICPh was detected by in situ hybridization of paraffin-embedded tissue sections. The phytoplasmas were restricted to phloem cells of the host plant species tested. The results were corroborated by observations under electron microscope. Short oligonucleotide-based in situ hybridization technique is capable of discriminating between targets with high sequence similarity (Webb et al. 1999).

Weed species exhibiting chlorosis and stunting symptoms belonging to 14 different taxonomic groups present in vineyards were indexed for the presence of phytoplasmas infecting grapevines. Nested PCR assays using primers specific for the phytoplasma 16S rDNA gene showed three of six *Calendula arvensis*, one of two *Solanum nigrum* and one of seven *Chenopodium* spp. tested positive. The RFLP analyses and sequencing of amplified 16S rDNA fragments identified a putative phytoplasma in the ribosomal subgroup 16Sr II-E. The presence of the stolbur phytoplasma causing "Bois noir" disease of grapevine was not detected by PCR assay in any of the weed species indexed (Tolu et al. 2006).

Ornamental and weed plant species exhibiting symptoms suggestive of phytoplasma infection were examined. PCR assay using universal phytoplasma primer pairs P1/P7 and R16F2n/R16F2 that directed amplification of rDNA sequences was applied for the detection of phytoplasmas in these weed species. All infected plants were PCR-positive. The phytoplasmas were characterized and differentiated by RFLP and sequence analysis of PCR-amplified rDNA. The phytoplasmas detected in *Asclepias curassavica* and *Celosia argentea* were identified as members of Clover proliferation phytoplasma group (16SrIV group). The phytoplasma detected in *Limonium sinuatum* showed restriction profiles identical to subgroup 16SrI-C. Phytoplasmas infecting *Gomphocarpus physocarpus*, *Tanatacetum partenium*, *Lactuca serriola*, *Tagetes patula* and *Coreopsis lanceolata* were identified as members of the subgroup 16SrI-B. *Catharanthus roseus* and *Rudbeckia hirta* were infected by the phytoplasmas belonging to subgroup 16srI-A as indicated by restriction profiles (Babaie et al. 2007).

The incidence of a new Zebra complex (ZC) disease affecting potato tubers was observed in the past few years and the association of '*Candidatus* Liberibacter' spp. with this disease was consistently demonstrated by a real-time PCR assay. A multiplex PCR assay using '*Ca. L. solanacearum*'-specific primers and primers specific for the ß-tubulin DNA regions from potato was developed to detect the pathogen in different solanaceous plant species. The ZC-associated bacterium was detected in silver leaf nightshade (*Solanum elaeagnifolium*), wolfberry (*Lycium barbarum*), black nightshade (*S. ptychanthum*) and jalpeno pepper (*Capsicum annuum*). 'Ca. Liberibacter' species detected in all samples divided into two clusters in all samples, sharing similarity of 98.8% in their partial 16S rRNA gene sequences and 99.3% in their partial intergenic spacer region (ISR)-23S rRNA gene sequences. The results indicated that these additional host plant species might have a role in the pathogen survival by serving sources of inoculum in the absence of potato crops (Wen et al. 2009).

3.2.2 Detection of Phytoplasmal Pathogens in Insect Vectors

Transmission through the insect vectors appears to be the only natural mode of transmission of phytoplasmas from infected plants to healthy plants. The insects that acquire the phytoplasmas after feeding on the infected plants for a definite period of time are able to transmit them to the healthy plant, after the completion of variable incubation period the length of which varies depending on the phytoplasma concerned. The phytoplasmas have to pass through different organs of the insect that has acquired the target phytoplasma from the infected plants and reach the salivary glands and then they are ejaculated into the healthy plants along with the saliva.

3.2.2.1 Immunoassays

Polyclonal antibodies (PAbs) were generated in the rabbits against the partially purified phytoplasma from grapevine Flavescence dorée (FD)-infected plant extracts. The PAbs were used in direct immunofluorescence staining technique for the detection of FD in the excised salivary glands in situ of the experimental leafhopper vector *Euscelidius variegatus*. The salivary glands were examined from the 4th to 10th week after the commencement of phytoplasma acquisition from infected plants. Detection of FD phytoplasma by immunofluorescence staining technique was comparable to ELISA test applied to insect corpses after removal of salivary glands (Lherminier et al. 1989).

3.2.2.2 Nucleic Acid-Based Assays

Nucleic Acid Hybridization Technique

Short nucleotides nonradioactively labeled (with biotin/digoxigenin) and designed as specific primers could be employed as probes for hybridization of phytoplasma in insect tissues. Biotinylated cloned DNA probes were employed for detection of aster yellows (AY) phytoplasma in the leafhopper vector *Macrosteles fascifrons* (Davis et al. 1992). Italian clover phyllody phytoplasma (ICPh) was detected in the leafhopper *Euscelidius variegatus* using oligonucleotide (20–24-mer) probes 5′ end labeled with either Cy5 fluorochrome, biotin or digoxigenin. Labeled probes were detected in insect tissue using a chromogenic alkaline phosphatase-nitroblue tetrazolium chloride/5 bromo-4-chloro-3-indolyl-phosphate reaction. The presence of the ICPh phytoplasma in the salivary glands and midgut was detected by the in situ hybridization technique (Webb et al. 1999).

Phytoplasma associated with bunchy top disease (BTD) of papaya occurring in Cuba was detected by a nonradioactive nucleic acid hybridization (nrNAH) assay. The 16S rDNA products amplified by PCR were purified, labeled with alkaline phosphatase and employed as a probe for the detection of the phytoplasma by the system of direct alkaline phosphatase labeling and chemiluminescent detection (Alkphos, Amersham, UK). The BTD phytoplasma was detected in 94% of the leafhopper *Empoasca papayae* collected from BTD affected papaya fields indicating that *E. papayae* was a potential vector capable of transmitting the disease to healthy plants. The probe yielded hybridization signals reacting with reference controls and insects carrying the BTD phytoplasma. The phytoplasma belonged to the 16SrII, '*Candidatus* Phytoplasma aurantifolia'. The nrNAH technique has the potential for use in phytosanitary surveillance for assessing disease incidence and estimating vector population carrying the BTD phytoplasma (Arocha et al. 2008).

Detection of phytoplasmas in insects using different methods may be a step forward toward the identification of a vector species, but it does not constitute an evidence for its vectoring ability. An alternative to biological transmission assays that can be used to determine the insect inoculativity was developed. This method consists of PCR detection of phytoplasmas in the insect feeding medium (sucrose) after allowing the test insect to feed. A correlation was established between the transmissibility of Flavescence dorée phytoplasma by *Euscelidius variegatus* and its detection by PCR in the insect feeding medium. PCR protocol for detection in the feeding medium appears to reflect vectoring ability of the test insect, probably by detecting it in the insect saliva. PCR assay was also able to detect the phytoplasma in the feeding medium after allowing the field-collected *Orosius albicinctus* and *Anaceratagallia laevis* (Tanne et al. 2001).

A real-time TaqMan PCR assay capable of detecting and quantifying a group 16SrVI phytoplasma in DNA extracts of beet leafhoppers was developed. The primers and probe were designed from the 16S rRNA gene of the Columbia Basin potato purple top phytoplasma which is closely related to the beet leafhopper (*Circulifer tenellus*)-transmitted virescence agent. The pathogen was readily detected in extracts from single or groups of five beet leafhoppers and the concentration of phytoplasma in individual leafhoppers was variable. Detection of phytoplasma DNA in beet leafhoppers by real-time PCR and nested PCR using the extracts of the same insect were in agreement, indicating that about 30% of the leafhoppers contained detectable concentrations of the phytoplasma. The real-time PCR was as sensitive as the commonly used, but more labor-intensive nested PCR for the detection of this phytoplasma. However, real-time PCR format can test large number of insects collected from commercial fields providing the results rapidly (Crosslin et al. 2006).

Chrysanthemum yellows (CY) phytoplasma is transmitted by *Macrosteles quadripunctatus* and *Euscelis incisus*. Two universal primer pairs designed on conserved sequences of 16S rRNA gene and one primer pair designed on extrachromosomal DNA of a severe strain of Western aster yellows phytoplasma were employed for the detection of CY phytoplasma in the leafhoppers. The chromosomal primer pairs were able to amplify phytoplasma-specific bands only when DNA was extracted by CTAB method. On the other hand, plasmid primer pairs were effective when Tris-EDTA buffer was also used for DNA extraction. Amplification of CY phytoplasma DNA was accomplished with as little as 1/10,000 of the total DNA extracted from a single leafhopper (Marazachi et al. 1998).

During survey conducted since 1998, over 200 leafhoppers and planthoppers captured on yellow traps were examined by employing the universal primer pair P1/P7 for the detection of phytoplasma DNA. Seven batches of insects tested positive for the presence of phytoplasmas in their body. RFLP analysis of the PCR products showed that five batches contained elm yellows (EY) group phytoplasma and two batches had a stolbur phytoplasma or a clover phyllody phytoplasma. Stolbur and clover phyllody phytoplasmas were ubiquitous that were hosted in many weeds and several insect species (Fig. 3.4; Boudon-Padieu et al. 2004). The leafhopper *Deltocephalus vulgaris* was shown to transmit sugarcane grassy shoot disease (Singh

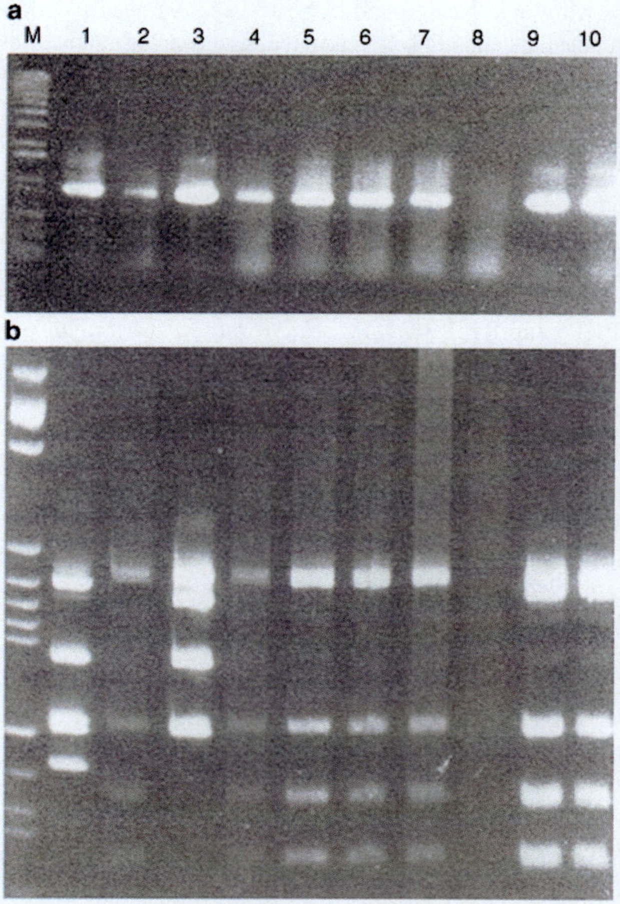

Fig. 3.4 Detection of elm yellows phytoplasma in different vector species by nested PCR assay and RFLP analysis of PCR products (**a**) Native amplimers; (**b**) RFLP patterns of amplimers shown in A after digestion with *Tru*91; Lane M: 1 kb DNA ladder; Lane 1: *Circulifer fenestratus*; Lane 2: *Cixius* sp. Lane 3: *Adarrus multinotatus*; Lane 4: *Iassus scutellarius*; Lane 5: *Allygidius furcatus*; Lanes 6 and 7: two specimens of undetermined species; Lanes 9 and 10: DNA from American EY (EY1) and European EY (ULW) respectively maintained in periwinkle (Courtesy of Boudon-Padieu et al. 2004; Invest Agrar: Sist Recur For, Dijon, France)

et al. 2002). Nymphs of *D. vulgaris* fed on infected sugarcane leaves for 15 days were tested by extracting DNA and subjecting it to nested PCR assay. The universal primer pair P1/P7 was employed for the first round and the amplified product of the first round was subjected to amplification by the primer pair P4/P7. A 500-bp product was detected in the DNA extract from the nymphs as well as in the sugarcane leaves infected by the grassy shoot phytoplasma (Srivastava et al. 2006).

In order to demonstrate the ability of leafhoppers *Orosius albicinctus* to transmit aster yellows (AY) phytoplasma, the insects were collected from and around phytoplasma-infected vineyards. Since the phytoplasmas are nonculturable pathogens, attempts to recover the phytoplasmal DNA from infected plants or insects resulted in preparations with large background of host DNA. A novel method of obtaining host-free phytoplasma, genomic DNA from the insect vector's saliva was developed. The phytoplasmal DNA present in the saliva was detected by nested PCR format which was more sensitive in detecting the pathogen DNA than standard PCR assay. The new protocol allowed isolation and characterization of 78 new putative phytoplasmal open reading frames and their deduced proteins (Melamed et al. 2003).

The vineyards in Canada were surveyed to assess the extent of incidence of phytoplasma diseases and the insect species that act as vectors of the phytoplasmas. Of the 22 leafhopper species in British Columbia, 11 species tested positive for phytoplasma DNA when nested PCR format was applied employing universal primer pairs P1/P6 and R16R2/R16F2. *Macrosteles quadrilineatus*, predominant species was the vector of aster yellows, stolbur and clover proliferation phytoplasmas. In Ontario, *Erthroneura* sp. carried phytoplasma strain 16SrI-C (AY group), whereas *Scaphoideus titanus* carried the elm yellows phytoplasma strain and also Flavescence dorée (FD) phytoplasama. The results suggested that high levels of infection risks of AY phytoplasma existed in the Canadian vineyards (Olivier et al. 2008).

Carrot crops are infected by the phytoplasma belonging to aster yellows (AY) 16SrI-A and 16SrXII-A (stolbur) subgroups. The leafhoppers and treehoppers captured in carrot fields were tested by employing nested PCR assay using the universal phytoplasma primer pair P1/P7 followed by F1/B6 primers to amplify amplicons obtained with P1/P7 primer pair. RFLP analyses were carried out using the restriction enzyme *Tru*I on F1/FB6 amplicons. The restriction fragments were resolved by electrophoresis on a 5% polyacrylamide gel. The bands were visualized after staining with ethidium bromide under UV transilluminator. Direct PCR did not result in clear recognition of amplified products which appeared as faint bands. On the other hand, nested PCR format yielded amplicons of expected size (about 1,700-bp) for eight of 43 insects. The three species *Macrosteles quadripunctulatus*, *M. sexnotatus* and *M. laevis* tested positive for phytoplasmas belonging to the subgroups 16SrI-B and 16SrXII-A subgroups. *M. quadripunctulatus* and *M. sexnotatus* specimens were positive for AY phytoplasma, whereas *M. laevis* specimen showed the presence of stolbur phytoplasma. Further, two more insect species *Psammotettix notatus* and *Stictocephala bisonia* carrying other phytoplasmas infecting pear and apple respectively were recorded as vectors of phytoplasmas for the first time. This study indicated the need for periodical monitoring the population of insects that may serve as potential vectors of phytoplasmal diseases (Duduk et al. 2008).

Spiroplasama kunkelii, causing corn stunt disease is one of the members of Mollicutes that has been cultured in cell-free synthetic medium. *S. kunkelii* is transmitted by the phloem-feeder *Dalbulus maidis* (leafhopper). The gene encoding a novel adhesion-like protein was identified in the genome of *S. kunkelii*. Adhesins of the sarpin family including SARP1 from *S. citri* and SKARP1 from *S. kunkelii* are considered to have a role in the adhesion of spiroplasma cells to the gut cells of the leafhopper vector during early stages of infection of vector tissues (Berg et al. 2001; Davis et al. 2005). A field-deployable real-time PCR technique was applied for the detection of *S. kunkelii* in the insect tissues as well as in corn plant tissues infected by corn stunt disease with a detection limit of 5 fg. This real-time PCR protocol was rapid, sensitive, specific and reliable, because the DNA from the culture of *S. citri*, healthy corn plant or aster yellows phytoplasma was not amplified by this assay (Wei et al. 2006).

Appendix 1: Detection of *Ralstonia solanacearum* Race 1 Strains in Soils by BIO-PCR Assay (Lin et al. 2009)

Bacterial Enrichment Using MSM-1 Broth

1. Prepare the MSM-1 medium without agar consisting of 10 g peptone, 5 g glucose, 1 g casein hydrolysate, 50 mg 2,3,5-triphenyl tetrazolium chloride (TTC) in 1 l sterile distilled water (SDW) (deleting 15 g agar) with additional antimicrobial compound – 5 mg chloramphenicol, 5 mg crystal violet, 5 mg cycloheximide, 100 mg polymyxin B sulphate and 20 mg tyrothricin.
2. Transfer 10 g soil sample to 90 ml SDW; shake at 180 rpm at room temperature for 30 min; add 1 ml sample suspension to 9 ml MSM-1 broth and incubate at 30°C at 160 rpm for enrichment.
3. Transfer 5 μl enriched suspension to 200 μl PCR tube; cover with one drop of sterile mineral oil; boil for 5 min and keep the tube in ice.
4. Perform the PCR amplification using primer pair AU759f/AU760r with a 25-μl reaction mixture [consisting of 1 × PCR buffer containing 10 mM Tris-HCl, pH 9.0, 50 mM KCl and 0.1% TritonX-100)], 1.5 mM $MgCl_{2,}$ 0.005 mM each dNTP, 1 pmol of each primer, 2 U Taq DNA polymerase (Promega, USA) and 5 μl boiled bacterial suspension or enriched cultures.
5. Provide the following conditions using a thermal cycler: denaturing at 94°C for 3 min, annealing at 53°C for 1 min, extension at 72°C for 1.5 min followed by 30 cycles of 94°C for 18 s, 60°C for 18 s and 72°C for 5 min.
6. Resolve the PCR product of 282-bp on 1.5% agarose gel and staining in ethidium bromide solution (1 μg/ml) and visualize under UV light.

Appendix 2: Detection of *Xylella fastidiosa* (Xf) in Vector Insects by PCR Assay (Krell et al. 2007)

Extraction of DNA

1. Use DNeasy Tissue Kit (Qiagen, USA) to extract DNA separately from heads of *Homalodisca liturata* (Hl) and heads, thoraxes and abdomens of *Diceroprocta apache* (Da) and grind the individual insects in phosphate buffered saline (PBS) in sample extraction pouches (Agadia).
2. Use Ready-To-Go PCR beads (Amersham Biosciences, UK) with a grape-specific primer pair XF2542-L and XF2542-R for PCR amplifications; for each reaction provide one Ready-To-Go PCR bead, 0.25 μl of each primer, 10 μl extracted DNA and 14.5 μl sterile water.
3. Perform PCR amplification using a thermal cycler programmed for one cycle of 95°C for 5 min followed by 35 cycles of 94°C for 40 s, 55°C for 40 s, 72°C for 1 min and finally one cycle of 72°C for 5 min and held at 4°C.
4. Resolve the amplicons by electrophoresis on 1.5% agarose-tris-borate-EDTA gel and stained with ethidium bromide (0.5 μg/ml of gel) to visualize the separated bands.

References

Álvarez B, López MM, Biosca EG (2007) Influence of native microbiota on survival of *Ralstonia solanacearum* phylotype II in river water microcosms. Appl Environ Microbiol 73: 7210–7217.

Arocha Y, Almeida R, Jones P (2008) Using ribosomal probe for the detection of the phytoplasma associated with bunchy top symptom of papaya (BTS) in plant and insect hosts. Rev Prot Veg 23: 16–20.

Babaie G, Khatabi B, Bayat H, Rastgou M, Hosseini A, SalekDeh GH (2007) Detection and characterization of phytoplasmas infecting ornamental and weed plants in Iran. J Phytopathol 155: 368–372.

Berg M, Melcher U, Fleltcher J (2001) Characterization of *Sprioplasma citri* adhesion related protein SARP1 which contains a domain of novel family designated sarpin. Gene 275: 57–64.

Bock CH, Parker PE, Gottwald TR (2005) Effect of simulated wind-driven rain on duration and distance of dispersal of *Xanthomonas axonopodis* pv. *citri* from canker-infected citrus trees. Plant Dis 89: 71–80.

Boudon-Padieu E, Larrue J, Clair D, Hourdel J, Jeanneau A, Sforza R, Collin E (2004) Detection and prophylaxis of elm yellows phytoplasmas in France. Invest Agrar Sist Recur For 13: 71–80.

Buzombo P, Jaimes J, Lam V, Cantrell K, Harkness M, McCullough D, Morano L (2006) An American hybrid vineyard in the Texas Gulf Coast: Analysis within a Pierce's disease of hot zone. Am J Enol Vitic 57: 347–355.

Caruso P, Palomo JL, Bertolini E, Alvarez B, López MM, Biosca EG (2005) Seasonal variation of *Ralstonia solanacearum* biovar 2 populations in a Spanish river: Recovery of stressed cells at low temperatures. Appl Environ Microbiol 71: 140–148.

Crockett AO, Wittwer CR (2001) Fluorescein-labeled oligonucleotides for real-time PCR: using the inherent quenching of deoxyguanosine nucleotides. Anal Biochem 290: 89–97.

Crosslin JM, Vandemark GJ, Munyaneza JE (2006) Development of a real-time, quantitative PCR for detection of the Columbia Basin potato purple top phytoplasma in plants and beet leafhoppers. Plant Dis 90: 663–667.

Damsteegt VD, Brlansky RH, Phillips PA, Roy A (2006) Transmission of *Xylella fastidiosa* causal agent of citrus variegated chlorosis by the glassy-winged sharpshooter, *Homalodisca coagulata*. Plant Dis 90: 567–570.

Davis RE, Dally EL, Jomantiene R, Zhao Y, Roe B, Lin SP, Shao J (2005) Cryptic plasmid pSKU146 from wall-less plant pathogen *Spiroplasma kunkelii* encodes an adhesion and components of a type IV translocation-related conjugation system. Plasmid 53: 179–190.

Davis RE, Sinclair WA, Lee I-M, Dally E (1992) Cloned DNA probes specific for detection of a mycoplasma-like organisms (MLOs) associated with ash yellows. Mol Plant Microbe Interact 5: 163–169.

Dittapongpitch V, Surat S (2003) Detection of *Ralstonia solanacearum* in soil and weeds from commercial tomato fields using immunocapture and polymerase chain reaction. J Phytopathol 151: 239–246.

Duduk B, Calarai A, Paltrinieri S, Duduk N, Bertaccini A (2008) Multigenic analysis for differentiation of aster yellows phytoplasmas infecting carrots in Serbia. Ann Appl Biol doi:10.1111/j.1774-7348.2008.00285.x

Elphinstone JG (1996) Survival and possibilities for extinction of *Pseudomonas solanacearum* (Smith) Smith in cool climates. Potato Res 39: 403–410.

Elphinstone JG, Stanford HM, Stead DE (1998) Survival and transmission of *Ralstonia solanacearum* in aquatic plants of *Solanum dulcamara* and associated with surface water in England. Bull OEPP 28: 93–94.

Fleischer SJ, de MacKiewicz D, Gildow FE, Lukezic FL (1999) Serological estimates of the seasonal dynamics of *Erwinia tracheiphila* in *Acalymma vittata* (Coleoptera: Chrysomelidae). Environ Entomol 28: 470–476.

Francis M, Lin H, Rosa JC, Doddapaneni H, Civerolo EL (2006) Genome-based PCR primers for specific and sensitive detection and quantification of *Xylella fastidiosa*. Eur J Plant Pathol 115: 203–213.

García-Salazar C, Gildow FE, Fleischer SJ, Cox-Foster D, Lukezic FL (2000) ELISA versus immunolocalization to determine the association of *Erwinia tracheiphila* in *Acalyamma vittata* (Coleoptera: Chrysomelidae). Environ Entomol 29: 542–550.

Gottwald TR, Irey M (2007) Post-hurricane analysis of citrus canker II: Predictive model estimation of disease spread and area potentially impacted by various eradication protocols following catastrophic weather events. Plant Mgt Network: doi: 10.1094/PHP-2007-0405-01-RS.

Graham J, Lloyd AB (1978) An improved indicator plant method for the detection of *Pseudomonas solanacearum* race 3 in soil. Plant Dis Rep 62: 35–37.

Graham J, Lloyd AB (1979) Survival of potato (strain race 3) of *Pseudomonas solanacerarum* in deeper soil layers. Aust J Agric Res 30: 489–496.

Griep RA, van Twisk C, van Beckhoven JRCM, van der Wolf JM, Schots A (1998) Development of specific recombinant monoclonal antibodies against lipopolysaccharide of *Ralstonia solanacearum* race 3. Phytopathology 88: 795–803.

Holmes A, Govan J, Goldstein R (1978) Agricultural use of *Burkholderia* (*Pseudomonas*) *cepacia*: a threat to human health. Emerg Infect Dis 4: 221–227.

Horita M, Yano K, Tsuchiya K (2004) PCR-based specific detection of *Ralstonia solanacearum* race 4 strains. J Gen Plant Pathol 70: 278–283.

Huang Q, Bentz J, Sherald JL (2006) Fast, easy and efficient DNA extraction and one-step polymerase chain reaction for the detection of *Xylella fastidiosa* in potential insect vectors. J Plant Pathol 88: 77–81.

Hyman LJ, Sullivan L, Toth IK, Perombelon MCM (2001) Modified crystal violet pectate medium (CVP) based on a new pectate source (Slendid) for the detection and isolation of soft rot erwinias. Potato Res 44: 265–270.

Janse JD, Schans J (1998) Experience with the diagnosis and epidemiology of bacterial brown rot (*Ralstonia solanacearum*) in the Netherlands. Bull OEPP 28: 65–67.

Khakvar R, Sijam K, Yun WM, Radu S, Lin TK (2008) Improving a PCR-based method for identification of *Ralstonia solanacearum* in natural sources of West Malaysia. Am J Agric Biol Sci 3: 490–493.

Khan IA, Turpin FT, Lister RM, Scott DH, Whitford F (1997) Using enzyme-linked immunosorbent assay (ELISA) for the detection of *Erwinia stewartii* (Smith) Dye in corn plants and corn flea beetles (*Chaetocnema pulicaria* Melsheimer). Sarhad J Agric 13: 391–397.

Krell RK, Boyd EA, Nay JE, Park Y-L, Perring TM (2007) Mechanical and insect transmission of *Xylella fastidiosa* to *Vitis vinifera*. Am J Enol Vitic 58: 211–216.

Kuo TT, Huang TC, Wu RY, Yang CM (1967) Characterization of three bacteriophages of *Xanthomonas oryzae* (Uyeda et Ishiyam) Dowson. Bot Bull Acad Sinica 8: 246–254.

Kuske CR, Banton KI, Adorada DL, Stark PC, Hill KK, Jackson PJ (1998) Small-scale DNA sample preparation method for field PCR detection of microbial cells and spores in soil. Appl Environ Mircrobiol 64: 2463–2472.

Lherminier J, Terwisscha van ST, Boudon-Padieu E, Caudwell A (1989) Rapid immunofluorescence detection of the grapevine Flavescence dorée mycoplasma-like organism in the salivary glands of the leafhopper *Euscelidius variegatus* Kbm. J Phytopathol 125: 353–360.

Lin C-H, Hsu S-T, Tzeng K-C, Wang J-F (2009) Detection of race 1 strains of *Ralstonia solanacearum* in field samples in Taiwan using BIO-PCR method. Eur J Plant Pathol 124: 75–85.

Manjunath KL, Halbert SE, Ramadugu C, Webb S, Lee RF (2008) Detection of '*Candiatus* Liberibacter asiaticus' in *Diaphorina citri* and its importance in the management of citrus huanglongbing in Florida. Phytopathology 98: 389–396.

Manome A, Kageyama A, Kurata S, Yokomaku T, Koyama O, Kanagawa T, Tamaki H, Tagawa M, Kamagata Y (2008) Quantification of potato common scab pathogens in soil by quantitative competitive PCR with fluorescent quenching-based probes. Plant Pathol 57: 887–896.

Marazachi C, Veratti F, Bosco D (1998) Direct PCR detection of phytoplasmas in experimentally infected insects. Ann Appl Biol 133: 45–54.

McGaha LA, Jackson B, Bextine B, McCullough D, Morana L (2007) Potential plant reservoirs for *Xylella fastidiosa* in south Texas. Am J Enol Vitic 58: 398–401.

Melamed S, Tanne E, Ben-Haim R, Edelbaum O, Yogev D, Sela I (2003) Identification and characterization of phytoplasmal genes, employing a novel method of isolating phytoplasmal genomic DNA. J Bacteriol 185: 6513–6521.

Miller DN (2001) Evaluation of gel filtration resins for the removal of PCR-inhibitory substances from soils and sediments. J Microbiol Meth 44: 49–58.

Olivier C, Lowery DT, Vincent C, Stobbs LM, Saguez J, Galka B, Bittner L (2008) Phytopalsma diseases in Canadian vineyards. Endure International Conference La Grand-Motte, France, 2008, pp. Q.46, 1–3.

Pirone L, Chiarini L, Dalmastri C, Bevivino A, Tabacchioni S (2005) Detection of cultured and uncultured *Burkholderia cepacia* complex bacteria naturally occurring in the maize rhizosphere. Environ Microbiol 7: 1734–1742.

Pooler MR, Myung IS, Bentz J, Sherald J, Hartung JS (1997) Detection of *Xylella fastidiosa* in potential insect vectors by immune magnetic separation and nested polymerase chain reaction. Lett Appl Microbiol 25: 123–126.

Pradhanang PM, Elphinstone JG, Fox RTV (2000) Sensitive detection of *Ralstonia solanacearum* in soil: a comparison of different detection techniques. Plant Pathol 49: 414–422.

Priou S, Aley P, Fernandez H, Gutarra L (1999) Sensitive detection of *Ralstonia solanacearum* (race 3) in soil by post-enrichment DAS-ELISA. Bacterial Wilt Newsletter No. 16, 10–13.

Priou S, Grutarra L, Aley P (2006) An imporoved enrichment broth for the sensitive detection of *Ralstonia solanacearum* (biovars 1 and 2A) in soil using DAS-ELISA. Plant Pathol 55: 36–45.

Qu X, Wanner LA, Christ BJ (2008) Using the *TxtAB* operon to quantify pathogenic *Streptomyces* in potato tubers and soil. Phytopathology 98: 405–412.

Redak RA, Purcell AH, Lopes JRS, Blua MJ, Mizell RF, Andersen PC (2004) The biology of xylem fluid-feeding insect vectors of *Xylella fastidiosa* and their relation to disease epidemiology. Annu Rev Entomol 49: 243–270.

Rodrigues JLM, Silva-Stenico ME, de Souza AN, Lopes JRS, Tsai SM (2006) *In situ* probing of *Xylella fastidiosa* in honey dew of a xylem sap-feeding insect using 16S rRNA-targeted fluorescent oligonucleotides. Environ Microbiol 8: 747–754.

Rodrgues JLM, Silva-Stenico ME, Gomes JE, Lopes JRS, Tsai SM (2003) Detection and diversity assessment of *Xylella fastidiosa* in field-collected plant and insect samples by using 16S rRNA and *gyrB* sequences. Appl Environ Microbiol 69: 4249–4255.

Romeiro RS, Silva HAS, Beriam LOS, Rodrigues Neto J, Carvalho MG de (1999) A bioassay for detection of *Agrobacterium tumefaciens* in soil and plant material. Summa Phytopathol 25: 359–362.

Salles JF, de Souza FA, van Elsas JD (2002) Molecular method to assess the diversity of *Burkholderia* species in environmental samples. Appl Environ Microbiol 68: 1595–1603.

Singh V, Baitha A, Sinha OK (2002) Transmission of grassy shoot disease of sugarcane by a leafhopper (*Deltocephalus vulgaris* Dash & Viraktamath). Ind J Sugar Technol 17: 60–63.

Srivastava S, Singh V, Gupta PS, Sinha OK, Baitha A (2006) Nested PCR assay for detection of sugarcane grassy shoot phytoplasma in the leafhopper vector *Deltocephalus vulgaris*: a first report. Plant Pathol 55: 25–28.

Tanne E, Boudon-Padieu E, Clair D, Davidovich M, Melamed S, Klein M (2001) Detection of phytoplasma by polymerase chain reaction of insect feeding medium and its use in determining vectoring ability. Phytopathology 91: 741–746.

Tolu G, Botti S, Garau R, Prota VA, Sechi A, Prota U, Bertaccini A (2006) Identification of a 16SrII-E phytoplasma in *Calendula arvensis*, *Solanum nigrum* and *Chenopodium* spp. Plant Dis 90: 325–330.

Tomlinson DL, Elphinstone JG, Hanafy MS, Shoala TM, El-Fatah HA, Agag SH, Kamal M, El-Aliem MMA, Fawzi FG, Stead DE, Janse JD (2009) Recovery of *Ralstonia solanacearum* from canal water in traditional potato-growing areas of Egypt but not from designated pest-free areas. Eur J Plant Pathol 125: 589–601.

Van der Wolf JM, Van Beckhoven JRCM, De Haan EG, Van den Bovenkamp GW, Leone GOM (2004) Specific detection of *Ralstonia solanacearum* 16S rRNA sequences by AmpliDet RNA. Eur J Plant Pathol 110: 25–33.

Van der Wolf JM, Vriend SGC, Kastelein P, Nijhuis EH, van Bekkum PJ, van Vuurde JWL (2000) Immunofluorescence colony-staining (IFC) for detection and quantification of *Ralstonia* (*Pseudomonas*) *solanacearum* biovar 2 (race 3) in soil and verification of positive results by PCR and dilution plating. Eur J Plant Pathol 106: 123–133.

Wakkimoto S (1957) A simple method for comparison of bacterial populations in a large number of samples by phage technique. Ann Phytopathol Soc Jpn 22: 159–163.

Webb DR, Bonfiglioli RG, Carraro L, Osler R, Symons RH (1999) Oligonucleotides as hybridization probes to localize phytoplasmas in host plants and insect vectors. Phytopathology 89: 894–901.

Wei W, Opgenorth DC, Davis RE, Chang CJ, Summers CG, Zhao Y (2006) Characterization of a novel adhesion-like gene and design of a real-time PCR for rapid, sensitive and specific detection of *Spiroplasma kunkelii*. Plant Dis 90: 1233–1238.

Wen A, Mallik I, Alavarado VY, Pasche JS, Wang X, Li W, Levy L, Scholthof HB, Rush CM, Gudmestad NC (2009) Detection, distribution and genetic variability of '*Candidatus* Liberibacter' species associated with Zebra complex disease of potato in North America. Plant Dis 93: 1102–1115.

Wenneker M, Verdel MSW, Groeneveld RMW, Kempennar C, van Beuningen AR, Janse JD (1999) *Ralstonia (Pseudomonas) solanacearum* race 3 (biovar 2) in surface water and natural weed hosts: First report on stinging nettle (*Urtica dioca*). Eur J Plant Pathol 105: 307–315.

Wicks TJ, Hall BH, Morgan BA (2007) Tuber soft rot and concentrations of *Erwinia* spp. in potato washing plants in South Australia. Aust Plant Pathol 36: 309–312.

Yamada T, Kawasaki T, Nagata S, Fujiwara A, Usami S, Fujie M (2007) New bateriophages that infect the phytopathogen *Ralstonia solanacearum*. Microbiology 153: 2630–2639.

Yung ANL, ChiChung W (2000) The design of specific primers for the detection of *Ralstonia solanacearum* in soil samples by polymerase chain reaction. Bot Bull Acad Sinica 41: 121–128.

Chapter 4
Assessment of Variability in Bacterial and Phytoplasmal Pathogens

Abstract Variability is the essential requirement of bacterial and phytoplasmal pathogens inducing diseases in plants/crops to survive and perpetuate in various environmental conditions and the agricultural practices followed in different ecosystems. Variations in the genetic constitution of bacteria are reflected in their phenotypic characteristics expressed in cultural, biochemical/physiological and immunological characteristics that may have an impact on the pathogenicity (pathogenic potential) of bacterial pathogens. Growth of and production of enzymes, toxins and antibiotics by bacterial pathogens may show variations in nutrient media. Several techniques have been applied to assess the variations in the nature of enzymes and toxins elaborated by bacterial pathogens, in addition to their reactivity to antibodies. Nucleic acid-based techniques have been shown to be more efficient in differentiating strains/races of bacterial pathogens more rapidly and reliably when compared to methods based on the biological, biochemical and immunological characteristics. The phytoplasmal pathogens have not yet been successfully cultured on artificial nutrient media. Hence, differences in their cultural and biological characteristics cannot be utilized as the basis for differentiation of phytoplamas. Application of nucleic acid-based techniques have been demonstrated to be more useful and precise in differentiating the strains of phytoplasmal pathogens than the biological methods depending on the pathogenicity of these pathogens.

Variations in cultural, biochemical, immunological and genetic characteristics of plant bacterial pathogens have been assessed. Phytoplasmal pathogens, on the other hand, have to be differentiated primarily based on the variations in genetic characteristics, since these pathogens are yet to be cultured on cell-free artificial media. It is important to relate the variations observed in different characteristics to the changes in the pathogen potential (virulence) of the bacterial pathogen(s) under investigation, since this information can form the basis for developing economically feasible disease management systems.

P. Narayanasamy, *Microbial Plant Pathogens-Detection and Disease Diagnosis: Bacterial and Phytoplasmal Pathogens, Vol. 2*, DOI 10.1007/978-90-481-9769-9_4,

4.1 Assessment of Variability in Bacterial Pathogens

4.1.1 Assessment of Variations in Pathogenicity

The International Society of Plant Pathology has laid down the criteria for applying the term 'pathovar' for those pathogenic bacteria that do not satisfy the criteria for species designation. Pathovars are distinguished within a bacterial species primarily based on the ability or inability of isolates to infect a set of host plant species or cultivars. The basic unit of variation in classifying physiologic specialization is physiologic race that is a combination of virulent and non-virulent reactions induced on a standard set of differential cultivar of a crop plant species. A differential set of host varieties that can distinguish the responses induced by pathogen isolates is essentially required for understanding the extent of variations in the pathogenic potential of a bacterial isolate/pathovar/race. Races of bacterial pathogens infecting different crop plants such as *Xanthomonas axonopodis* pv. *malvacearum* (*Xam*) infecting cotton (Brinkerhoff 1970), *X. campestris* pv.*vesicatoria* (*Xcv*) infecting pepper (Cook and Stall 1969) and *X. axonopodis* pv. *glycines* (*Xag*) infecting soybean (Cross et al. 1966) have been identified using respective set of differential varieties. Seven distinct pathogenic races of *X. axonopodis* pv.*glycines* (*Xag*) were differentiated by using seven differential soybean cultivars. Soybean cv. Acme considered as universally susceptible to *Xag* races was found to be resistant to races 5 and 6, whereas the cv. Flambeau considered as resistant was susceptible to race 4 (Cross et al. 1966). Races of *Xanthomonas oryzae* pv. *oryzae* (*Xoo*), causative agent of rice bacterial blight disease could be differentiated by inoculating the differential rice genotypes with different resistance genes (Table 4.1) (Vera Cruz and Mew 1989).

4.1.2 Assessment of Variations in Phage Sensitivity

Bacteriophages cause lysis (dissolution) of bacterial cells resulting in the formation of 'plaques' of clear areas in the bacterial culture grown in the appropriate growth medium. The number of plaques formed is proportional to the phage concentrations. Phages have been used to detect the presence of bacterial pathogens in infected leaves, irrigation water and soil. On the basis of sensitivity to phages, bacterial strains may be divided into different lysotypes. The relationship between sensitivity of bacterial pathogens to specific bacteriophages and their virulence was studied by Choi et al. (1981) using *Xanthomonas oryzae* pv. *oryzae* (*Xoo*) and the phage OP2. The virulence of *Xoo* strains was closely correlated with sensitivity to the phage OP2. The correlation was not distinct when other phages were used for the test. Twenty one *Xoo* isolates belong to six races based on similarity of reaction to a set of phages (10^6) were tested. The phages that showed clear plaque formation were multiplied from a purified single plaque of each phage isolate on the most sensitive

Table 4.1 Identification of races of *Xanthomonas oryzae* pv. *oryzae* (*Xoo*) using differential rice cultivars (Vera Cruz and Mew 1989)

Races of Xoo	Differential cultivars (resistance gene)[a] IR 24 (0)	IR20 (Xa-4)	Cas 209 (Xa-10)	IR1545-339 (Xa-5)	DV 85 (Xa-5, Xa-7)
Race 1 (PXO61)	S[b] (40–45)	R (3–5)	S (40–45)	R (3–5)	R (1–3)
Race 2 (PXO86)	S (30–35)	S (30–35)	R (1–3)	R (5–10)	R (1–3)
Race 3 (PXO79)	S (35–40)	S (30–35)	S (35–40)	R (3–5)	R (1–3)
Race 4 (PXO71)	S (40–45)	MR (15–20)	S (40–45)	S (30–35)	R (3–5)
Race 5 (PXO112)	S (40–45)	R (1–3)	R (1–3)	R (3–5)	R (1–3)
Race 6 (PXO99)	S (40–45)	S (40–45)	S (45–50)	S (40–45)	MS (25–30)

[a] Plant age at inoculation – 40 days after sowing
[b] Reactions of differentials: S susceptible; MS moderately susceptible; MR moderately resistant; R resistant. Numbers in parenthesis are ranges of % lesion area

bacterium of each race: PXO 13 (race1) for group A, IRN801 (race2) for group B, IRN702 (race3) for group C, PXO71 (race4) for group D and PXO111 (race5) for group E. None of the strains of race 6 formed plaque from any of the phages tested. Based on the distribution of lysotype, some races could be differentiated. All isolates of race 4 (100%) gave the same reaction to all phage groups (A–E), whereas all isolates of race 6 were resistant to all *Xoo* phages. The majority of race1 isolates (77.2%) was resistant to group E phages (Vera Cruz and Mew 1989). The pathogenic potential (virulence) of two races of *X. axonopodis* pv. *malvacearum* (*Xam*) was shown to be related to the phage sensitivity. The phage sensitivity of these two races of *Xam* exhibited distinct differences, when six phages were used for typing the pathogen isolates. Race1 was sensitive to three or rarely four of the six phages used for typing. On the other hand, all six phages lysed the isolates of race 2 of *Xam* tested (Freigoun et al. 1994).

4.1.3 Assessment of Variations in Cultural Characteristics

Variations in cultural or physiological characteristics of *Xanthomonas oryzae* pv. *oryzae* (*Xoo*) isolates belonging to pathotypes I, II, III, IV and V existing in Japan and Indonesia did not have any relationship with pathotypes tested (Hifini et al. 1975). Strains of *Xoo* were compared by computer-assisted numerical taxonomy, testing 132 unit characters such as cell morphology, standard biochemical tests, acid production from carbohydrates, growth on various carbon sources, growth under different conditions and resistance to antibiotics. The races showed variations in the utilization of trehalose and sodium aconitate. Strains of race 1 (84.3%) could not utilize trehalose as a carbon source, while all strains (100%) of races 4 and 6 grew luxuriantly on the medium containing trehalose. The majority of races (94.3%), 3 (92.8%) and 5 (96.4%) also could grow on the medium with trehalose as carbon source. Sodium aconitate inhibited the growth of races 4 and 6 and 98.2% of race 1. However, the differentiation based on the utilization of these two compounds was not consistent and reliable, as the reactions among the strains overlapped in other races (Vera Cruz and Mew 1989).

Metabolic fingerprinting using API 20E, API 50CH, ATBG-5 and API-ZYM tests and BIOLOG metabolic fingerprinting procedure were adopted to assess the intraspecific diversity of a collection of 63 Spanish strains of *Erwinia amylovora* (*Ea*), the fire blight pathogen infecting apple, pear and other rosaceous plants. The API 20E tests showed that all strains of *Ea* were positive for acetoin production and acid production from glucose and sucrose. All Spanish strains of *Ea* utilized the carbohydrates in API 50CH tests. Most of the strains (75 of 78) were sensitive to all antibiotics at concentrations tested in the ATBG-5 format. API-ZYM analysis revealed similar hydrolytic patterns for the *Ea* stains. The Spanish strains were identified as members of *Ea* with 100% probability in all cases. The results indicated that these tests did not show significant variations in the Spanish strains tested. Variability detected among some strains was not reproducible. Hence, the utility of these biochemical tests to assess variability in bacterial plant pathogens seems to be limited (Donat et al. 2007).

4.1.4 Assessment of Variations in Biochemical Characteristics

4.1.4.1 Fatty Acid Analysis

Bacterial whole-cell fatty acid methyl esters (FAMEs) are extracted and fatty acid profiles are generated for the bacterial strains to be differentiated. All pathovars of *Pseudomonas syringae*, *P. viridiflava* and the pear and radish strains were considered to be closely related, as revealed by the dendrogram based on the fatty acid composition. Lauric acid and palmitoleic acid were major fatty acids present in these two bacterial pathogens. Strains of *P. syringae* were pathogenic on pear blossoms, whereas strains of *P. viridiflava* infected radish leaves causing rotting symptoms (Khan et al. 1999). Strains of *Xanthomonas axonopodis* pv. *citri*, *Xac*-A, *Xac*-A^W, *Xac*-A* and *X. axonopodis* pv. *citrumelo* strains formed distinct clusters, based on the multivariate analysis of the fatty acid profiles of each strain. A dendrogram constructed with representative strains showed that *Xac*-A strains were separated from *Xac*-A^W and *Xac*-A* (Sun et al. 2004).

4.1.4.2 Polyacrylamide Gel Electrophoretic Analysis

By applying polyacrylamide gel electrophoresis (PAGE) technique and silver staining of SDS-lysed cells of *X. campestris* pv. *vesicatoria* (*Xcv*), two major phenotypes could be differentiated. The presence of broad dark gray bands with MW of 32.35 and 25.57 kDa designated α- and ß was recognized in different strains of *Xcv*. The α-band was detected in 192 of 197 tomato race and the ß-band was noted in all 55 strains of tomato race of *Xcv.* The differentiation of strains of race1 could be achieved by assessing the ability of the strains to hydrolyze starch (Amy$^+$) or degrade pectate (Pec$^+$). Race1 strains expressing the α-band could not hydrolyze starch (Amy$^-$) or degrade pectate (Pec$^-$) In contrast, most strains of race 2 were Amy$^+$ and Pec$^+$. In addition, silver staining of protein profiles and testing for amylolytic activity may form a reliable basis for the differentiation of strains of *Xcv* (Bouzar et al. 1994). Generation of envelope protein profiles has shown to be a reliable basis of differentiation of pathovars of *Pseudomonas syringae.* Three protein bands 60, 65, and 150 kDa present in *P. syringae* pv. *pisi* strains could be used to differentiate them from strains (29) of *P. syringae* pv. *syringae* which lacked these three proteins (Malandrin et al. 1996).

An advanced version of PAGE, pulsed-field gel electrophoresis (PFGE) is able to overcome some problems associated with conventional PAGE format. The strains of *Erwinia amylovora* (*Ea*), causing fire blight disease in apple and other rosaceous plants from the Mediterranean and European countries were analyzed by PFGE procedure. The PFGE profiles were determined by assaying the bacterial genomic DNA after digestion with the restriction enzyme *Xba*I. The pathogen strains from Austria and Czechia showed similarity to the strains included in Central European type Pt1, whereas the group Pt2 included the strains from Eastern Europe and Mediterranean region. Italian strains exhibited patterns of both types (Zhang et al. 1998). The genomic DNAs of 63 Spanish strains of *Ea*, after digestion with *Xba*I were analyzed by PFGE procedure. Three distinct patterns Pt1, Pt3 and Pt4 were observed among strains of *Ea.* Pt3 and Pt4 could be distinguished by a single band of 130-kb present in Pt4, but absent in Pt3. The only difference noted between Pt1 and Pt4 was the slightly greater distance between the pair of bands situated between 194 and 242.5-kb. Among the Spanish strains, 51% were identified as Pt4, while 30% and 17% of the strains belonged to Pt1 and Pt3 respectively. All strains present in imported plant materials were identified as Pt3. The PFGE pattern identified for each strain could be reproduced in all cases. Characterization of strain diversity in *E. amylovora* is essential for tracking long or short distance pathogen dispersal and to recognize the possible sources of infection (Donat et al. 2007).

Strains of *Xanthomonas axonopodis* pv. *citri* (*Xac*), *Xac*-A, *Xac*-A^W and *Xac* A* were analyzed by PFGE procedure. The DNA derived from the pathogen strains were digested with the restriction enzymes *Xba*I or *Spe*I and the large quantities of digests were separated by PFGE. These strains clustered based on the genetic difference derived from similarity coefficients of DNA fragments after digestion with restriction enzymes *Xba*I or *Spe*I. The clustering of strains was very similar with both restriction enzymes. Strains *Xac*-A, *Xac*-A^W and *Xac*-A* formed separate distinct clusters. However, the *Xac*-A^W and *Xac*-A* strains were more related to *Xac*-A than *X. axonopodis aurantifolii* (*Xaa*) strains (Sun et al. 2004).

The DNA fingerprinting profiles of *Acidovorax avenae* subsp. *citrulli* (*Aac*), causing bacterial fruit blotch (BFB) in cucurbits in Israel, were generated by PFGE technique and rep-PCR assays. The isolates of *Aac* showed 23 unique bands and five different profiles, each one containing 9–13 bands. Based on the profiles, the isolates of *Aac* were classified into two groups one of which included strains that were more associated with watermelon (group II). The group I enclosed the isolates that were associated with non-watermelon cucurbits. PFGE analysis provided a distinct advantage of giving reproducible results, in addition to possibility of assessing the genetic diversity of bacterial pathogens more reliably (Burdman et al. 2005).

Clavibacter michiganensis subsp. *michiganensis* (*Cmm*), the causal agent of tomato bacterial wilt and canker disease, is a quarantine organism under the European Union Public Health legislation. Strains (58) of *Cmm* collected for 10 years in Israel and 18 strains from other sources were analyzed by macrorestriction PFGE technique. The genomic DNA after digestion with restriction enzymes *Vsp*I or *Dra*I produced several fragments. The pathogenic strains of *Cmm* were differentiated into 11 haplotypes by either *Vsp*I or *Dra*I. The strains from Israel formed four distinct groups. Group A included 16 strains, while group B enclosed 32 strains which constituted the major clusters. These two groups of strains were present in the region where tomatoes were grown under cover. The results of macrorestriction PFGE analysis were confirmed by rep-PCR assay employing ERIC or BOX primers (Kleitman et al. 2008) [Appendix 1].

4.1.5 Assessment of Variations in Immunological Characteristics

Enzyme-linked immunosorbent assay (ELISA) has been applied more frequently to differentiate the strains of bacterial pathogens than any other immunoassays. Monoclonal antibodies (MAbs) have been shown to be efficient in differentiating the strains of bacterial pathogens. *Xanthomonas oryzae* pv. *oryzae* (*Xoo*) strains (178) were classified into four groups (I–IV) based on their reactivity with four MAbs in ELISA tests (Benedict et al. 1989). In a later study employing both MAbs and polyclonal antibodies (PAbs), 63 strains of *Xoo* were placed in nine reaction types consisting of four serovars and seven subserovars (Huang et al. 1993). Indirect ELISA format was employed to differentiate two strains of *X. axonopodis* pv. *citri* (*Xac*) with unique host specificity designated *Xac*-A (Asiatic strains) and *Xac*-A^W (Wellington strain). Commercially available MAb A1 specific for *Xac*-A strain used in the assays, reacted specifically with *Xac*-A strain, but not with *Xac*-A^W strain (Sun et al. 2004). The specific MAbs 7AH10, 5HB3 and 4AD2 were used to distinguish the strains *of X. axonopodis* pv. *vesicatoria* (*Xav*). The *Xav* strains could be differentiated into two groups (i) those that were capable of hydrolyzing starch (Amy^+) and (ii) those that were unable to hydrolyze starch (Amy^-). Most of *Xav* strains (97%) collected from different countries were recognized by MAb 5HB3 in ELISA tests. Japanese collections of *Xav* strains could be differentiated into Amy^+ and Amy^- by employing the three MAbs (Tsuchiya et al. 2003).

4.1.6 Assessment of Variations in Nucleic Acid Characteristics

Nucleic acid-based techniques have been applied to detect and differentiate several bacterial pathogens with greater precision and reliability. Restriction fragment length polymorphism (RFLP), polymerase chain reaction (PCR) either alone or in conjunction with random amplified polymorphic DNA (RAPD) and amplified fragment length polymorphism (AFLP) procedures have been successful in differentiating strains of bacterial pathogens to varying degrees.

4.1.6.1 Restriction Fragment Length Polymorphism Method

By using specific restriction enzyme(s), the bacterial DNA is cleaved at specific sites, resulting in formation of DNA fragments that migrate to different distances in the gels, depending on their sizes. The resultant RFLP patterns may be unique to different species/strains of the bacterial species under investigation. The RFLP patterns generated by digestion with *Pst* restriction enzyme from *X. oryzae* pv. *oryzae* (*Xoo*) strains of race 2 commonly prevalent in the Philippines could be used as the basis of differentiation (Leach and White 1991). Based on the band patterns generated, the strains of *X. campestris* were divided into three RFLP groups (1, 2, and 3). These RFLP groups were correlated well with results of biochemical and physiological tests and host range studies. RFLP procedure could provide results much earlier than biological tests for distinguishing strains of *X. campestris* causing bacterial leaf streak in cereal crops (Alizadeh et al. 1999). Bacterial pathogens infecting rice *Burkholderia glumae*, *B. gladioli* and *B. vandii* were differentiated by using RFLP patterns generated, following digestion of genomic DNA with specific restriction enzymes (Ura et al. 1998). It was possible to differentiate 14 strains of *X. axonopodis* pv. *citri* (*Xac*) and five pathovars of *X. campestris* (*Xc*) based on the RFLP patterns which were distinctly different for each strain/pathovar tested. The incidence of *Xc* in citrus nurseries could be reliably confirmed based on the RFLP pattern generated (Hartung and Civerolo 1989; Kanamori et al. 1999).

Erwinia amylovora (*Ea*), causing fire blight disease in apple and other fruit trees, induces hypersensitive reaction (HR) (production of necrotic lesions) on nonhosts like tobacco. HR as well as pathogenicity are governed by a gene cluster designated *hrp* (30-kb). The *dsp* region close to the *hrp* gene cluster codes for disease-specific functions (Bauer and Beer 1991). The genetic variability was examined by applying RFLP analysis. The *3′ hrpN* gene and/ or a fragment of 1,341-bp of the *dspA/E* region of the *Ea* strains (73) collected from 13 Maloideae host species and *Rubus* spp. were subjected to RFLP analysis. An RFLP pattern that was able to differentiate *Ea* strains from *Rubus* spp. and *Amelanchier* sp. from all other strains, was identified enabling reliable identification of *Ea* strains (Giorgi and Scortichini 2005). Sugarcane leaf scald disease is caused by *Xanthomonas albilineans* and it produces a toxic compound albicidin, the production of which is governed by the 49-kb albicidin biosynthesis gene. RFLP analysis was performed on 137 strains of *X. albilineans*. Fourteen haplotypes and two major genetic groups [albicidin (ALB-RFLPA and ALB-RFLPB] were identified. Albicidin genetic diversity was very similar to the genetic diversity of the pathogen based on the whole genome. However, no relationship could be established between variations in albicidin biosynthesis and pathogenicity of *X. albilineans* strains (Champoiseau et al. 2006a, b).

4.1.6.2 Polymerase Chain Reaction-Based Techniques

By employing universal primers, an array of DNA amplified products are generated from the DNA of target bacterial species and these amplicons constitute the genomic fingerprints for the bacterial species concerned. Repetitive sequence based rep-PCR procedures are more commonly used to fingerprint plant pathogenic bacteria. They generate several genomic fragments via PCR which are resolved as banding patterns that are useful for classifying the bacteria. The specific conserved repetitive sequences [repetitive extragenic palindromic (REP)] sequences, enterobacterial repetitive intergenic consensus (ERIC) sequences and BOX elements] are distributed in the genomes of bacterial species of diverse origin. Rep-PCR assays are employed using primer sets that can amplify the specific conserved repetitive sequences, leading to the differentiation of bacterial strains within a species.

The rep-PCR fingerprint analysis of isolates of different bacterial pathogens has been preferred for characterization of pathogenic bacterial species because of the simplicity, robustness and relatively high resolution power of the technique. Rep-PCR after gel electrophoresis was shown to be effective in differentiating five subspecies of *Clavibacter michiganensis* subsp. *michiganensis* (*Cmm*), *nebraskensis*, *sepedonicus* and *insidiosum* (Opgenorth et al. 1996). Genetic diversity of the genome of isolates of *Xanthomonas oryzae* pv. *oryzae* (*Xoo*) was investigated using the primers based on the sequences of the repetitive elements IS1112 commonly present in *Xoo* (George et al. 1997; Gupta et al. 2001). The geographic origin of strains of *X. axonopodis* pv. *citri* (*Xac*) introduced into Florida, USA was established by employing rep-PCR using ERIC and BOX primers (Cubero and Graham 2002).

The interrelatedness, genetic diversity and geographical distribution of the common bean bacterial blight (CBB) pathogen *Xanthomonas axonopodis* pv. *phaseoli* (*Xap*) and *X. axonopodis* pv. *phaseoli* var. *fuscans* were studied by employing RFLP analysis of PCR amplified 16S ribosomal gene including the 16S–23S intergenic spacer region and rep-PCR. The DNAs of *Xap* and *Xap* var. *fuscans* produced six and five fragments respectively. *Xap* isolates showed the presence of 300- and 400-bp fragments which were absent in isolates of *Xap* var. *fuscans*, but they had a 700-bp fragment instead. In addition, polymorphic bands and fragments characteristic of these two strains were produced due to the action of the restriction enzyme *Mbo*I. It was possible to precisely differentiate the isolates of *Xap* and *Xap* var. *fuscans* by employing rep-PCR analyses (Mahuku et al. 2006).

Xanthomonad-like bacteria associated with common bacterial blight (CBB) of the common bean were characterized on the basis of carbohydrate metabolism, brown pigment production, genetic analysis by rep-PCR and RAPD and pathogenicity on cultivars representing the two common bean gene pools (Andean and Middle American). *X. axonopodis* pv. *phaseoli* (*Xap*) was more prevalent (80%) than *X. phaseoli* var. *fuscans*. Genetic analyses and pathogenicity tests indicated that the East African strains represent distinct xanthmonads that independently evolved to be pathogenic on common bean. *Xap* var. *fuscans* strains were more closely related and genetically distinct from *Xap* strains. But two distinct clusters of *Xap* var. *fuscans* strains were identified and one of the cluster enclosed the most New World strains and the other cluster included the most African strains. Spanish strains were placed in both clusters, but irrespective of the cluster grouping, all strains were highly pathogenic on bean cultivars of both gene pools. The results consistently pointed to the multiple introductions of CBB bacteria into Spain (López et al. 2006).

Pseudomonas syringae pv. *syringae* (*Pss*) is associated with pear blast disease occurring in many countries. Relationships among pathogen population levels, quantity of ice nuclei and incidence of dormant flower bud blast, as well as the variability of *Pss* strains in virulence, production of phytotoxic and biocidal compounds have been investigated (Montesinos and Vilardell 1991; Zeller et al. 1997). The genetic diversity of *Pss* strains in Italy associated with certain specific symptoms of pear blast disease was assessed by employing repetitive sequence PCR with the BOX A1 primer. Eighty nine isolates were putatively identified as *Pss*. Four genomic patterns A, B, C and D were recognized by BOX-PCR analysis. Patterns A and C were identical to those *Pss* strains earlier isolated from pear cv. Coscia. The *syrB* gene governing the synthesis of syringopeptides was detected in 38.8% of the isolates, while ice-nucleation activity was noted in 77.7% of the isolates. The isolates possessing the *syrB* gene were more virulent in pathogenicity tests. Isolates producing C and D patterns were more aggressive than A and B on lemon fruits and pear leaves. On the other hand, A, B, and C isolates were more aggressive than D isolates on lilac leaves, indicating in variation in the susceptibility levels of hosts to isolates producing different genomic patterns assessed by BOX-PCR assay (Natalini et al. 2006).

Six pathogenic races of *Xanthomonas campestris* pv. *campestris* (*Xcc*) infecting *Brassica* spp. were differentiated based on their virulence on differential plant species. The genetic diversity of isolates of *Xcc* was assessed by applying rep-PCR, PFGE and amplified fragment length polymorphism (AFLP) methods. Analyses by rep-PCR, AFLP and PFGE distinguished 13, 12 and 11 differential genotypes respectively. High level of heterogeneity with *Xcc* than that was estimated earlier, was revealed by these techniques, as different genotypes of *Xcc* were detected in all collections of isolates made in Israel and other geographical locations. Two race 3 isolates HRI5212 and HRI6412 were placed in different rep-PCR, PFGE and AFLP types indicating that DNA fingerprinting might not be effective in fixing race affiliation. There was no relationship between genotypes determined by the molecular technique employed and the pathogenic potential assessed by inoculation on differential host plants (Valverde et al. 2007).

The PCR-based suppression subtractive hybridization (SSH) has been shown to be effective in identifying differences between prokaryotic genomes with differing phenotypes including those of pathogenic and nonpathogenic strains of the same species and between different closely related species. In order to identify genomic differences among plant pathogenic *Erwinia* spp. strains, SSH was employed to generate six subtractive libraries to compare the genomes of tree-infecting *Erwinia amylovora* (*Ea*) strains with those of *E. pyrifoliae*, a Japanese *Erwinia* sp. strains and a *Rubus*-infecting strain of *Ea*. Strain-specific sequences including genes enclosing a putative type III secretion system (T3SS) effector, a T3SS apparatus component and several putative proteins were evaluated. The SSH in combination with dot blot hybridization screening enabled identification of a number of genomic differences between *Erwinia* strains with differing host ranges including components of two novel type III secretion systems in *E. amylovora*, a putative tyrosine phosphatase effect and several sequences related to membrane transport or polysaccharide biosynthesis (Triplett et al. 2006).

The genetic diversity of Spanish strains of *Erwinia amylovora* (*Ea*) causing fire blight disease of apple and pear was studied by applying PFGE, PCR-ribotyping and minisatellite-primed (MSP)-PCR and RAPD assays. Amplification with the primers based on the spacer regions between the 16S and 23 S ribosomal genes produced several bands with a size range between 700- and 1,100-bp. All strains of *Ea* (78) produced five bands corresponding to a unique profile that fit into PCR-ribotype 1, regardless of the source, geographic location, year of isolation and PFGE pattern. This was the dominant profile among New Zealand and eastern North America strains. In MSP-PCR analysis, clustering of 29 of 30 strains into a very homogeneous group with identical fingerprints was observed. RAPD analysis also showed the similarity of fingerprints of these strains of *E. amylovora.* PFGE analysis together with MSP-PCR and RAPD assays provided highly reproducible results which were discriminative enough to suggest that several *E. amylovora* introductions might have occurred in Spain (Donat et al. 2007).

Pear bacterial shoot blight (BSB) agent *Erwinia pyrifoliae* was studied to establish the identity by investigating phylogenetic relationship with *Erwinia* sp. The 16S rRNA gene sequence was similar to that of *E. amylovora*. The 16S–23S rRNA ITS regions differentiate *E. pyrifoliae* from *E. amy*lovora (Kim et al. 2001). In a later investigation, the phylogenetic relationships among *E. amylovora* bv. 4 and other biovars of *E. amylovora* and *E. pyrifolia* were examined on the basis of the nucleotide sequences of the genes for 16S rRNA, DNA gyrase B subunit (*gyrB*) and *rpoD*. These genes are essential for bacterial cell survival. The isolates of *E. amylovora* and *E. pyrifoliae* could be divided into two major groups. One group contained isolates of *E. amylovora* bv. 1, 2, 3 and other group included isolates of *E. amylovora* bv. 4 and *E. pyrifoliae*. The results suggested that phylogenetic analysis based on the *gyrB* and *rpoD* sequences might provide higher resolution of relationship between pathogen isolates than when sequences of 16S rRNA were used (Matsurra et al. 2007).

Plasmids are present abundantly in plant pathogenic bacteria and they are considered to contribute to the fitness or even virulence of bacterial pathogens. The indigenous plasmid pEA29 was assumed to be present in all strains of *Erwinia amylovora* (*Ea*). As this plasmid carries genes involved in thiamine biosynthesis, lack of pEA29 may reduce the fitness of *Ea* for colonization of plant tissues (Falkenstein et al. 1989; McGhee and Jones 2000). The strains of *Ea* from various geographical regions were screened for the presence of pEA29. By employing novel primers designed from the sequences of the chromosomal *ams* region and TaqMan probe from pEA29, the absence of the plasmid pEA29 in one strain from Iran, one strain from Spain and two strains from Egypt was recorded. The strains with pEA29 were virulent in assays with immature pears or apple seedlings, but showed reduced growth level in minimal medium without amino acids and thiamine. The newly designed chromosomal primers, used for conventional and for real-time PCR formats, identified *Ea* strains lacking pEA29 in field samples. Although these variants were apparently rare, they could be detected in isolates from different countries/different regions in the world with fire blight disease incidence (Mohammadi et al. 2009).

Ralstonia solanacearum (*Rs*) is capable of infecting a wide range of plant species belonging to more than 50 families causing serious diseases of tomato, potato, banana etc. The strains of *Rs* are classified into races and biovars mainly based on their pathogenicity. The strains are divided into four phylotypes (I–IV) based on sequence analysis of the ITS region. Strains of *Rs*, *Pseudomonas szygii* and banana blood disease bacterium (BDB) from different countries were examined by PCR amplification using the primer pair 759 and 760 and the resultant 282-bp fragments were sequenced. Strains of *Rs* were classified into three groups. Group I strains consisted of strains belonging to biovars 3, 4, 5 and biovar N2 from Japan. Most of these strains were of Asian origin except for two strains from Australia and Guyana. Group II contained strains belonging to biovar 1 and 2 and biovar N2 from Brazil. Group III included strains belonging to biovar N2 from Japan and the Philippines. All strains of *P. szygii* and BDB clustered in group III (Villa et al. 2003). The sequences of polymorphic bands derived from rep-PCR fingerprinting wee used to design two primer sets that amplified specifically the DNA of Rs race 4 strains from ginger, moga and curcuma. One primer set (AKIF/AKIR) amplified a single band (165-bp) from genomic DNA obtained from all mioga andcurcuma and some ginger isolates. The other set 21F/21R amplified a 125-bp fragment from other ginger isolates. The specificity and discriminating capacity of the primer sets was indicated by the absence of amplification from genomic DNAs of other Rs strains or other related bacteria (Horita et al. 2004).

In a further study, partial sequencing of 16S rDNA, endoglucanase and *hrpB* genes of Asian strains of *Rs* complex including 31 strains of *Rs* and two strains each of BDB bacterium and *P. szygii* was performed. Various levels of polymorphisms were observed in each of these DNA regions. The highest polymorphism (ca. 25%) was seen in the endoglucanase gene sequence. The *hrpB* sequence showed about 22% polymorphism. Four clusters were recognized based on the phylogenetic analysis. Cluster 1 enclosed all strains of *Rs* from Asia, which belong to biovars 3, 4, 5 and N2. Cluster 2 contained the Asian strains of *Rs* (biovars N2 and 1) isolated from potato and clove as well as BDB and *P. szygii*. Cluster 3 included race 3 biovar 2 strains from potato, race 2 biovar 1 strains from banana and race 1 biovar 1 strains isolated from America, Asia and other countries. African strains of *Rs* formed a separate cluster 4 (Villa et al. 2005).

Common scab disease caused by gram-positive actinomycete *Streptomyces scabies* infects potato and other root crops such as radish, beet, carrot, turnip and sweet potato. The genetic diversity of *Streptomyces* spp. isolated from field-grown potatoes in six states in the US was examined. The isolates were classified into species based on the sequence of variable regions in the 16S rRNA gene and the genes associated with pathgenicity island (PAI) of *S. turgidiscabies*. About 50% of the isolates belonged to *S.scabies* or *S. europaeiscabies* based on 16S rDNA sequence and had characteristic features of PAI. These isolates were pathogenic to potato and radish and present in all locations. Some of the other isolates (about 10%) were pathogenic, but lacked one or more genes characteristic of PAI, although they had genes for biosynthesis of the pathogenicity determinant thaxtomin. Approximately 40% of the isolates were nonpathogens with diverse morphological characteristics. Four common scab-causing species *S. scabies*, *S. europaeiscabies*, *S. stelliscabies* and the new group X were identified based on sequences of variable regions in the 16S rRNA gene. The results of the survey suggested that different populations of scab-causing species could contribute to the differences in the incidence and severity of potato common scab disease in different locations and years (Wanner 2006).

Banana bacterial wilt diseases are due to the members of *R. solanacearum* (*Rs*) species complex. *R. solancearum* is a heterogeneous species which has been divided into four genetic groups known as haplotypes. Strains causing Moko and Bugtok diseases belong to phylotype II, while the blood disease bacterium (BDB) belongs to phylotype IV. Phylogenetic analysis of partial endoglucanase gene sequences was applied to further assess the evolutionary relationships between *Rs* strains. Moko disease strains were found to be polyphyletic, forming four related, but distinct strain clusters. Bugtok disease-strains could not be differentiated from strains causing Moko disease in the Philippines. Based on phylogenetic analysis of partial endoglucanase gene sequences of strains causing BDB confirmed a close relationship of these strains to *Rs* strains within phylotype IV of the species complex (Fegan and Prior 2006).

Strains of *Clavibacter michiganensis* subsp. *michiganensis* (*Cmm*) were analyzed by rep-PCR assay employing either ERIC or BOX primers. The strains of *Cmm* were placed in four groups A, B, C and D. Similar results were obtained by PFGE analyses also. The Israel *Cmm* strains were highly homogeneous (>92% similarity) within each of the four groups (Fig. 4.1). PCR-based detection of three genes located on different sites of the pathogenicity island revealed that all tested pathogenic strains of *Cmm* were positive for the presence of *ppaA*, *chpC and tomA*, whereas nonpathogenic strains lacked *ppaA*, *chpC*, but not *tomA*, indicating that *ppA* and *chpC* might be essential for virulence *of Cmm* (Kleitman et al. 2008).

Genetic variability of '*Candidatus* Liberibacter' species associated with Zebra disease complex of potato occurring in North America, was investigated. '*Ca. Liberibacter solanacearum*' and '*Ca. Liberibacter psyllaurous*' were detected in potato by applying conventional PCR and real-time PCR assays using pathogen-specific primers and probes. '*Ca.* Liberibacter' species were divided into two clusters showing similarity of 99.8% in their partial 16S rRNA gene sequences and 99.3%

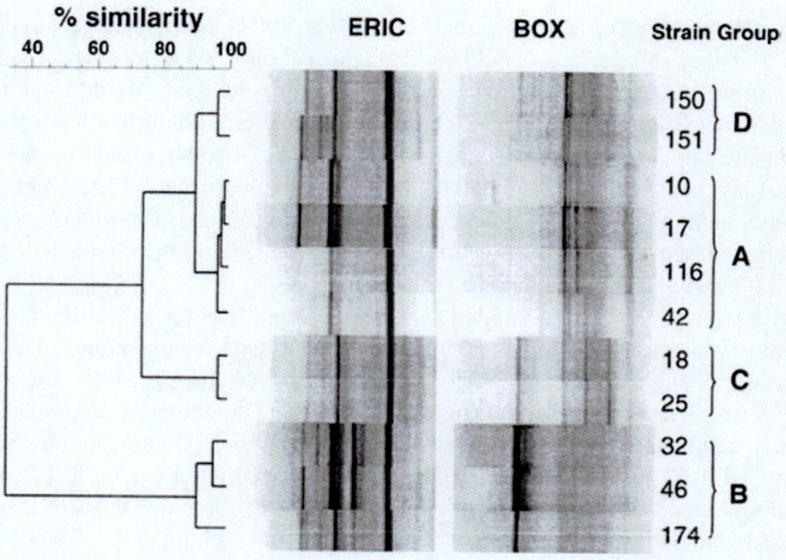

Fig. 4.1 Differentiation of isolates of *Clavibacter michiganensis* subsp. *michiganensis* (*Cmm*) by rep-PCR assay. Strain number and groups differentiated using pulsed field gel electrophoresis (PFGE) are indicated in columns on the right hand side (Courtesy of Kleitman et al. 2008 and with kind permission of Springer Science and Business Media, Heidelberg, Germany)

in their partial intergenic spacer region ISR-23S RNA gene sequences. Genetic variation in the 16S rDNA consistently matched that of the ISR-23S rDNA region. In this partial 16S-ISR-23S rDNA region, eight single nucleotide polymorphisms were detected among '*Ca. L. psyllaurous*' and '*Ca. L. solanacearum*' strains examined. These two species of '*Ca.* Liberibacter' were found to be very closely related. The results were useful for clarifying the current taxonomic status of '*Ca. L. solanacearum*' and '*Ca. L. psylloaurous*' associated with the Zebra disease complex affecting potatoes (Wen et al. 2009).

4.1.6.3 Random Amplified Polymorphic DNA Technique

Differentiation and establishment of relationship between bacterial pathogen species and strains/pathovars have been achieved by applying the random amplified polymorphic DNA (RAPD) procedure. Detection of polymorphism throughout the entire bacterial genome may be more effective by employing RAPD in conjunction with PCR, than other nucleic acid-based methods. RAPD markers have been demonstrated to be useful for the reliable strain identification and differentiation as well as for other genetic analysis of microbial plant pathogens.

Different *Erwinia* spp. and pectolytic pseudomonads are associated with soft rot diseases of vegetables. RAPD procedures have been found to be effective in differentiating *E. carotovora* subsp. *carotovora* (*Ecc*) and *atroseptica* (*Eca*) (Parent et al.

1996) and the strains of *E. amylovora* (Momol et al. 1997). *Ea* causing bacterial soft rot of mulberry (*Morus* spp.) could be distinguished into two types. Type 1 strains were similar to *Ecc*. On the other hand, type 2 strains were distinctly different from *Ecc* and other *E. carotovora* strains, when the RFLP patterns obtained after digestion of *pel* gene with *Sau*3 restriction enzyme (Seo et al. 2003, 2004). Genetic variability in *Ea* may be due to recombination and/or mutation in the bacterial genome. Sequence analysis of the *hrpN* gene revealed that strains of *E*a from *Rubus* spp. and *Amelanchier* sp. had variations in the noncoding region (60-bp fragment) downstream (Giorgi and Scortichini 2005).

RAPD analysis has been applied for efficient identification and differentiation of strains of bacterial pathogens belonging to the genera *Xanthomonas*. The rice bacterial leaf blight (BLB) pathogen *X. oryzae* pv. *oryzae* (*Xoo*) is known to occur in the form of several strains in different rice growing countries. The *Xoo* isolates (16) from different locations in India, were analyzed by RAPD technique. Seven primers OPA-03, OPA-04, OPA-10, OPA-11, OPK-7, OPK-12 and OPK-17 generated simple, specific and reproducible fingerprints from genomic DNAs of the *Xoo* isolates tested. However, these primers were not useful in grouping the *Xoo* isolates. The primers PJEL1 and PJEL2 with an insertion element IS1112 occurring in relatively high copy number (about 80 copies) were employed to generate fingerprint patterns for the same set of *Xoo* isolates. Combination of the RAPD-PCR (seven primers) and IS1112-PCR (two primers) resulted in production of fingerprints with which five groups could be recognized. This procedure was found to be effective for rapid identification of *Xoo* isolates and assessment of genetic variation in the *Xoo* populations present in India (Gupta et al. 2001).

Xanthomonas axonopodis pv. *diffenbachiae* (*Xad*) causes Anthurium blight disease of *Anthurium* and other aroids. The genetic diversity of *Xad* was assessed by using 25 strains from different hosts and geographical locations. The fingerprints of the strains were generated by employing RAPD-PCR procedure and analyzed by the National Taxonomy System Software (NTSYS). The specificity of some of the RAPD fragments selected from PCR profiles was tested by Southern analyses of PCR products. RAPD markers (209) were generated in eight individual DNA profiles. The serotypes and RAPD profiles of some groups of isolates were found to be correlated. Likewise, a possible link between the host range of isolates tested and their RAPD profiles for strains was inferred. Cluster analysis by unweighted pair group method arithmetic average (UPGMA) confirmed that the pathovar is genetically diverse with some strains that were clustered together showing similar host preferences (Khoodo and Jaufeerally-Fakim 2004).

Xylella fastidiosa (*Xf*), a xylem-limited bacterial pathogen, causes serious losses in citrus, grapevine, plum, peach, almond and oak. Application of RAPD, variable number of tandem repeats (VNTR) and rep-PCR procedures have indicated the existence of genetic diversity within the strains of *Xf* isolated from citrus (Rosato et al. 1998; Coletta-Filho and Machado 2002). A total of 360 *Xf* strains were used for assessment of genetic variation among them by employing 15 VNTR and 58 RAPD markers. The VNTR markers were found to be more efficient for discrimination of *Xf* strains isolated from citrus, compared with RAPD markers. The results suggested a genetic structure according to the region of host origin (Coletta-Filho and Machado 2003). In a later investigation, isolates from Pierce's disease (PD)-infected grapevine, phony peach (PP)-affected peach, leaf scald (PLS)-affected plum, oak (OAK) and oleander (OLS) were analyzed. A higher resolution DNA sequence approach was adopted to examine the evolutionary relationships, geographic variation and divergence lines among *Xf* isolates. Two of three genetically divergent clades corresponded to *X. fastidiosa* subsp. *piercei* (PD and some ALS isolates) and *X. fastidiosa* subsp. *multiplex* (OAK, PD, PLS and some ALS isolates). All OLS isolates were placed in the third clade and this group was named as *X. fastidiosa* subsp. *sandyi* (Schuenzel et al. 2005).

4.1.6.4 Amplified Fragment Length Polymorphism Technique

Bacterial spot disease of stone fruits is caused by *Xanthomonas arboricola* pv. *pruni* (*Xap*). Fluorescent amplified fragment length polymorphism (FAFLP) procedure was applied to compare *Xap* populations from Italy, France and USA on the basis of sequences of four housekeeping genes *atpD*, *dnaK*, *efp* and *glnA* and the ITS regions. Only a low level of diversity was observed among the 23 strains of *Xap* tested. Population diversity as indicated by AFLP was lower for the west European population than for the American population (Boudon et al. 2006). *X. albilineans* causing sugarcane leaf scald disease is a systemic xylem-invading bacterial pathogen. Several genetic variants of *X. albilineans* (haplotypes) and albicidin production groups (albivars) were identified within this bacterial species. AFLP method was applied to determine genetic variations which could not be identified by other nucleic acid based procedures. Selected strains (19) which exhibited difference in pathogenicity, were tested. Banding patterns were observed for all strains and these patterns varied according to the strain of *X. albilineans* following PCR amplification of total genomic DNA with all four selective primer combinations. Six combined haplotypes were differentiated. The combined haplotypes were distributed into two major genetic groups – AFLPA and AFLPB – by cluster analysis. In each of these two groups, strains of *X. albilineans* originated from different locations and belonged to different groups showing different extent of stalk colonization were present. Similarly, strains belonging to a single haplotype (GA2 or GB3) showed variations in their virulence (Champoiseau et al. 2006a,b).

Pseudomonas savastanoi pv. *savastanoi* (*Pss*), causing olive knot disease known since ancient times, also infects oleander and ash trees. The classification of the causal bacterium remains unclear and controversial. The genetic diversity of 71 *Pss* strains isolated from different host species and from diverse geographical regions was determined by fluorescent amplified fragment length polymorphism (FAFLP) analysis. Three different selective primer combinations were employed for amplification reactions. A single band of about 1,450-bp was detected in all strains of *Pss* isolated from olive, oleander and ash following amplification and sequencing of their 16S rDNA fragment. All similarity groupings in a cluster, forming a taxon clearly separates from outgrowing strains. *Pss* strains formed subclusters that correlated with host species. Strains identified within these subclusters were related to the geographical regions where the strains were isolated. Strains from olive were divided into two subclusters. Likewise, strains from oleander were differentiated from those from ash. Three strains from jasmine exhibited a high level of similarity among them. The results demonstrated the discriminative ability of AFLP technique in obtaining detailed analysis of the genetic variations between strains of *P. savastanoi* pv. *savastanoi* both from different host species and from diverse geographical locations (Sisto et al. 2007).

The intraspecific diversity of a collection of 63 Spanish strains of *Erwinia amylovora* was assessed by applying AFLP technique, in addition to other biochemical and molecular techniques with a view to grouping them based phenotypic and genotypic criteria. AFLP amplification reactions were performed using *Eco*RI/*Mse*I enzymes and two pairs of primers EOO/MO1-g and E3/M1. The reaction yielded a fingerprint of 41 DNA bands including seven differential bands (A to G with a size ranging from 71 to 847-kb), whereas the remaining bands were shared by all strains analyzed. Band A was present in eight of 30 strains tested, while band B was detected in 21 fingerprints and the remaining five bands (C to G) were uniquely present in only one strain. The distinct introductions of fire blight pathogen *E. amylovora* into Spain, suggested earlier by results of PCR and RAPD techniques, were confirmed by the results of AFLP method, employing two primer pairs in this investigation (Donat et al. 2007).

4.1.6.5 DNA Microarray Technology

Microarrays have been demonstrated to be a powerful tool for genomic comparisons and analysis of microbial populations and their diversity. As thousands of probes can be used simultaneously, it is possible to compare and differentiate microorganisms on the basis of several characteristics. Microarray technology was employed to identify individual genes and gene profiles that were associated with strains of *Pseudomonas syringae* (Sarkar et al. 2006). The gene contents of 18 strains of *Ralstonia solanacearum* were determined and about 50% of the genes were found to be conserved, including many pathogenicity associated genes (Guidot et al. 2007).

Erwinia carotovora subsp. *carotovora* (*Ecc*) (renamed as *Pectobacterium carotovorum*) and *E. carotovora* subsp. *atroseptica* (*Eca*) (*P. atrosepticum*) cause blackleg and soft rot diseases in potatoes. A microarray format was designed to distinguish bacterial pathogens of potatoes. Gene-specific probes were designed for all genes of *Eca*, about 50% of genes of *Streptomyces scabies* (common scab disease) and 30% of genes of *Clavibacter michiganensis* ssp. *sepedonicus* (*Cms*) (ring rot disease), utilizing whole genome sequence information available. In the case of *S. turgidisacbies* (associated with common scab disease), 226 probes were designed based on the sequences of the pathogenicity island containing important virulence genes. In addition, probes were designed for the virulence-associated *nip* (necrosis-inducing protein) genes of *Eca* and *Ecc* and for the intergenic spacer (IGS) sequences of the 16S–23S rRNA gene region. The probes contained about 40 target specific nucleotides and they were synthesized on the array in situ, organized as eight subarrays with an identical set of probes which could be used for hybridization with different samples. All bacterial species were readily distinguished using a single channel system for signal detection. Almost all of about 1,000 probes designed for *Cms*, about 50% and 40% of about 4,000 probes designed for the genes of *S. scabies* and *Eca* respectively and over 100 probes for *S. turgidiscabies* showed significant signals only with the respective species. *Eca* and *Ecc* could be detected with 110 common probes. In contrast, the strains of these species differed in their signal profiles. Probes targeting IGS region and *nip* genes could be employed for differentiating the strains which showed differences in their virulence levels. The results demonstrated that custom-designed genome-wide microarrays might provide a robust means for distinguishing the bacterial pathogens infecting potatoes (Aittamaa et al. 2008).

Huanglongbing (HLB), also known as citrus greening disease is considered to be caused by a phloem-limited fastidious alaphproteobacterium, '*Candiatus* Liberibacter' spp. Attempts to culture this pathogenic entity in artificial medium has been unsuccessful so far. A combination of high-density phylogenetic 16S rRNA gene microarrays and 16S rRNA gene clone library sequencing was applied to determine the bacterial diversity associated with leaf midribs in symptomatic and asymptomatic citrus plants. PhyloChip arrays indicated that nine taxa of bacteria were significantly more abundant in symptomatic midribs. '*Ca. Liberibacter asiaticus*' was detected at a very low level in asymptomatic plants but was over 200 times more abundant in symptomatic plants. The results of PhyloChip analysis were confirmed by sequencing 16S rRNA gene clone libraries which indicated the abundance of '*Ca. Liberibacter asiaticus*' in symptomatic citrus leaves. The PhyloChip arrays may be useful in discriminating the putative pathogenic bacteria from other bacteria belonging to different phyla associated with citrus plants (Sagaran et al. 2009).

4.1.7 Assessment of Variability in Resistance to Chemicals

Copper compounds have been applied for the control of plant bacterial pathogens for several decades. But copper-based bactericides became ineffective in reducing the loss due to the bacterial diseases. Hence, such a situation due to development of copper-resistant bacterial pathogen strains, has necessitated revisiting of the approach of bacterial disease management through chemicals. Copper resistance genes were located on 188–200 kb self transmissible plasmids in the strains of *Xanthomonas campestris* pv. *vesicatoria* (*Xcv*) occurring in Florida and Oklahoma (Bender et al. 1990). Later, chromosome-encoded copper resistance was identified in a strain of *Xcv* from Taiwan (Basim et al. 1999). In a further study, copper resistance genes in strain XvP26 of *Xcv* were found to be localized to 7,652-bp region. These genes hybridized only weakly, as determined by Southern hybridization analysis to other copper resistance genes in *Xanthomonas* and *Pseudomonas* strains. The size of the 5.5-kb region that is functional to copper resistance in XvP26 was similar to the sizes of *P. syringae* pv. *tomato*, *X. arboricola* pv. *juglandis* and *X. c.* pv. *vesicatoria.* PCR amplification analysis of strains from Taiwan showed that none of the strains collected from other locations contained the uniquely oriented copper gene cluster found in XvP26 strain of *Xcv.* This probably indicated that there was a rare introduction of this strain into Taiwan or chromosomal transfer from another organism would have occurred (Basim et al. 2005).

Burkholderia glumae causes bacterial seedling and grain rot disease in rice crops and seeds. Oxolinic acid (OA) has been shown to be effective, when used for seed treatment and for spraying at heading stage in controlling the disease in seeds and seedlings. However, an OA-resistant strain of *B. glumae* was isolated from rice seedlings grown from OA-treated seeds (Hikichi et al. 1998). Amino acid substitutions with arginine (Arg) and isoleucine (Ile) at position 83 in GyrA (GyrA83) were considered to be responsible for OA-resistance in field strains of *B. glumae* (Maeda et al. 2004a). Another technique, mismatch amplification mutation assay (MAMA)-PCR was developed by employing specific primers based on GyrA83, to detect OA-resistant strains of *B. glumae* in infested rice seeds (Maeda et al. 2004b). In a later investigation, rep-PCR analysis using BOX and ERIC as primers was applied. A highly divergent phylogeny was revealed among field strains of *B. glumae*. The amino acid at position 83 of GyrA (GyrA83) involved in OA resistance and the DNA patterns from the rep-PCR and partial nucleotide sequences of *gyrB* and *rpoD* from various strains were analyzed. Based on the band patterns in the rep-PCR analysis and the *gyrB* and *rpoD* sequences two highly OA-resistant Toyama strains, Pg-13 and Pg-14 were identified. The results suggested that the bacteria might acquire OA resistance faster than phylogenetic diversity as determined with the repetitive sequences BOX and ERIC and with *gyrB* and *rpoD* (Maeda et al. 2007).

For the management of the fire blight disease affecting apple, pear and other rosaceous plants caused by *Erwinia amylovora* (*Ea*), the registered antibiotic available was streptomycin only. Resistance to streptomycin in *Ea* was reported as early as in 1971 and the resistant strains of *Ea* have been found to be endemic in many parts of the United States, the exception being the Northeast apple-growing region till 2002. During a routine survey, highly resistant isolates were observed in two orchards in New York State. The molecular analysis showed that resistance was caused by the acquisition of the *strA-strB* gene pair, inserted into the ubiquitous nontransmissible *Ea* plasmid pEA29. The resistant isolates from Michigan also showed similar mechanism of resistance to streptomycin earlier. The results revealed how unintentional movement of nursery material could nullify the attempts to check the spread of antibiotic-resistant strains of *E. amylovora* (Russo et al. 2008).

4.2 Assessment of Variability in Phytoplasmal Pathogens

As the phytoplasmas have defied all attempts to culture them on artificial media, information on the biological characteristics and biochemical/physiological functions of axenically cultured phytoplasmas is not available. Hence, differentiation and identification of phytoplasmas has to be accomplished primarily based on the variations in the immunological characteristics and genomes of phytoplasmas isolated from infected plants.

4.2.1 Assessment of Variability in Immunological Characteristics

The possibility of differentiating the phytoplasmas based on serological characteristics has been indicated in a few cases. In many phytoplasmas, a single reasonably abundant surface membrane protein has been shown to be immunodominant. In other phytoplasmas, several proteins in the membrane preparations are strongly antigenic. Flavescence dorée and elm yellows phytoplasmas have three and four proteins respectively and they have been shown to be serologically related to members of the same subclade (Lee et al. 2000). These proteins appear to have a role in binding the phytoplasmas to cells of insect vectors. The antigenic membrane proteins (AMPs) are serologically variable between phytoplasmas. Aster yellows (AY) and related phytoplasmas form the largest division of the phytoplasmas in cladograms based on 16S rRNA gene sequences. Chlorante AY (AY-C) and clover phyllody (CP) are two serologically distinct phytoplasmas that have been placed in separate subclades IB and IC. The predicted properties of AMPs of AY-C and CP phytoplasmas were consistent with their membrane proteins. The large difference between them, despite their being in the same rRNA clade, might suggest that they are under relatively strong divergent selective pressure (Barbara et al. 2002).

4.2.2 Assessment of Variability in Genomic Characteristics

Availability of quality DNA isolated from infected plant tissues is a critical factor in studying the genomic characteristics of phytoplasmal pathogens. Differentiation of various phytoplasmas and their strains has been possible based on comprehensive restriction fragment analyses with sufficient restriction enzymes. The results have been consistent with phylogenetic groups delineated based on sequence data. With RFLP analyses of phytoplasma 16S rDNA sequences, major phytoplasma groups and subgroups have been classified. Finer subgroup differentiation within each major group has been accomplished by using less-conserved sequences such as ribosomal protein gene clusters or the 16S–23S rRNA intergenic spacer (IGS) region. Universal and group-specific primers may be developed on the basis of 16S–23S IGS region ribosomal protein (*rp*) gene, elongation factor EF-Tu (*tuf*) and other gene sequences. Analyses of PCR-amplified sequences of the conserved genes or a specific DNA may be used as the basis for a simple and rapid method of discriminating phytoplasmas. More than 800 phytoplasma 16S rRNA gene sequences have been deposited in the nucleotide database of the National Center for Biotechnology Information's (NCBI). Availability of high quality sequence data facilitates simulation of restriction digestion in silico and generation of virtual RFLP patterns that are consistent with those obtained by actual RFLP analysis. This computer simulated procedure allows high throughput analysis for identification, differentiation and classification of diverse phytoplasmas. Further, the computer-simulated RFLP analysis system is compatible with traditional RFLP analysis system (Lee et al. 1998, 2007). Classification scheme for identification and differentiation of phytoplasmas based on the above mentioned characteristics has been adopted (Chapter 2). Phytoplasmas may infect two or more plant species in the same or different locations and different phytoplasmas may induce overlapping symptoms. In such situations, it becomes mandatory to apply more discriminatory techniques that can assist in differentiation of various isolates within a 16S rRNA groups. Various molecular techniques have been developed to resolve the strains with greater precision and consistency.

4.2.2.1 Polymerase Chain Reaction-Dependent Techniques

Restriction fragment length polymorphism (RFLP) analyses were performed on PCR amplicons of phytoplasmal DNA from eight samples from *Ulmus* spp. (elms) affected by elm yellows (EY) disease in Italy and the USA, from *Catharanthus roseus* infected with strain EY1 and from five other plant species infected with phytoplasmas of EY group (16SrV). RFLP profiles obtained with *Taq1* restriction enzyme from ribosomal DNA amplified with primer pair P1/P7 differentiated elm-associated phytoplasmas from strains originally detected in *Apocynum cannabinum*, *Prunus* spp., *Rubus fruticosus*, *Vitis vinifera* and *Ziziphus jujuba*. RFLP profiles obtained with *Bfa*I differentiated strains from *A. cananbinum* and *Vitis vinifera* from other phytoplasms of group 16SrV. Elm-associated phytoplasmas strains from Italy could be distinguished from those of American origin by RFLP profiles obtained with *Mse*I in the same fragment of non-ribosomal DNA (Griffiths et al. 1999). Maize bushy stunt (MBS) (red stunt) caused by phytoplasma and corn stunt (pale stunt) induced by the spiroplasma, *Spiroplasma kunkelii* are economically important maize diseases. The genetic variability of the phytoplasma and spiroplasma from infected maize plants in Brazil was assessed by employing multiple PCR-RFLP technique. A DNA fragment of 500-bp was amplified from the spiralin gene in *S. kunkelii* and the fragment (1,200-bp) was amplified from the 16S rRNA gene in the phytoplasma (Fig. 4.2). The partial sequences of the spiralin gene exhibited similarity of 98% among the isolates of *S. kunkelii*, whereas isolates of the phytoplasma showed similarity up to 98%, indicating a very narrow genetic variability in the gene sequences examined (Gomes et al. 2004).

'*Candiatus Phytoplasma australiense*' is associated with Australian grapevine yellows (AGY or AUSGY), papaya dieback (PDB), strawberry lethal yellows (SLY) and strawberry green petal (SGP) diseases occurring in New Zealand (NZ) and Australia. Variations in the *tuf* gene in a diverse samples of NA isolates (36) of *Ca. P. australiense*' was investigated to determine intraspecific variations among the isolates. Partial *tuf* gene sequences of isolates from four different host genera revealed nine different variants which clustered into two distinct groups without any obvious correlation with host or geographic origin. Three distinct clades, one enclosing isolates prevalent entirely in Australia, one including the isolates only from NZ and the third containing isolates from both countries, were recognized. These divisions were consistent with differences detected in the 16S–23S rRNA ITS region. These distinct subgroups *tuf1*, *tuf2* and *tuf3* could be distinguished by diagnostic PCR assay (Table 4.2) (Anderson et al. 2006).

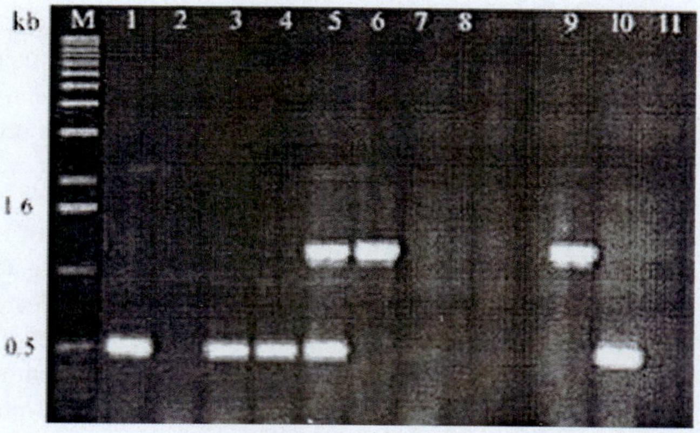

Fig. 4.2 Differentiation of isolates of *Spiroplasma kunkelii* and phytoplasma from maize plants showing stunting symptoms by PCR assay employing primers targeting spiralin gene of *Spiroplasma* and 16S rDNA gene of phytoplasma isolates. M: 1 kb DNA ladder; Lanes 1, 3 and 4: spiroplasma-specific fragment; Lane 5: both spiroplasma + phytoplasma fragments; Lanes 2, 7 and 8: absence of both pathogens; Lane 9: phytoplasma-positive control; Lane 10: spiroplasma-positive control; Lane 11: negative control (Courtesy of Gomes et al. 2004; Pesq agropec bras Brasilia, Brazil)

Table 4.2 Differentiation of *tuf* gene clades 1 and 2 of '*Candidatus Phytoplasma australiense* (Anderson et al. 2006)

	Clade	Position	Sequence 5′–3′
Pa TC1F	1	306–324	CATTCTAGTTGTTTCTGGT
PaTC1R	1	858–843	GTCTTCTCGGTTAATC
PaTC2F	2	306–324	TATTTTAGTTGTTTCTGGA
PaTC2R	2	858–843	GTCTTCGCGGTTAATA

Terminal-restriction fragment length polymorphism (T-RFLP) analysis is the most widely applied molecular method for bacterial ecological studies. The T-RFLP analysis is a direct DNA profiling method that targets usually rRNA gene. This genetic fingerprinting method uses a fluorescently labeled oligonucleotide primer for PCR amplification and the digestion of the PCR products with one or more restriction enzymes. This generates labeled terminal restriction fragments (TRFs) of various lengths depending on the DNA sequence of the target phytoplasma present and the enzyme used to cut the sequence. The results of T-RFLP are obtained through TRF separation by high-resolution gel electrophoresis on automated DNA sequences. The laser scanning system of the DNA sequences detects the labeled primer and from this signal the sequencer can record corresponding fragment sizes and the consequent abundances. The resulting data are easy to analyze, as they are presented as figures for statistical analysis and graphically for rapid visual interpretation (Hodgetts et al. 2007).

The T-RFLP technique was employed using the primers based on the 23S ribosomal genes. The selected primers amplified part of the 23 S rRNA gene. The T-RFLP procedure provided improved resolution between almost all of the 16Sr groups of phytoplasmas compared to the conventional restriction enzyme analyses of the 16S rRNA. By employing the restriction enzymes *Bsh* 12361 and *Mse*I on PCR products and fragment analysis in the range of 68–640-bp, 37 isolates from ten of the 16Sr groups were tested. Distinct and unambiguous T-RFLP profiles were produced for nine of the ten taxonomic groups. Almost all isolates tested within a group shared the same profile and it was possible to differentiate the isolates of one group from the isolates of the other taxonomic groups. In addition, the presence of mixtures of phytoplasmas from different groups in the same sample could be detected and differentiated. PCR amplification of appropriate sequences by the selected primers was verified by including a primer that amplified a terminal restriction fragment (TRF) product of a specific defined size (461-bp) from the host plant chloroplast DNA, as a built-in internal control. The T-RFLP protocol has the potential for simultaneous detection and taxonomic grouping of phytoplasmas in test samples using a single PCR (Hodgetts et al. 2007).

In another investigation, *rp* operon gene *rpl V* (*rpl 22*) and *rpsC* (*rps3*) were adapted to amplify these genes from a wide range of phytoplasmas. These sequences were used to construct a phylogenetic tree for 87 phytoplasma strains belonging to 12 16Sr groups which resulted in a finer resolution of lineages within groups. Furthermore, these sequences were useful to design group-specific primers for phytoplasma diagnostics (Martin et al. 2007). This approach was further extended by designing a new set of primers from the non-ribosomal *secA* gene which encodes SecA, the ATP-dependent force generator in the bacterial precursor protein translocation cascade system. These primers were employed in a semi-nested PCR assay that generated a fragment of expected size (about 480-bp) from all 34 phytoplasmas examined, including strains of representative of 12 16Sr groups. The results of this protocol divided 16SrII group into two distinct clusters, while phytoplasmas associated with coconut lethal yellowing-type diseases were differentiated into three distinct groups. The ability to differentiate 16Sr groups and subgroups by virtual RFLP analysis of *secA* gene sequences suggested that this gene may provide an informative alternate molecular marker for pathogen identification and differentiation and diagnosis of phytoplasmal diseases (Hodgetts et al. 2008).

Grapevine plants are seriously infected by phytoplasmas causing flavescence dorée (FD) and Bois noir (BN or Vergilbungskrankhiet, VK) diseases. Three subtypes of '*Candidatus phytoplasma solani*' causative agent of BN have been described and shown to be associated with different host plant species. A novel and rapid real-time PCR allelic discrimination assay was developed for distinguishing the two major BN phytoplasma subtypes, VK type I and II. TaqMan probes carrying different fluorescent dyes were designed to specifically bind to a polymorphism characteristic for two BN phytoplasma subtypes, thereby allowing discriminative amplification in a single-tube and single-step assay. The conventional PCR-RFLP and the new protocol were applied to analyze 259 BN-positive grapevine samples collected over 5 years. With TaqMan procedure, 257 out of 259 samples could be typed compared with 200 out of 259 samples that could be typed using conventional procedure. The overall concordance of these two methods was 100%. The results demonstrated the superiority of the TaqMan procedure for reliable and rapid determination of Bois Noir phytoplasma subtypes in infected grapevine plants (Berger et al. 2009).

Aster yellows (AY) phytoplasmas '*Candidatus Phytoplasma asteris*', beet leafhopper-transmitted virescence agent (BLTVA) and *Spiroplasma citri* (SC) cause indistinguishable overlapping symptoms on carrot. Genetic variability of AY phytoplasmas detected in carrot fields under non-epidemic conditions was investigated by including four additional genes in addition to 16S rDNA sequences and RFLP analysis. PCR assays followed by RFLP analyses and/or sequencing of phytoplasma16S rDNA and ribosomal protein genes *l22* and *s33*, *tuf*, a putative aa kinase plus ribosomal recycling factor genes and DNA helicase gene were performed. This approach allowed the first molecular identification of phytoplasmas infecting carrot in Serbia, using several molecular markers. Use of more genes, in addition to 16S rDNA for phytoplasma differentiation and classification will be useful for the epidemiological investigations (Duduk et al. 2009).

Spiroplasma citri, a phloem-limited mollicute causes citrus stubborn disease and *S. citri* also infects carrot and several weed species. Genetic diversity among *S. citri* strains from various locations was assessed by performing RAPD-PCR assay using 35 strains cultured from 1980 to 1993 and 35 strains cultured during 2005–2006. A total of 20 primer pairs were employed and considerable diversity among strains was observed. However, no unique genetic signatures could be inferred for differentiating strains collected 15–18 years earlier (historic) from the strains collected recently (new) from California orchards. This showed that genetic patterns of *S. citri* within an orchard could be similar or very different. Genetic homogeneity in some citrus groves could be a consequence of dispersion by nursery propagation. One RAPD primer OPA-13 differentiated between carrot and citrus strains of *S. citri*. The lack of major differences among historic and new strains suggested that genetic changes in *S. citri* genome would not lead to any reemergence of CSD in California orchards (Mello et al. 2008).

4.2.2.2 Subtractive Suppression Hybridization

Flavescence dorée (FD) phytoplasma, a quarantine pathogen of grapevine occurring in France was investigated to evaluate the genetic diversity and genetic relationship with other 16SrV group phyoplasmas in Europe. The variability of two newly characterized FD phytoplasma genes *map* and *degV* isolated by subtractive suppression hybridization (SSH) procedure was assessed. Sequence typing of these housekeeping genes *map* and *degV* encoding a protein of unknown function allowed consistent differentiation of three major FD groups of isolates occurring in France or Italy. This investigation confirmed that the three groups of FD phytoplasmas constitute consistent lineages. Variants could be distinguished on the basis of new single nucleotide polymorphisms (SNPs) in *map* and *uvrB-degV* genetic loci in two of three strain clusters, FD1 and FD3. The FD2 strain cluster was the most widespread (83%) and FD1 strain cluster showed an incidence of 17% in France. The FD3 strain cluster was restricted to Italy. All three strain clusters could be transmitted to grapevine by the leafhopper *Scaphoideus titanus*. The results provided strong genetic evidence of a common origin between alder yellows (AldY) and FD phytoplasmas (Aranaud et al. 2007).

Stolbur phytoplasmas (StolP) infect a wide range of crop plants including grapevine, strawberry, sugar beet and celery. StolP belongs to the 16SrXII-A group with the suggested species designation '*Candidatus Phytoplasma solani*'. The survey of StolP genome identified a partial coding sequence SRO1H10 having no homologue in the '*Candidatus Phytoplasma asteris*'. The *vmp1* gene encodes a protein of 557 amino acids predicted to possess a putative signal peptide and a potential C-terminal transmembrane domain. Southern blotting experiments detected multiple sequences homologous to *vmp1* in the genomes of nine StolP isolates. The *vmp1* gene is variable in size and eight different *vmp1 Rsa*I RFLP profiles could be distinguished among 12 StolP isolates. Comparison of *vmp1* sequences revealed that insertions in the largest forms of the gene encode an additional copy of a repeated domain of 81 amino acids, while variations in 11-bp repeats resulted in gene disruption in two StolP isolates. The results suggested that *vmp1* may exhibit more variability than three housekeeping genes involved in protein translation, maturation and secretion (Cimerman et al. 2009).

4.2.2.3 Single-Strand Conformation Polymorphism Analysis

Single-strand conformation polymorphism (SSCP) analysis is a sensitive and rapid method for detection of polymorphism in PCR-amplified fragments. After chemical and/or heat denaturation, ssDNA of different primary structure fold into different conformations as a result of self-complementarity and intramolecular interactions under appropriate electrophoretic conditions. Different ssDNA species migrate to different distances and the presence of potential mutations is detected as a band-shift or as a change in number of bands. Hence, conditions for the SSCP analysis should be formulated empirically for each set of fragments.

The applicability of SSCP analysis as a tool for detection of molecular variability in isolates of phytoplasmas infecting grapevines was evaluated. Phytoplasma isolates (70) from grapevines belonging to the 16SrXII-A ribosomal subgroups (stolbur) were examined. Routine PCR-RFLP analyses of phytoplasma 16S rRNA gene fragments (580-bp) and *tuf* gene (850-bp) revealed no polymorphisms among the analyzed samples. At least two different restriction profiles and even mixed profiles were observed, when 1,270-bp amplimer of phytoplasma *dnaB* pseudogene were analyzed by PCR-RFLP assay. DNA fragments amplified by PCR were subjected to the SSCP analysis under conditions optimized for each fragment length. SSCP analyses revealed the presence of polymorphisms in the phytoplasma 16S rRNA gene, *tuf* gene and *dnaB* pseudogene that was undetectable in routine RFLP analysis. SSCP analyses revealed several different profiles in 16S rRNA gene fragment, two different patterns in *tuf* gene fragment and five profiles in *dnaB* pseudogene fragment of the isolates. Reliability of the SSCP analysis was confirmed by the multiple alignments of representative sequences showing different SSCP profiles, followed by the phylogenetic analyses using various bioinformatics tools (Music et al. 2007).

Apple proliferation (AP) is one of the most important phytoplasmal diseases in Europe. Several genes and procedures have been adopted to differentiate '*Candidatus Phytoplasma mali*' strains on molecular basis and to demonstrate genetic variability (Seemüller and Schneider 2007; Martini et al. 2008). The 16S rRNA sequences are almost identical and hence, unsuitable for strain differentiation. But other non-ribosomal DNA fragments display higher sequence variability. Based on the sequence analysis of the immunodominant membrane protein (*imp*) gene of *Ca. P. mali*', 16 strains from apple and periwinkle from Germany, Italy, Austria and France, were differentiated into five groups (Danet et al. 2008). Ribosomal protein genes and the *pnp* gene were employed to type pathogen strains and to assess the frequency of '*Ca. P. mali*' subtype combinations occurring in the field conditions (Danet et al. 2008; Martini et al. 2008).

The association of the *hflB* gene with variations in the strain virulence of *Ca. P. mali* was examined. HflB is an ATP-dependent membrane-associated Zn^{2+} protease associated with protein degradation and it is directly associated with membrane transport considered to play an important role in bacterial virulence. The *hflB* gene is present as a single gene in most bacteria. On the other hand, in phytoplasma up to 24 copies of this gene are present. A 530-bp fragment of this gene was amplified for analysis of strain variability by SSCP analysis from 42 accessions originating from apple prolifieration (AP)-infected plants using the primer pair of HflB3–1/rHflB. More than 20 different profiles were generated. The PCR amplimers were sequenced to corroborate SSCP results. The nucleic acid homology within '*Ca. P. mali*' strains ranged from 94.2% to full identity, whereas the identity to the *hflB* gene of '*Ca. Phytoplasma pyri*' and '*Ca. Phytoplasma prunorum*' was at least 85.9% and 84.6% or higher respectively. The molecular analysis of the *hflB* gene in combination with SSCP typing allowed high throughput analysis of samples and enabled differentiation of '*Ca. P. mali* strains to a higher degree than published markers (Schneider and Seemuller 2009).

4.2.2.4 Heteroduplex Mobility Assay

Heteroduplex mobility assay (HMA) is a fast and inexpensive method for determining relatedness between DNA sequences, when two non-identical, but closely related single-stranded DNA fragments anneal. Such molecules show structural distortions at mismatched base pair and at unpaired bases where an insertion or a deletion in the nucleotide sequence has occurred. Heteroduplexes migrate more slowly than a homoduplex in PAGE. The extent of this retardation is proportional to the degree of divergence between the two DNA sequences. The presence of an unpaired base affects the mobility of a heteroduplex more than a mismatched nucleotide (Upchurch et al. 2000). The HMA procedure has been adopted for differentiating phytoplasma isolates causing witches' broom disease in *Populus nigra* cv. Italica and stolbur or big bud symptoms on tomato (Cousin et al. 1998), for differentiation of phytoplasmas in the aster yellows group and clover proliferation group (Wang and Hiruki 2001), determination of genetic variability among isolates of Australian grapevine phytoplasamas (Constable and Symons 2004) and for assessing genetic diversity of phytoplasama isolates from North America, Europe and Asia (Wang and Hiruki 2005).

The HMA procedure was adopted to determine genomic diversity among African isolates of coconut lethal yellowing phytoplasmas causing Cape St. Paul wilt disease (CSPD, Ghana), lethal disease (LD, Tanazania) and lethal yellowing (LYM, Mozambique), in comparison with the Caribbean lethal yellowing (LY). A DNA fragment of 1,850-bp covering the 16S rRNA gene of each isolate was amplified with primer pair P1/P7 followed by HMA analysis for sequence variation. A PCR product from GH5D (CSPD) isolate as a reference was combined with each PCR product and electrophoresed on polyacrylamide gels. Three groups of phytoplasma associated with various coconut lethal yellowing diseases were identified by HMA. The samples from Mozambique (LYM) and Ghana (CSPD) constituted one group which was different from the second group LD from Tanzania. These two groups were different from the third group of Caribbean isolates. The HMA protocol developed in this investigation has the potential to provide a simple and rapid means of identifying and establishing the diversity of isolates within the coconut lethal yellowing disease group, because of the combination of 16S rRNA gene PCR-HMA procedures. This test required only 24–36 h to provide results (Marinho et al. 2008).

Appendix 1: Differentiation of Strains of *Clavibacter michiganensis* subsp. *michiganensis* (*Cmm*) by Macrorestriction Pulsed Field Gel Electrophoresis (PFGE) Technique (Kleitman et al. 2008)

1. Grow the bacteria (*Cmm*) overnight in LB broth at 28°C; transfer 1 ml each of the culture into an Eppendorf tube; centrifuge at 14,000 rpm for 1 min; resuspend the pellet in 1 ml of suspension buffer containing 100 mM Tris-HCl, pH 7.5 and 100 mM EDTA, pH 8 and adjust the OD of the suspension to 1.33.
2. Mix 250 μl of the suspension with 250 μl of 1% Seakem Gold agarose (SKG) (FMC Rockland, USA) in TE buffer (consisting of 10 mM Tris, 1 mM EDTA, pH 8.0) at 50°C; dispense the mixture immediately into the wells of reusable plug molds (Bio-Rad Laboratories, USA) and allow the mixture to solidify at room temperature for 15 min.
3. Cut the agarose plugs into three pieces with a sterile spatula; transfer one piece into a 2-ml Eppendorf tube containing 1 ml of 2 mg lysozyme (Sigma-Aldrich, Switzerland) in suspension buffer; incubate overnight at 37°C in an orbital shaker; remove the lysozyme and add 1 ml of lysis solution containing 1 ml of NDS buffer (containing 0.5 M EDTA, pH 8 and 1% sarcosyl) and 50 l of 20 mg/ml proteinase K.
4. Allow lysis to continue overnight at 50°C with agitation in an orbital shaker; wash the agarose plugs twice with double-distilled water and then four times

with TE buffer at 50°C; prewarm the water and buffer at 50°C before use and use the plugs either for restriction digestion or store at 40°C for up to 6 months.

5. Prepare a 3 mm wide slice from each plug; use 30 units of either *Vsp*I or *Dra*I restriction enzyme in 150 l buffer for digestion at 37°C for 16 h; load the plug slices in appropriate wells of a 1% SKG agarose gel prepared in 0.5 × TBE (Tris-borate-EDTA buffer, Sigma) and use lambda concatemer (Bio-Rad Laboratories, USA).
6. Perform the electrophoresis providing the following conditions: an initial switch time of 0.2 s and a final switch time of 54.2 s (with a gradient of 6 V/cm and an angle of 120° at 14°C) for 22 h and stain the gels with ethidium bromide for further analysis.
7. Analyze the PFGE patterns with the Molecular Analysts Fingerprinting II software package, version 3 (Bio-Rad).

References

Aittamaa M, Somervuo P, Pirhonen M, Mattinen L, Nissinen R, Auvinen P, Valkonenn JPT (2008) Distinguishing bacterial pathogens of potato using a genome wide microarray approach. Mol Plant Pathol 9: 705–717.

Alizadeh A, Arlat M, Sarrafi A, Boucher CA, Barrault G (1999) Restriction fragment length polymorphism analysis of Iranian strains of *Xanthomonas campestris* from cereals and grasses. Plant Dis 81: 31–35.

Anderson MT, Newcomb RD, Liefting LW, Beever RE (2006) Phylogenetic analysis of '*Candidatus Phytoplasma australiense*' reveals distinct populations in New Zealand. Phytopathology 96: 838–845.

Aranaud G, Malembic-Maher S, Salar P, Bonnet P, Maxiner M, Marcone C, Boudon-Padieu E, Foissac X (2007) Multilocus sequence typing confirms the close genetic interrelatedness of three distinct flavescence dorée phytoplasma strain clusters and group 16SrV phytoplasmas infecting grapevine and alder in Europe. Appl Environ Microbiol 73: 4001–4010.

Barbara DJ, Morton A, Clark MF, Davies DL (2002) Immunodominant membrane proteins from two phytoplasmas in the aster yellows clade (chlorante aster yellows and clover phyllody) are highly divergent in the major hydrophilic region. Microbiology 148: 157–167.

Basim H, Minsavage GV, Stall RE, Wang J-F, Shanker S, Jones JB (2005) Characteriztion of a unique chromosomal copper resistance gene cluster from *Xanthmonas campestris* pv. *vesicatoria*. Appl Environ Microbiol 71: 8284–8291.

Basim H, Stall RE, Minsavage GV, Jones JB (1999) Chromosomal gene transfer by conjugation in the plant pathogen *Xanthomonas axonopodis* pv. *vesicatoria*. Phytopathology 89: 1044–1049

Bauer DW, Beer SV (1991) Further characterization of an *hrp* gene cluster of *Erwinia amylovora*. Mol Plant Microbe Ineract 4: 493–499.

Bender CL, Malvick DK, Conway KE, George S, Cooksey DA (1990) Characterization of pXv10A, a copper resistance plasmid in *Xanthomonas campestris* pv. *vesicatoria*. Appl Environ Microbiol 56: 170–175.

Benedict A, Alvarez AM, Berestecky J, Inanka W, Minzumoto C, Pollard L, Mew T, Gonzalez C (1989) Pathovar-specific monoclonal antibodies for *Xanthomonas campestris* pv. *oryzae*. Phytopathology 79: 322–328.

Berger J, Dalla-Via J, Baric S (2009) Development of a TaqMan allelic discriminations assay for the distinction of two major subtypes of the grapevine yellows phytoplasma Bois Noir. Eur J Plant Pathol 124: 521–526.

Boudon S, Manceau C, Nolteghem JL (2006) Structure and origin of *Xanthomonas arboricola pv. pruni* populations causing bacterial spot of stone fruit trees in Western Europe. Phytopathology 96: 1081–1088.

Bouzar H, Jones JB, Minsavage GV, Stall RE, Scott JE (1994) Proteins unique to phenotypically distinct groups of *Xanthomonas campestris* pv. *vesicatoria* revealed by silver staining. Phytopathology 84: 39–44.

Brinkerhoff LA (1970) Variation in *Xanthomonas malvacearum* and its relation to control. Annu Rev Phytopathol 8: 85–110.

Burdman S, Kots N, Kritzman G, Kopelowitz J (2005) Molecular, physiological and host range characterization of *Acidovorax avenae* subsp. *citrulli* isolates from watermelon and melon in Israel. Plant Dis 89: 1339–1347.

Champoiseau P, Daugrois J-H, Girard D-C, Royer M, Rott PC (2006a) Variation in albicidin biosynthesis genes and in pathogenicity of *Xanthomonas albilineans*, the sugarcane leaf scald pathogen. Phytopathology 96: 33–45.

Champoiseau P, Daugrois J-H, Pieretti S, Cocciancich PS, Royer M, Rott P (2006b) High variation in pathogenicity of genetically closely related strains of *Xanthmonas albilineans*, the sugarcane leaf scald pathogen in Guadeloupe. Phytopathology 96: 1081–1091.

Choi JE, Tsuchiya K, Matsuyama N, Wakimoto S (1981) Bacteriological properties related to virulence in *Xanthmonas campestris* pv. *oryzae*. Ann Phytopathol Soc Jpn 47: 668–675.

Cimerman A, Pacifico D, Salar P, Marazachi C, Foissac X (2009) Striking diversity of *vmp1*, a variable gene encoding a putative membrane protein of the stolbur phytoplasma. Appl Environ Microbiol 75: 2951–2957.

Coletta-Filho HD, Machado MA (2002) Evaluation of genetic structure of populations of *Xylella fastidi*osa from different *Citrus sinensis* varieties. Appl Environ Microbiol 68: 3731–3736.

Coletta-Filho HD, Machado MA (2003) Geographical genetic structure of *Xylella fastidiosa* from citrus in Sao Paulo State, Brazil. Phytopathology 93: 28–34.

Constable FE, Symons RH (2004) Genetic variability amongst isolates of Australian grapevine phytoplasmas. Austr Plant Pathol 33: 115–119.

Cook AA, Stall RE (1969) Differentiation of pathotypes among isolates of *Xanthomonas vesicatoria*. Plant Dis Rept 53: 617–619.

Cousin M-T, Roux J, Boudon-Padieu E, Berges R, Seemuller E, Hiruki C (1998) Use of heteroduplex mobility analysis (HMA) for differentiating phytoplasma isolates causing witches' broom disease on *Populus nigra* cv. Italica and stolbur or big bud symptoms on tomato. J Phytopathol 146: 97–102.

Cross JE, Kennedy BW, Lambert JW, Cooper RL (1966) Pathogenic races of the bacterial blight pathogen of soybean, *Pseudomonas glycinea*. Plant Dis Rept 50: 557–560.

Cubero J, Graham JH (2002) Genetic relationship among worldwide strains of *Xanthomonas* causing canker in citrus species and design of new primers for their identification by PCR. Appl Environ Microbiol 68: 1257–1264.

Danet JL, Bahriz H, Cimerman A, Foissac X (2008) New molecular typing tools to monitor fruit tree phytoplasma variability in the 16SrX taxonomic group. Acta Hortic 4: 343–349.

Donat V, Biosca EG, Peñalver J, López MM (2007) Exploring diversity among Spanish strains of *Erwinia amylovora* and possible infection sources. J Appl Microbiol 103: 163–1649.

Duduk B, Calari A, Paltrinieri S, Duduk N, Bertaccini A (2009) Multigenic analysis for differentiation of aster yellows phytoplasmas infecting carrots in Serbia. Ann Appl Biol 154: 219–229.

Falkenstein H, Zeller W, Geider K (1989) The 29 kb plasmid common in strains of *Erwinia amylovora* modulates development of fire blight symptoms. J Gen Microbiol 135: 2643–2650.

Fegan M, Prior P (2006) Diverse members of the *Ralstonia solanacearum* species complex cause bacterial wilts of banana. Aust Plant Pathol 35: 93–101.

Freigoun SO, ElFaki HI, Gelie B, Schirmer M, Lemattre M (1994) Phage sensitivity in relation to pathogenicity and virulence of the cotton bacterial blight pathogen of Sudan. Plant Pathol 43: 493–497.

George MLC, Bustaman M, Cruz WT, Leach JE, Nelson RJ (1997) Movement of *Xanthomonas oryzae* in southeast Asia detected using PCR-base DNA fingerprinting. Phytopathology 87: 302–309.

Giorgi S, Scortichini M (2005) Molecular characterization of *Erwinia amylovora* strains from different host plants through RFLP analysis and sequencing of *hrpN* and *dspA/E* genes. Plant Pathol 54: 789–798.

Gomes EA, Jardim SN, Guimarães CT, de Souza IRP, de Oliveira E (2004) Genetic variability of Brazilian phytoplasma and sprioplasma isolated from maize plants. Pesq agrospec bras Brasilia 39: 61–65.

Griffiths HM, Sinclair WA, Boudon-Padieu E, Daire X, Lee I-M, Sfalanga A, Bertaccini A (1999) Phytoplasmas associated with elm yellows: molecular variability and differentiation from related organisms. Plant Dis 83: 1101–1104.

Guidot A, Prior P, Schoenfeld J, Carrere S, Gemin S, Boucher C (2007) Genomic structure and phylogeny of plant pathogen *Ralstonia solanacearum* inferred from gene distribution analysis. J Bacteriol 189: 377–387.

Gupta VS, Rajebhosale MD, Sodhi M, Sukhwinder S, Gnanamanickam SS, Dahliwal HS, Ranjekar PK (2001) Assessment of genetic variability and strain identification of *Xanthomonas oryzae* pv. *oryzae* using RAPD-PCR and IS 1112-based PCR. Curr Sci 80: 1043–1049.

Hartung JS, Civerolo EL (1989) Restriction fragment length polymorphisms distinguish *Xanthomonas campestris* strains isolated from Florida citrus nurseries from *X. campestris* pv. *citri*. Phytopathology 79: 793–799.

Hifini HR, Nishiyama K, Izuka A (1975) Bacteriological characteristics of some isolates of *Xanthomonas oryzae* different in their pathogenicity and locality. Contr Cent Res Inst Agric Bogor (Indonesia) 16: 1–18.

Hikichi Y, Egami H, Oguri Y, Okuno T (1998) Fitness for survival of *Burkholderia glumae* resistant to oxolinic acid in rice plants. Ann Phytopathol Soc Jpn 64: 147–152.

Hodgetts J, Ball T, Boonham N, Mumford R, Dickinson M (2007) Use of terminal restriction fragment length polymorphism (T-RFLP) for identification of phytoplasmas in plants. Plant Pathol 56: 357–365.

Hodgetts J, Boonham N, Mumford R Harrison N, Dickinson M (2008) Phytoplasma phylogenetics based on anlysis of secA and 23 S rRNA gene sequences for improved resolution of candidate species of '*Candidatus* Phytoplasma'. Int J Syst Evol Microbiol 58: 1826–1837.

Horita M, Yano K, Tsuchiya K (2004) PCR-based specific detection of *Ralstonia solanacearum* race 4 strains. J Gen Plant Pathol 70: 278–283.

Huang BC, Zhu H, Hu GG, Li QX, Gao JL, Goto M (1993) Serological studies on *Xanthomonas campestris* pv. *oryzae* with polyclonal and monoclonal antibodies. Ann Phytopathol Soc Jpn 59: 123–127.

Kanamori H, Sugimoto H, Ochiar K, Kaku H, Tsuyumu S (1999) Isolation of *hrp* cluster from *Xanthomonas campestris* pv. *citri* and its application for RFLP analysis of xanthomonads. Ann Phytopathol Soc Jpn 65: 110–115.

Khan AA, Furuya N, Matsumoto M, Matsuyama N (1999) Trial for rapid identification of pathogens from blighted pear blossom and rotted radish leaves by direct colony and whole cell fatty acid analysis. J Fac Agric Kyushu Univ 43: 327–335.

Khoodo MHR, Jaufeerally-Fakim Y (2004) RAPD-PCR fingerprinting and Southern analysis of *Xanthomonas axonopodis* pv. *diffenbachiae* strains isolated from different aroid hosts and locations. Plant Dis 88: 980–988.

Kim WS, Hildebrand M, Jock S, Geider K (2001) Molecular comparison of pathogenic bacteria from pear trees in Japan and the fire blight pathogen *Erwinia amylovora*. Microbiology 147: 2951–2959

Kleitman F, Barash I, Burger A, Iraki N, Falah Y, Sessa G, Weinthal D, Chalupowicz L, Gartemann K, Eichenlaub R, Manulis-Sasson S (2008) Characterization of a *Clavibacter michiganensis* subsp. *michiganensis* population in Israel. Eur J Plant Pathol 121: 463–475.

Leach JE, White FE (1991) Molecular probes for disease diagnosis and monitoring. In: Khush GS, Toenniessen GH (eds) Rice Biotechnology. CAB International, UK and International Rice Research Institute, Philippines, pp. 281–301.

Lee I-M, Davis RE, Gunderson-Rindal DE (2000) Phytoplasma: phytopathogenic mollicutes. Annu Rev Microbiol 54: 221–155.

Lee I-M, Gunderssen-Rindal DE, Bertaccini A (1998) Phytoplasma: ecology and genomic diversity. Phytopathology 88: 1359–1366.

Lee I-M, Zhao Y, Davis RE, Wei W, Martin M (2007) Prospects of DNA-based systems for differentiation and classification of phytoplasmas. Bull Insectol 60: 239–244.

López R, Asensio C, Gilbertson RL (2006) Phenotypic and genetic diversity in strains of common blight bacteria (*Xanthomonas campestris* pv. *phaseoli* and *X. campestris* pv. *phaseoli* var.*fuscans*) in a secondary center of diversity of the common bean host suggests multiple introduction events. Phytopathology 96: 1204–1213.

Maeda Y, Horita M, Shinohara H, Kiba A, Ohnishi K, Taushima S, Hikichi Y (2007) Analysis of sources of oxolinic acid resistant field strains of *Burkholderia glumae* based on rep-PCR analysis and nucleotide sequences of *gryB* and *rpoD*. J Gen Plant Pathol 73: 46–52.

Maeda Y, Kiba A, Ohnishi K, Hikichi Y (2004a) Implications of amino acid substitutions in *GyrA* at position 83 in terms of oxolinic acid resistance in field isolates of *Burkholderia glumae*, a causal agent of bacterial seedling rot and grain rot of rice. Appl Environ Microbiol 70: 5613–5620.

Maeda Y, Kiba A, Ohnishi K, Hikichi Y (2004b) New method to detect oxolinic acid-resistant *Burkholderia glumae* infesting rice seeds using a mismatch amplification mutation assay polymerase chain reaction. J Gen Plant Pathol 70: 215–217.

Mahuku GS, Jara C, Henriquez MA, Castellanos G, Cuasquer J (2006) Genotypic characterization of the common bean bacterial blight pathogens, *Xanthomonas axonopodis* pv. *phaseol*i and *Xanthomonas axonopodis* pv. *phaseoli* var. *fuscans* by rep-PCR and PCR-RFLP of the ribosomal genes. J Phytopathol 154: 35–44.

Malandrin L, Huard A, Samson R (1996) Discriminant envelope profiles between *Pseudomonas syringae* pv. *pis*i and *P. syringae* pv. *syringae*. FEMS Microbiol Lett 41: 11–17.

Marinho VLA, Fabre S, Dollet M (2008) Genetic variability among isolates of coconut lethal yellowing phytoplasmas determined by heteroduplex mobility assay (HMA). Trop Plant Pathol 35: 377–380.

Martini M, Ermacora P, Flaginella L, Loi N, Carraro L (2008) Molecular differentiation of *Candidatus Phytoplasma mali* and its spreading in Friuli VeneziaGiulia region (north-east Italy). Acta Hortic 781, 395–402.

Martin M, Lee I-M, Bottner KD, Zhao Y, Botti S, Bertaccini A, Harrison NA, Carraro L, Marconi C (2007) Ribosomal protein gene-based phylogeny-based phylogeny for finer differentiation and classification of phytoplasmas. Internat Syst Evol Microbiol 57: 2037–2051.

Matsurra T, Shinohara H, Inoue Y, Azegami K, Tsushima S, Tsukamoto T, Mizuno A (2007) *Erwinia* isolates from the bacterial shoot blight of pear in Japan are closely related to *Erwinia pyrifoliae* based on phylogenetic analyses of *gyrb* and *rpoD* genes. J Gen Plant Pathol 73: 53–58.

McGhee G, Jones AL (2000) Complete nucleotide sequence of ubiquitous plasmid pEA29 from *Erwinia amylovora* strain Ea 88: Gene organization and intraspecies variation. Appl Environ Microbiol 66: 4897–4907.

Mello AFS, Yokomi RK, Melcher U, Chen JC, Wayadande AC, Fletcher J(2008) Genetic diversity of *Spriroplasma citri* strains from different regions, hosts and isolation dates. Phytopathology 98: 960–968.

Mohammadi M, Moltmann E, Zeller W, Geider K (2009) Characterization of naturally occurring *Erwinia amylovora* strains lacking the common plasmid pEA29 and their detection with real-time PCR. Eur J Plant Pathol 124: 293–302.

Momol MT, Momol EA, Lamboy WF, Norelli JL, Beer SV, Aldwinckle HS (1997) Characterization of *Erwinia amylovora* strains using random amplified polymorphic DNA fragments (RAPDs). J Appl Microbiol 82: 389–398.

Montesinos E, Vilardell P (1991) Relationships among population levels of *Pseudomonas syringae*, amount of ice nuclei and incidence of blast of dormant flower buds in commercial pear orchards in Catalunya, Spain. Phytopathology 81: 113–119.

Music MS, Krajačić M, Škorić D (2007) Evaluation of SSCP analysis as a tool for detection of phytoplasma molecular variability. Bull Insectol 60: 245–246.

Natalini E, Rossi MP, Barionovi D, Scortichini M (2006) Genetic and pathogenic diversity of *Pseudomonas syringae* pv. *syringae* isolates associated with bud necrosis and leaf spot of pear in a single orchard. J Plant Pathol 88: 219–223.

Opgenorth DC, Smart CD, Louwas PJ, Bruijn FJ de, Kirkpatrick BC (1996) Identification of *Xanthomanas fragariae* field isolates by rep-PCR genomic fingerprinting. Plant Dis 80: 868–873.

Parent JG, Lacroix M, Page D, Vezina L, Vegiard S (1996) Identification of *Erwinia carotovora* from soft rot diseased plants by random amplified polymorphic DNA (RAPD) analysis. Plant Dis 80: 494–499.

Rosato YB, Neto JR, Miranda VS, Carlos EF, Manfio GP (1998) Diversity of *Xylella fastidiosa* population isolated from *Citrus sinensis* affected by citrus variegated chlorosis in Brazil. Syst Appl Microbiol 21: 593–598.

Russo NL, Burr TJ, Breth DI, Aldwinckle HS (2008) Isolation of streptomycin-resistant isolates of *Erwinia amylovora* in New York. Plant Dis 92: 714–718.

Sagaran US, De Angelis KM, Trivedi P, Andersen GL, Lu S-E, Wang N (2009) Bacterial diversity analysis of Huanglongbing pathogen-infected citrus using PhyloChip arrays and 16S rRNA gene clone library sequencing. Appl Environ Microbiol 75: 1566–1574.

Sarkar SF, Gordon JS, Martin GB, Guttman DS (2006) Comparative genomics of host-specific virulence in *Pseudomonas syringe*. Genetics 174: 1041–1056.

Schneider B, Seemuller E (2009) Strain differentiation of '*Candidatus Phytoplasma mali*' by SSCP and sequence analysis of the *hflB* gene. J Plant Pathol 91: 103–112.

Schuenzel EL, Scally M, Stouthamer R, Nunney L (2005) A multigenic phylogenetic study of clonal diversity and divergence in North American strains of the plant pathogen *Xylella fastidiosa*. Appl Environ Microbiol 71: 3832–3839.

Seemüller E, Schneider B (2007) Differentiation in virulence and genomic features of strains of '*Candiatus* Phytoplasma mali', the apple proliferation agent. Phytopathology 97: 964–970.

Seo ST, Furuya N, Lim CK, Takanami Y (2003) Phenotypic and genetic characterization of *Erwinia carotovora* from mulberry (*Morus* spp.) Plant Pathol 52: 140–146.

Seo ST, Tsuchiya K, Murakami R (2004) Characterization and differentiation of *Erwinia carotovora* strains from mulberry trees. J Gen Plant Pathol 70: 120–123.

Sisto A, Cipriani MG, Tegli S, Cerboneschi M, Stea G, Santilli E (2007) Genetic characterization by fluorescent AFLP of *Pseudomonas savastanoi* pv. *savastanoi* strains isolated from different host species. Plant Pathol 56: 366–372.

Sun X, Stall RE, Jones JB, Cubero J, Gottwald TR, Graham JH, Dixon WN, Schubert TS, Chaloux PH, Stromberg VK, Lacy GH, Sutton BD (2004) Detection and characterization of a new strains of citrus canker bacteria from Key/Mexican lime and Alamow in South Florida. Plant Dis 88: 1179–1188.

Triplett LR, Zhao Y, Sundin GW (2006) Genetic differences between blight-causing *Erwinia* species with differing host specificities identified by suppression subtractive hybridization. Appl Environ Microbiol 72: 7359–7364.

Tsuchiya K, D'Ursel CCM, Horita M, Nozu Y (2003) Relation of Japanese *Xanthomonas campestris* pv. *vesicatoria* with worldwide strains revealed with three specific monoclonal antibodies. J Gen Plant Pathol 69: 310–315.

Upchurch DA, Shankarappa R, Mullins JI (2000) Position and degree of mismatches and the mobility of DNA heteroduplexes. Nucleic Acids Res 28: E69–e69.

Ura H, Matsumoto M, Iiyama K, Furuya N, Matsuyama N (1998). PCR-RFLP analysis for distinction of *Burkholderia* species, causal agents of various rice diseases. Proc Assoc Plant Protect Kyushu 44: 5–8.

Valverde A, Hubert T, Stolov A, Dagar A, Kopelowitz J, Burdman S (2007) Assessment of genetic diversity of *Xanthomonas campestris* pv. *campestris* isolates from Israel by various DNA fingerprinting techniques. Plant Pathol 56: 17–25.

Vera Cruz CM, Mew TW (1989) How variable is *Xanthomonas campestris* pv. *oryzae*? In: Bacterial Blight of Rice, Proceedings of the International Workshop, International Rice Research Institute, Manila, Philippines, pp. 153–166.

Villa J, Tsuchiya K, Horita M, Natural M, Opina N, Hyakumachi M (2003) DNA analysis of *Ralstonia solanacearum* and related bacteria based on 282-bp PCR amplified fragment. Plant Dis 87: 1337–1343.

Villa JE, Tsuchiya K, Horita M, Natural M, Opina N, Hyakumachi M (2005) Phylogenetic relationships of *Ralstonia solanacerum* species complex strains from Asia and other countries based on 16SrDNA, endoglucanase and *hrpB* gene sequences. J Gen Plant Pathol 71: 39–46.

Wang K, Hiruki C (2001) Use of heteroduplex mobility assay for identification and differentiation of phytoplasmas in aster yellows group and the clover proliferation group. Phytopathology 91: 546–552.

Wang K, Hiruki C (2005) Distinction between phytoplasmas at the subgroup level detected by heteroduplex mobility assay. Plant Pathol 54: 625–633.

Wanner LA (2006) A survey of genetic variation in *Streptomyces* isolates causing potato common scab in the United States. Phytopathology 96: 1363–1371.

Wen A, Mallik I, Alvarado VY, Pasche JS, Wang A, Li W, Levy L, Scholth HB, Rush CM, Gudmestad NC (2009) Detection, distribution and genetic variability of '*Candidatus* Liberibacter' species associated with Zebra complex disease of potato in North America. Plant Dis 93: 1102–1115.

Zeller W, Xie X-L, Bereswill S, Geider K (1997) Taxonomy and virulence of bacterial blight (*Pseudomonas syringae pv. syringae*) from poem fruit trees. In: Rudolph K, Burr TJ, Mansfield JW, Stead DE, Vivian A, Von Kietzell J (ed) *Pseudomonas syringae* and related pathogens, Kluwer Academic Publishers, Dordrecht, The Netherlands, pp. 456–459.

Zhang Y, Merighi M, Bazzi C, Geider K (1998) Genomic analysis by pulsed-field gel electrophoresis of *Erwinia amylovora* from the Mediterranean region including Italy. J Plant Pathol 80: 225–232.

Chapter 5
Diagnosis of Bacterial Diseases of Plants

Abstract Bacterial pathogens infect various crop plants from seedling to heading stages in the field conditions as well as after harvest and during storage. As the basic requirement, the pathogenicity of the putative bacterial pathogen(s) isolated from the diseased plants, has to be proved and all steps of Koch's postulates have to be fulfilled. Diagnosis of the bacterial diseases based on the symptoms induced, host range and reactions on differential host plant species or crop cultivars has been accomplished. Biochemical and physiological characteristics like utilization of carbon sources using Biolog system have also been used as the basis for identification of bacterial species. However, these tests require long time and often provide inconsistent results which have to be confirmed by other tests. Tests based on immunological and genomic nucleic acid characteristics have been demonstrated to be more efficient and reliable for detection and identification of the varieties and strains within a bacterial species. Phytoplasmal diseases may be distinguished based on the symptoms commonly induced in many cases. Pathogenicity of the putative phytoplasma has to be proved by grafting or by using the insects present on the crop plants species affected by the disease. Diagnosis of phytoplasmal diseases by nucleic acid-based techniques has been shown to be more accurate and reliable in establishing the presence of the causative agent and etiology of the disease, even in the asymptomatic plants and the propagative plant materials that do not exhibit disease symptoms.

Crops cultivated in different ecosystems are exposed to various bacterial plant pathogens, in addition to different kinds of other biotic and abiotic agents that can adversely affect their growth and yield potential to varying magnitudes, depending on the duration for which the disease-inducing agents have access to the susceptible plant species. Any deviation from the normal functions and development leads to disease in the affected plant species. The nature of the agent(s) involved in observed diseased condition has to be ascertained immediately. Microbial plant pathogens are able to cause distinct macroscopic symptoms that can be visually seen in most cases and hence, these diseases can be separated with some experience from the problems caused by insect pests or adverse environmental condition or industrial pollutants injected into the atmosphere. Disease diagnosis pertains to the identification of the nature and cause of the disease, while detection relates to establishment

P. Narayanasamy, *Microbial Plant Pathogens-Detection and Disease Diagnosis: Bacterial and Phytoplasmal Pathogens, Vol. 2*, DOI 10.1007/978-90-481-9769-9_5,

of the identity of the causative agent present in the infected plants. Plant pathogen detection and disease diagnosis are inseparable processes, the results of one process complimenting the other resulting in reliable identification of the disease-causing agent precisely for initiating follow up action for application of the effective measures to minimize disease spread and the consequent yield losses to the growers.

Disease diagnosis is based on the postulates enunciated by Robert Koch and it is followed essentially to recognize the primary pathogen causing the disease. In a disease complex, two or more pathogens may be involved, making it difficult to identify the primary pathogen and the secondary invaders of the affected plant tissues. Symptom expression following infection by the bacterial pathogens may not be clear or may take long time for the development of symptoms in tissues far away from the site of infection in some cases as in bacterial wilt diseases. In such cases the infection may take place in the roots or basal portions of the stem and the symptoms are expressed only after lapse of appreciable time. By the time symptoms become observable, the pathogen might have well established in the internal tissues of the plants, making it difficult, if not impossible, to apply any effective measures to reduce disease spread. However, reliability on the development of external symptoms of infection alone may lead to erroneous identification of the pathogen and consequently the control methods that are suggested may prove to be ineffective. Hence, accurate identification of the disease-causing bacterial species/strains is the basic requirement for planning effective disease management strategies. In addition, precise identification and differentiation of the cause of the disease-causing bacterial species/subspecies/strains enables more informed decision to be made about the choice of cultivar (showing resistance, if available) and application of appropriate chemicals in time.

Bacterial plant diseases may be diagnosed based on the general symptom picture which differs from diseases caused by fungal and viral pathogens. The absence of any kind of spores or spore-bearing structures on the plant surface and failure of sporulation when the infected tissues are incubated in moist chamber may indicate that the disease under investigation may not be due to a fungal pathogen. Presence of water-soaked areas around the lesions in infected leaves and stem and bacterial ooze from the cut ends of stems or petioles of infected plants is a common feature of bacterial diseases (Birch and Patil 1987). The absence of the above mentioned features may form the basis for eliminating the virus/viroid as the possible cause of the disease.

5.1 Choice of Diagnostic Tests for Bacterial Diseases

5.1.1 Conventional Techniques

Phytopathogenic bacterial species are less numerous when compared with fungal plant pathogens. As the first step in adopting Koch's postulates to provide the evidence that the bacterial species is the cause of the disease to be diagnosed, the bacterium has to be isolated in pure cultures using appropriate nutrient medium that may favor the development of the test bacterial species. Isolation of the bacterial species suspected to be the cause of the disease may not always be easy and straight forward. The presence of saprophytic bacteria which can overgrow the slow-growing pathogen often hampers the attempt to isolate the putative pathogen in pure culture. Selective media amended with antibiotics have been used to inhibit the saprophytes and to encourage differential development of the putative pathogenic bacterial species. *Xanthomonas campestris* pv. *translucens* from the seeds of wheat and barley (Schaad and Forster 1985), *Pseudomonas syringae* pv. *phaseolicola* from bean (Naumann et al. 1988), *P. viridiflava* from onion (Gaitaitis et al. 1997) and *X. campestris* pv. *manihotis* (Fessehaie et al. 1994) have been successfully isolated in pure cultures using appropriate selective media.

The bacterial isolates from pure cultures are inoculated onto the natural host plant species to verify their pathogenicity (ability to infect the susceptible plants). *Brenneria* (*Erwinia*) *quercina* was shown to be the incitant of the bark canker and drippy nut disease of oaks (*Quercina* spp.) by inoculation of the isolated bacterial culture on this host plant species. The Spanish isolates were able to induce internal symptoms of necrosis and acorn exudation in *Q. ilex* and *Q. pyrenaica* (Biosca et al. 2003). The isolates of *Acidovorax avenae* subsp. *citrulli* was proved to be the causative agent of bacterial fruit blotch (BFB) disease of watermelon and melon by pathogenicity tests by inoculating the putative pathogen from the culture onto watermelon and melon plants and fruits (Burdman et al. 2005). A new leaf scorch disease was observed on highbush blueberry plants in Georgia, USA. The putative bacterial pathogen was isolated on nutrient media and it was inoculated on three cultivars of blueberry. Symptoms of the disease appeared at 54 days postinoculation. The causal agent was identified as *Xylella fastidiosa* (*Xf*). Positive reisolations of *Xf* from inoculated symptomatic plants fulfilled the Koch's postulates (Chang et al. 2009). Satisfying the different steps of Koch's postulates is considered as essential for accepting the results of the diagnosis of new diseases occurring now and then in different geographical locations. The colony morphology of the bacterial cultures isolated in the nutrient media and the bacterial cell characteristics are determined and the variations, if any, are recorded. As these characteristics do not show much variation within a morphologic species, identification of subspecies, varieties or strains has to be accomplished by employing several biochemical/physiological, immunological or nucleic acid-based techniques.

5.1.2 Biochemical Tests

Utilization of carbon sources by the bacteria efficiently is one of the important physiological functions of bacteria as the carbon sources are required obtaining necessary energy for all vital functions. The bacteria differ in their patterns of utilization of various carbon sources. Metabolic fingerprints characteristic of a bacterial species are generated by studying the pattern of utilization of a defined set of carbon sources. API system and Biolog assay have been employed to identify

Table 5.1 Identification of plant bacterial species using Biolog database system (Harris-Baldwin and Gudmestad 1996)

	Strains correctly identified (%)[a]		
	Genus	Species	Subspecies/pathovar
Bacterial species (No. of strains tested)	Trial 1/trial 2	Trial 1/trial 2	Trial 1/trial 2
Clavibacter michiganensis subsp. *sepedonicus* (22)	77/45	68/41	59/36
*C.m.*subsp. *michiganensis* (15)	60/47	33/27	20/0
*C.m.*subsp. *insidiosus* (15)	40/53	40/47	13/13
*C.m.*subsp. *nebraskensis* (13)	27/40	20/33	0/0
*C.m.*subsp. *tessellarius* (12)	42/33	33/25	0/0
Curtobacterium flaccumfaciens pv. *poinsettiae* (9)	100/100	33/33	0/0
C.f. pv. *flaccumfaciens* (10)	100/90	20/20	10/20
C.f. pv. *betae* (9)	100/100	20/33	11/0
C.f. pv. *oortii* (5)	100/100	60/80	0/0

[a] Correctness of identification compared to identifications made by conventional methods

Gram-negative bacterial species such as *Xanthomonas axonopodis* (*campestris*) pv. *citri* causing citrus canker disease (Bochner 1989; Verniere et al. 1993), *Xanthomonas oryzae* pv. *oryzae* and *X. oryzae* pv. *oryzicola* (Griffin et al. 1991) and *Pseudomonas syringae* pathovars (Gardan et al. 1984). The usefulness of the Biolog automated microbial identification system for identifying the slow-growing Gram-positive *Clavibacter* spp. and *Curtobacterium* spp. reliably and efficiently was demonstrated (Table 5.1) (Harris-Baldwin and Gudmestad 1996). Isolates of *Brenneria quercina*, causative agent of bark canker disease of oaks were tested by API20E, API20NE and API50CH systems. The Spanish isolates produced homogeneous profiles comparable to the reference strain (Biosca et al. 2003). In a later investigation, the API system was found to be effective in identifying the isolates of *Brenneria nigrifluens* causing bark canker disease of Persian walnuts, based on the metabolic fingerprints generated by the API system (Moretti et al. 2007).

Bacterial cell protein composition may form the basis for differentiation and identification of a bacterial species and its strains. The whole-cell protein analysis performed with *Xylella fastidiosa* (*Xf*), the causative agent of grapevine Pierce's disease by employing sodium dodecyl sulfate (SDS)-polyacrylamide gel electrophoresis (PAGE) technique revealed variations in the protein banding patterns among the strains of *Xf.* The presence or absence and intensity of ten protein bands indicated the identity of the strains of *Xf* which could be assigned to six different groups (Wichman and Hopkins 2002). An advanced variant of gel electrophoresis known as pulsed field gel electrophoresis (PFGE) is useful for separating larger fragments of bacterial DNA. The isolates of *Acidovorax avenae* subsp. *citrulli* (*Aac*) infecting cucurbits were identified by employing PFGE procedure. This method showed that the isolates of *Aac* could be divided into two classes based on the DNA profiles generated from their genomic DNAs (Burdman et al. 2005).

5.1.3 Immunoassays

Bacterial cell walls have different kinds of proteins that may function as specific antigenic determinants that can differentiate the strains of a bacterial species reliably. Polyclonal and monoclonal antibodies (PAbs and MAbs) have been used to diagnose infections by bacterial pathogens in plants, seeds and propagative materials which may or may not show any visible symptoms. Specific components or metabolites of bacteria can be detected in plant tissues leading to correct diagnosis of the disease newly observed.

Ralstonia solanacearum (*Rs*), causing brown rot of potato tubers and wilts of several crops, occurs in the form different strains. Double antibody sandwich (DAS)-ELISA format was applied to detect and differentiate 168 typical *Rs* strains. The MAb 8B-IVIA reacted positively only with the strains of *Rs* infecting potato, but not with 174 other pathogenic bacterial species or unidentified bacteria isolated from potato, indicating the specificity of the MAb in the identification of potato strains of *Rs*. Adoption of an internal enrichment step consisting of shaking the samples in modified Wilbrink broth at 29°C for 72 h followed by testing by DAS-ELISA was effective in diagnosing the disease in seed potato lots, even when the tubers did not exhibit any visible symptom of infection (Caruso et al. 2002). Indirect ELISA format was employed to detect different isolates of *Brenneria* (*Erwinia*) *quercina* infecting oaks in Spain, using PAbs generated against the Spanish isolate 1467a. The antiserum reacted positively with all Spanish isolates of *B. quercina*. Hence, it may be possible to diagnose infection by all isolates of *B. quercina* by employing the indirect ELISA test (Biosca et al. 2003).

Importation of maize seeds from the United States was prohibited by several countries, because of the possible introduction of *Erwinia stewartii* (*Pantoea stewartii*) causing the Stewart's wilt disease. DAS-ELISA test has been recommended for testing the maize seeds by the National Seed Health System as the standard method for phytosanitary seed health testing for *E. stewartii*. By using a commercially available ELISA reagent set (Agadia, Inc., Elkhart, USA) samples of 100-kernel lots could be tested for the presence of *E. stewartii* in the maize seed lots. This ELISA test was able to successfully detect *E. stewartii* even in seed lots treated with fungicides and insecticides (Lamka et al. 1991; Pataky et al. 2004). *Xylella fastidiosa* (*Xf*) has a wide host range of which some plant species are symptomless carriers of this pathogen. The PathoScreen Kit (Agadia, Inc.) was used to test the presence of *Xf* in various plant species. The ELISA test was found to be easier, less expensive and less-time consuming for screening various host plant species that might serve as sources of inoculum of *Xf* for citrus and grapevine crops (Costa et al. 2004).

Bacterial leaf scorch disease affecting blueberry (*Vaccinium virgatum*) in Georgia, USA, was shown to be caused by *Xylella fastidiosa* (*Xf*) by isolation-based conventional method. ELISA tests were performed using the commercially available kits (Agadia, Inc., Elkhart, IN) as per the recommendations of the manufacturer. All ELISA tests conducted with the tissues directly from the infected plants were positive and confirmed the results of conventional biological method. Agadia ELISA kits are widely used for detection of *Xf* in a wide range of host plant species because of absence of cross-reaction with the closest relatives of *Xanthomonas* spp. Hence, the results obtained in this investigation were considered to be conclusive and the pathogen infecting blue berry was identified as *Xylella fastidiosa* (Chang et al. 2009).

5.1.4 Nucleic Acid-Based Techniques

Diagnostic tests based on the characteristics of the genomic nucleic acids of the bacterial pathogens can provide more reliable results which are not dependent on the symptom expression and influence of environmental conditions. In addition, comparatively they are more precise and sensitive and specific, as they need very small amounts of pathogen population that may be below the detectable levels of other diagnostic methods. Tests based on hybridization or amplification of the target DNA fragments of the bacterial pathogen have been extensively used for specific detection and identification of subspecies, pathovars or strains of the bacterial species under investigation. However, the limitations of the nucleic acid-based methods such as the requirement of high quality DNA, presence of inhibitors of PCR amplification in the plants that can adversely affect the results and inability to differentiate the live and dead bacterial cells have to be borne in mind while interpreting the results obtained from these techniques.

Combination of biological and enzymatic amplification (BIO-PCR) technique enhances the sensitivity of the PCR assay significantly, because of the enrichment step to concentrate the bacterial populations present in the samples. Diagnosis of infection of bean seeds by *Pseudomonas syringae* pv. *phaseolicola* (*Psp*) was accomplished with a greater sensitivity level than the standard PCR assay. In addition, false positives due to the presence of dead bacterial cells and false-negatives due to the presence of potential PCR inhibitors in the seed extracts could be eliminated by applying the BIO-PCR format for the diagnosis of seed infection by *P. syringae* pv. *phaseolicola* (Schaad et al. 1995). A combined agar absorption and BIO-PCR assay was shown to be effective in diagnosing diseases caused by slow-growing pathogens like *Xylella fastidiosa* (*Xf*). This bacterial pathogen requires 10–14 days to produce visible colonies on the nutrient media. To overcome the need for obtaining the pure culture of *Xf*, the petioles of grapes and citrus leaves with symptoms were spotted onto the agar media, followed by washing after various time intervals and assaying by real-time PCR assay. The presence of *Xf* in 97% and 100% of the spots was detected respectively in grapes and citrus leaves infected by *Xf*, at 2 days and 4 h of absorption in agar. On the other hand, standard PCR format did not detect *Xf* in any of the samples, when tested simultaneously with the real-time PCR format, indicating the high level of sensitivity and specificity of the real-time PCR assay (Fatmi et al. 2005).

Bacterial pathogens elaborate several enzymes and toxins that have important role in their pathogenicity/virulence. The sequences of the *hrp* gene cluster are targeted by employing specific primers that amplify the unique DNA fragments based on which the pathovars/strains of bacterial pathogens can be detected and the diseases induced

Table 5.2 Detection of *Erwinia amylovora* in field samples by standard and modified REDExtract-N-Amp™ Plant PCR Kit (Stoger et al. 2006)

Nature of samples	No. of samples	No. of positive samples detected		
		Both methods	Standard method	Modified PCR Kit
Field samples	951	211	211	341
Symptomless samples	26	5	5	15

by them can be reliably diagnosed. Specific primer pair to amplify the *hrpF* gene of *Xanthomonas campestris* pv. *campestris* (*Xcc*), infecting *Brassica* spp. was employed to diagnose the seed infection by *Xcc*. The primers amplified a 619-bp fragment of the *hrpF* gene specifically and no amplimers were produced from other *Xanthomonas* spp. tested. Seedborne *Xcc* infection could be detected even if one seed was infected in a lot of 10,000 seeds by including primers targeting a 360-bp section of the ITS region of *Brassica* spp. in a multiplex PCR format (Berg et al. 2005).

Diagnosis of the fire blight disease of apple and pear caused by *Erwinia amylovora* (*Ea*) in asymptomatic plant materials is a major challenge for certification programs. The suitability of the REDExtract-N-Amp™ Plant PCR Kit and a modified protocol of this standard procedure developed in this investigation was compared with the existing methods that were found to be either time-consuming or insensitive or unsuitable for handling large number of samples routinely. The standard method and modified protocol were employed to diagnose 951 field samples. From a total of 341 positive samples, 187 tested positive for the standard method and modified protocol. A total of 154 samples were tested positive with the modified method (Table 5.2). Under greenhouse conditions, the modified method detected *Ea* in inoculated plants (with or without symptoms) 6 weeks after inoculation. On the other hand, the standard method was unable to detect *Ea* in asymptomatic plants (Stoger et al. 2006).

Potato tubers are infected by bacterial and fungal pathogens. *Clavibacter michiganensis* subsp. *sepedonicus* (*Cms*) causing bacterial ring rot disease that warrants serious concern, because of its potential to cause significant economic losses. The pathogen can readily spread through seed potato tubers, particularly by latent infection in which a low incidence of inoculum persists asymptomatically in potato for several generations. Hence, phytosanitary measures are considered to be the feasible approach to reduce the spread of the pathogen through certification of seed tubers. In order to prevent false-negative results that arise because of reaction failure, an internal reaction control in the real-time TaqMan PCR assay was included for the diagnosis of ring rot disease in the potato tuber samples. The superiority of the real-time PCR assay was revealed by the positive detection of *Cms* in as many as 20% (25 of 125) of the ELISA-negative samples. Asymptomatic tuber samples which were ELISA-negative may prove to be potential sources of inoculum, if more sensitive test is not employed for disease diagnosis (Smith et al. 2008). The effectiveness of a modified TaqMan PCR assay for the diagnosis of infection by *Ralstonia solanacearum* (*Rs*) in tomato and potato was demonstrated. Validation of negative results by inclusion of an internal control and confirmation of positive results in a closed-tube system were achieved. Identification of post-reaction products using melt peak analysis was possible because of the distinct melt peaks at 90°C for reaction control product (94-bp) and 83°C for diagnostic target amplicon (68-bp). Presence of *Rs* in infected tomato stems and leaves as well as in potato tubers and stems was diagnosed by the protocol developed by Smith and De Boer (2009).

The possibility of detecting two or more plant pathogens simultaneously is being explored, since simultaneous detection and identification of more than one pathogen at a time can be expected to save time and costs of testing them individually. *Acidovorax avenae* subsp. *citrulli*, the causal agent of bacterial fruit blotch (BFB) and *Didymella bryoniae*, the fungal pathogen causing stem blight disease are transmitted through the cucurbit seeds. A combination of magnetic capture hybridization (MCH) and multiplex real-time PCR was applied for the detection of these bacterial and fungal pathogens simultaneously. The MCH real-time PCR protocol facilitated detection of both target pathogens in watermelon and melon seed samples (n = 5,000 seeds/sample) in which 0.02% (two seeds in 1,000 seeds) were infested with *A.avenae* subsp *citrulli* and 0.02% of the seeds were infested with *D. bryoniae* (Ha et al. 2009).

5.2 Choice of Diagnostic Tests for Phytoplasmal Diseases

5.2.1 Biological Methods

Phytoplasmas induce certain common symptoms such as chlorosis, dwarfing, multiple branching, virescence and partial or total sterility of the affected plants. Newly observed disease suspected to be due to a phytoplasma, is transmitted to a healthy plant of that species and/or to the universal host plant species *Cantharanthus roseus* by grafting or using a parasitic dodder (*Cuscuta* spp.). The symptoms induced on the artificially inoculated plants are compared with those caused by other known phytoplasmas to have a clue on the identity of the new phytoplasma. As the phytoplasmas are known to be vectored by different leafhopper species, the insects belonging to this group are tested for their vector competence. If any insect species could be identified as the vector capable of transmitting the new phytoplasma from infected plants to healthy plants, the nature of relationship between the phytoplasma and the vector is determined. Phytoplasmas show a circulative (persistent) type of relationship with their vector and in some cases their ability to multiply in the vector has also been demonstrated. As the phytoplasmas have not been cultured on any cell-free media, further tests based on the genomic characteristics have to be performed for the precise identification of the putative phytoplasma causing the newly observed disease.

5.2.2 Nucleic Acid-Based Techniques

Detection and identification of phytoplasmas based on the biological characteristics is laborious and time-consuming and the results obtained are often inconsistent.

The characteristics of the axenic cultures of phytoplasmas are not available. Hence, the phytoplasma taxonomy has to heavily rely on the molecular characteristics. Nucleic acid-based analyses have been shown to be more accurate and reliable than the identification based on biological characteristics (Bertaccini 2007). PCR-based assays provide a better basis for classification of phytoplasmas. Use of phytoplasma group-specific universal (generic) or phytoplasma group-specific oligonucletoide primers based on conserved 16S rRNA gene sequences enables amplification of 16S rDNA sequences from strains belonging to a broad spectrum of phytoplasma groups and from diverse strains within a given group. Analysis of PCR-amplified sequences of the conserved genes or specific DNA fragment offers a simple and rapid means of classification of phytoplasmas (Lee et al. 2007).

Apple proliferation (AP) phytoplasama is classified as a quarantine organism in Europe and North America. However, a sensitive, rapid and specific diagnostic technique was lacking. In order to overcome this limitation, a PCR-based method was developed involving the use of an endogenous reference serving as an internal positive control. By simultaneous amplification and individual detection of pathogen and host DNA, not only the pathogen-specific fragment is amplified, but also uninfected plant material and PCR inhibition can be distinguished. The real-time PCR assay has been shown to be highly sensitive, rapid and reliable for the diagnosis of apple proliferation phytoplasma in plant materials especially when the infection level is low, facilitating application of possible preventive measures to reduce subsequent spread of the disease (Baric and Dall-Via 2004). Although the nucleic acid-based techniques are sensitive and precise, the cost-effectiveness and requirement of high technical expertise to perform the tests, limit their utilization for large scale routine diagnosis.

5.3 Agencies Involved in Disease Diagnosis

Disease diagnosis may be the primary responsibility of the personnel of diagnostic centers, plant quarantines and certification programs. In addition, diagnosis of disease may be required for epidemiological and breeding for disease resistance programs to some extent. Precision, sensitivity, rapidity, reproducibility, simplicity and cost-effectiveness of diagnostic methods may determine the quality and utility of diagnosis. The speed of providing reliable results may have high priority, while choosing the method suitable for epidemiological investigations, since large number of samples may have to be analyzed during field surveys. As the conventional methods, although simple, are labor-intensive and require large greenhouse space and longer time to give the results. The conventional methods are being largely replaced by more sensitive methods involving the use of antibody-dependent immunological and nucleic acid-based techniques which can provide results rapidly.

Disease diagnostic centers (DDCs) have the responsibility of analyzing the samples received from the growers and communicating the results with necessary advice for follow up action by the growers. Plant pathology departments attached to the universities/research stations are entrusted with this function of offering advisory service to the clientale group. Plant quarantines, both international and domestic, have the primary the responsibility of preventing the introduction of new pathogens/pests into the country or other regions by enforcing specific regulations for the import/movement of plants and plant materials. They are able to employ serodignostic methods more frequently compared to nucleic acid-based techniques. Certification personnel have the responsibility of certifying seeds and other propagative plant materials for freedom from all microbial plant pathogens. It may suffice to employ generic diagnostic methods for pathogen detection for plant quarantine and certification personnel, whereas the epidemiologist may need more specific and sensitive diagnostic methods that can identify the bacterial pathogens up to subspecies or varieties or pathovars of the bacterial species causing epidemics in one area. Such specific information may be needed to verify whether the same pathogenic strain already exists in other areas of the country and to take up necessary measures to restrict the spread of the disease to other parts of the country. The higher level of sensitivity and specificity of detection methods will be useful to assess the distribution of different strains of a pathogen species with varying levels of pathogenic potential (virulence). In addition, evolution of new strains of indigenous pathogens has to be monitored by epidemiologist by employing discriminative methods. Information on the population structure of the pathogen gathered will be useful for developing disease forewarning system to alert the farmers on the possible occurrence of a more aggressive pathogen strain, if necessary.

5.3.1 Disease Diagnostic Centers

Accurate and reliable identification and differentiation of the bacterial pathogens and their strains with different pathogenic potential is the basic requirement of the development of effective disease management systems. Plant disease diagnostic centers (DDCs) are expected to provide organized, systematic and professional service for rapid detection and precise identification of plant pathogens as the primary cause(s) of the disease problems observed in the region where DDC is located. Any disease management system will be ineffective, if it is not based on the proper diagnosis of the disease to be controlled. Hence, DDCs have the responsibility of making accurate identification of the disease problems and suggesting suitable changes in the disease management strategies with the aim of reducing both qualitative and quantitative losses that have to be otherwise incurred by the cultivator. The credibility of the DDCs depends on the availability of adequate facilities to analyze large number of samples within short time and to offer corrective measures without any loss of time, based on the results obtained.

Adoption of molecular techniques for the detection of pathogens and diagnosis of diseases caused by them, will enhance the sensitivity and reliability of results and reduce the false positive or negative results. Among the immunodiagnostic methods, enzyme-linked immunosorbent assay (ELISA) formats have been extensively employed to obtain results with greater reliability and reduction in time required for analysis compared with isolation-based methods. Polymerase chain reaction (PCR)-based techniques have been demonstrated to provide improved detection and differentiation of the bacterial pathogen isolates which exhibit differences in their virulence and other characteristics. The results from PCR-based techniques are more precise and can be obtained much earlier compared to the immunodiagnostic tests. Many DDCs in North America and Europe are capable of applying highly technical diagnostic procedures. Access to computers for maintenance and retrieval of information has significantly enhanced the quality of diagnostic services offered by these DDCs. On the other hand, DDCs in the developing countries have to be upgraded and they should be established in as many locations as possible to cater to the needs of more areas, so as to bring larger number of farmers under the protective umbrella. Thus the aim of establishing the DDCs to record the extent of disease incidence, diagnose the disease problems accurately and to offer effective suggestions to keep the pathogens at bay, can be satisfactorily realized. The DDCs have also been involved in field surveys on regional or national basis to assess the distribution of crop diseases and monitor the 'pathogen path' to indicate the vulnerability of adjacent states/regions for possible attack by the disease(s).

A plant biosecurity system was visualized for early detection, precise diagnosis and rapid follow up action to minimize disease spread. The National Plant Diagnostic Network (NPDN) has been established in the USA, with specific the objective of (i) establishing a national communications system linking plant diagnostic centers; (ii) upgrading infrastructure for effective diagnosis; and (iii) providing training to facilitate rapid reporting of outbreaks of pathogens/diseases. A nationwide network aims to function as a cohesive system to rapidly detect and diagnose high consequence pathogens in the agricultural and natural ecosystems. The network ensures that all participating diagnostic facilities are alerted to possible outbreaks and are technologically equipped to rapidly detect and identify high risk pathogens. NPDN has established an effective communication network and developed standardized diagnostic and reporting protocols. Cataloguing of diseases and pests occurrence has been taken up to widen the national database (Stack et al. 2006). The concept of establishing similar national network with suitable modifications may be considered by other countries for exchange of information on the occurrence of new pathogens in the respective countries.

5.3.2 Plant Quarantines

Exclusion of plant pathogens by adopting different approaches is one of the principles of crop disease management. Plant quarantines play a vital role in excluding plant pathogens by preventing their introduction through infected plants and plant materials. Domestic and international plant quarantines have been established by most countries to exercise statutory control of plant health through legislation under which appropriate actions can be taken by the country concerned to prevent introduction or spread of the notified plant pathogen(s) that are not present or insignificant in that country concerned. The principles of establishing plant quarantines recognize the sovereignty of the country which has the right to impose the conditions and phytosanitary measures to be taken by the exporting country. The need for establishing modern quarantine laboratories has been well realized by several nations, after the implementation of the General Agreement on Tariffs and Trade (GATT), since there is a spectacular increase in the movement of people and plant materials that are likely to carry and introduce the plant and human pathogens into another country where they may be absent. This situation necessitated the enforcement of sanitary and phytosanitay measures at global level.

The International Plant Protection Convention (IPPC) was established in 1991, following the acceptance of GATT by majority countries. Basic principles required for formulating standards for plant quarantine procedures in relation to the international trade have been formulated by an expert committee (FAO 1991). Accordingly, an organism is considered to be of quarantine significance (QS), if its exclusion is perceived as important enough to protect agriculture and natural vegetation of the importing country. Sixteen basic principles have been suggested by the expert committee with the expectation that judicious application of these principles might lead to reduction or elimination of the use of unjustified phytosanitary measures as a barrier to trade. Eight regional plant protection organizations have been set up to advise and assist member governments on the technical, administrative and legislative measures concerned with plant quarantine activities. The QS organisms have been divided into two groups designated List A and List B organisms. List A includes dangerous organisms and List B encloses organisms of importance for individual countries. List A is subdivided into List A1 containing pathogens absent in all member countries (exotic pathogens) and List A2 enclosing pathogens present only in some countries in the region (Narayanasamy 2002). Some of the representative bacterial and phytoplasmal pathogens included in the List A1 and List A2 by European and Mediterranean Plant Protection Organization (EPPO) are as follows:

Bacterial pathogens – *Citrus huanglongbing* (citrus greening), *Liberibacter africanum*, *Liberibacter asiaticum*, *Xanthomonas axonopodis* pv. *citri*, *X. oryzae* pv. *oryzae*, *X. oryzae* pv. *oryzicola* and *Xylella fastidiosa*.

Phytoplasmal pathogens – '*Candidatus Phytoplasma ulmi*', Palm lethal yellowing phytoplasma, Peach rosette phytoplasma, Peach yellows phytoplasma, Potato purple-top phytoplasma, and Western-X phytoplasma.

The other approach of excluding fungal pathogens is through certification of seeds and asexually propagated plant materials such as tubers, suckers, corms, setts and budwood materials. In addition, the postharvest produce, fruits and vegetables that are exported, may carry the fungal pathogens which may be in the quiescent phase. Rapid and reliable methods have to be employed to detect and identify the fungal pathogens which may induce symptoms after the clearance of the consignments from the ship (Narayanasamy 2006). The imported plants and plant materials are checked carefully by the plant quarantine personnel by placing them in isolation in post-entry quarantines (PEQs). Disease-free materials are released after testing the plants and plant materials using appropriate techniques. In order to check the spread of the bacterial and phytoplasmal pathogens within the country, the certification personnel inspect the seed lots and mother plants from which propagative materials are taken and multiplied. Necessary tests are performed before certifying them for freedom from plant pathogens. Various kinds of detection techniques based on the biological, immunological and genomic nucleic acid characteristics have been employed for detection and differentiation of bacterial and phytoplasmal pathogens present in test materials. The relative usefulness and applicability for large scale use of the detection and identification methods have been indicated in Chapter 2 to enable the technical staff to choose the ones that suit their requirements.

References

Baric S, Dall-Via J (2004) A new approach to apple proliferation detection: a highly sensitive real-time PCR assay. J Microbiol Meth 57: 135–145.

Berg T, Tesoriero L, Hailstones DL (2005) PCR-based detection of *Xanthomonas campestris* pathovars in *Brassica* seed. Plant Pathol 54: 416–426.

Bertaccini A (2007) Phytoplasmas: diversity, taxonomy and epidemiology. Front Biosci 12: 673–689.

Birch RG (2001) *Xanthomonas albilineans* and the antipathogenesis approach to disease control. Mol Plant Pathol 2: 1–11.

Birch RG, Patil SS (1987) Correlation between albicidin production and chlorosis induction by *Xanthomonas albilineans*, the sugarcane pathogen. Physiol Mol Plant Pathol 30: 199–206.

Bochner BR (1989) Sleuthing out bacterial identities. Nature (London) 399: 157–158.

Biosca EG, González R, López-López MJ, Soria S, Montón C, Pérez-Laorga E, López MM (2003) Isolation and characterization of *Brenneria quercina*, causal agent of bark canker and drippy nut of *Quercus* spp. in Spain. Phytopathology 93: 485–492.

Burdman S, Kots N, Kritzman G, Kopelowitz J (2005) Molecular, physiological and host range characterization of *Acidovorax avenae* subsp. *citrulli* isolates for watermelon and melon in Israel. Plant Dis 89: 1339–1347.

Caruso P, Gorris MT, Cambra M, Palomo JP, Collar J, López MM (2002) Enrichment double antibody sandwich indirect enzyme-linked immunosorbent assay that uses a specific monoclonal antibody for sensitive detection of *Ralstonia solanacerum* in asymptomatic potato tubers. Appl Environ Microbiol 68: 3634–3638.

Champoiseau P, Daugrois J-M, Girard J-C, Royer M, Rott P (2006) Variation in albicidin biosynthesis genes and in pathogenicity of *Xanthomonas albilineans*, the sugarcane leaf scald pathogen. Phytopathology96: 33–45.

Chang C-J, Donaldson R, Brannen P, Krewer G, Boland R (2009) Bacterial leaf scorch, a new blueberry disease caused by *Xylella fastidiosa*. HortScience 44: 413–417.

Costa HS, Raetz E, Pinkard TR, Gispert C, Hernandez-Martinez R, Dumenyo CK, Cooksey DA (2004) Plant hosts of *Xylella fastidiosa* in and near southern California vineyards. Plant Dis 88: 1255–1261.

Fatmi M, Damsteegt VD, Schaad NW (2005) A combined agar-absorption and BIO-PCR assay for rapid sensitive detection of *Xylella fastidiosa* in grape and citrus. Plant Pathol 54: 1–7.

Fessehaie A, Wydra K, Rudolph K (1994) Development of a new semi-selective medium for isolating *Xanthomonas campestris pv. manihotis* from plant material and soil. Phytopathology 89: 591–597.

Food and Agriculture Organization (FAO) (1991 Annex I: Quarantine Principles as Related to International Trade. Report of Expert Consultation on Harmonization of Plant Quarantine Principles, 6–10 May 1991, Food and Agriculture Organization, Rome, pp.15–20.

Gaitaitis R, Summer D, Gay D, Smittle D, McDonald G, Maw B, Johnson WC III, Tolner B, Hung Y (1997) Bacterial streak and bulb rot of onion. I. A diagnostic medium for semiselective isolation and enumeration of *Pseudomonas viridifalva*. Plant Dis 87: 897–900.

Gardan L, Moreno B, Guinebertiere JP (1984) The use of a micromethod of auxanongram and data analysis for identification of *Pseudomonas.* Proceedings of the 2nd Working Group on *Pseudomonas syringae* pathovars, Athens, Greece, pp. 37–39.

Griffin DE, Dowler WM, Hartung JS, Bonde MR (1991) Differences in substrate utilization among isolates of *Xanthomonas oryzae* pv. *oryzae* and *X. campestris* pv. *oryzicola* from several countries. Phytopathology 81: 1222.

Ha Y, Fessehaie A, Ling KS, Wechter WP, Keinath AP, Walcott RR (2009) Simultaneous detection of Acidovorax avenae subsp. citrulli and Didymella bryoniae in cucurbit seedlots using magnetic capture hybridization and real-time polymerase chain reaction. Phytopathology 99: 666–678.

Harris-Baldwin A, Gudmestad NC (1996) Identification of phytopathogenic coryneform bacteria using the Biolog automated microbial identification system. Plant Dis 80: 874–878.

Lamka GL, Hill JH, McGee DC, Braun EJ (1991) Development of an immunosorbent assay for seedborne *Erwinia stewartii* in corn seeds. Phytopathology 81: 839–846.

Lee I-M, Zhao Y, Davis RE, Wei W, Martini M (2007) Prospects of DNA-based systems for differentiation and classification of phytoplasmas. Bull Insectol 60: 239–244.

Moretti C, Sivestri FM, Rossini E, Natalini G, Buonaurio R (2007) A protocol for rapid identification of *Brenneria nigrifluens* among bacteria isolated from bark cankers in Persian walnut plants. J Plant Pathol 89: 211–218.

Narayanasamy P (2002) Microbial Plant Pathogens and Crop Disease Management. Science Publishers, Enfield, NH.

Narayanasamy P (2006) Postharvest Pathogens and Disease Management, Wiley, Hoboken, NJ.

Naumann K, Karl H, Zielke R, Schmidt A, Griesbach E (1988) Comparative analysis of bean seeds on the incidence of causal agent of halo blight *Pseudomonas syringae* pv. *phaseolicola* by different identification methods. Zentr Mikrobiol 143: 487–498.

Pataky JK, Block CC, Michener PM, Shepherd LM, McGee DC, White DG (2004) Ability of an ELISA-based seed health test to detect *Erwinia stewartii* in maize seed treated with fungicides and insecticides. Plant Dis 88: 633–640.

Schaad NW, Cheong SS, Tamaki S, Hatziloukas E, Panopoulos NJ (1995) A combined biological and enzymatic amplification (BIO-PCR) technique to detect *Pseudomonas syringae* pv. *phaseolicola* in bean seed extracts. Phytopathology 85: 243–248.

Schaad NW, Forster RL (1985) A semiselective agar medium for isolating *Xanthomonas campestris* pv. *translucens* from wheat seeds. Phytopathology 75: 260–263.

Smith DS, De Boer SH (2009) Implementation of an artificial reaction control in a TaqMan method for PCR detection of *Ralstonia solanacearum* race 3 biovar 2. Eur J Plant Pathol 90: 128–136.

Smith DS, De Boer SH, Gourley J (2008) An internal reaction control for routine detection of *Clavibacter michiganensis* subsp. *sepedonicus* using a real-time TaqMan PCR-based assay. Plant Dis 92: 684–693.

Stack J, Cardwell K, Hammershmidt R, Byrne J, Loria R, Snover-Clift K, Baldwin W, Wisler G, Beck H, Bostock R, Thomas C, Luke E (2006) The national plant diagnostic network. Plant Dis 90: 128–136.

Stoger A, Scaffer J, Ruppitsch W (2006) A rapid and sensitive method for direct detection of *Erwinia amylovora* in symptomatic and asymptomatic plant tissues by polymerase chain reaction. J Phytopathol 154: 469–473.

Verniere C, Pruvost O, Civerolo EL, Gambin O, Jaquemoud-Collet JP, Luisetti J (1993) Evaluation of the Biolog substrate utilization system to identify and assess metabolic variation among strains of *Xanthomonas campestris* pv. *citri*. Appl Environ Microbiol 59: 243–249.

Wichman RL, Hopkins DL (2002) Differentiation of pathogenic groups of *Xylella fastidiosa* with whole cell protein profiles. Plant Dis 86: 875–879.

Index

P

T